STARTING CATEGORY THEORY

STARTING CATEGORY THEORY

PAOLO PERRONE
University of Oxford, UK

World Scientific

NEW JERSEY • LONDON • SINGAPORE • BEIJING • SHANGHAI • HONG KONG • TAIPEI • CHENNAI • TOKYO

Published by

World Scientific Publishing Co. Pte. Ltd.
5 Toh Tuck Link, Singapore 596224
USA office: 27 Warren Street, Suite 401-402, Hackensack, NJ 07601
UK office: 57 Shelton Street, Covent Garden, London WC2H 9HE

Library of Congress Cataloging-in-Publication Data
Names: Perrone, Paolo (Researcher at the University of Oxford), author.
Title: Starting category theory / Paolo Perrone, University of Oxford, UK.
Description: New Jersey : World Scientific, [2024] | Includes bibliographical references and index.
Identifiers: LCCN 2024000546 | ISBN 9789811286001 (hardcover) |
ISBN 9789811286018 (ebook for institutions) | ISBN 9789811286025 (ebook for individuals)
Subjects: LCSH: Categories (Mathematics)
Classification: LCC QA169 .P386 2024 | DDC 512/.62--dc23/eng/20240223
LC record available at https://lccn.loc.gov/2024000546

British Library Cataloguing-in-Publication Data
A catalogue record for this book is available from the British Library.

For any available supplementary material, please visit
https://www.worldscientific.com/worldscibooks/10.1142/13670#t=suppl

Desk Editors: Sanjay Varadharajan/Rok Ting Tan

Typeset by Stallion Press
Email: enquiries@stallionpress.com

To Alessandro Meani, my first math teacher

Preface

> The purpose of abstracting is not to be vague, but to create a new semantic level in which one can be absolutely precise.
>
> *E. W. Dijkstra,*
> *Turing Award lecture* [Dij72]

Mathematics is the science of patterns. Sometimes, in our everyday lives, at work, or in research, we have an intuitive "feeling" about something technical or conceptual, which can be hard to express or communicate. This "feeling" may, for instance, be a sudden understanding of *how something works*, or that *something new is similar to something we've seen before*, or that *something does not, or should not, depend on how it is presented*. Very often, listening to this "feeling" gives us insight and guides us in finding solutions to problems we may have.

However, when we are experiencing this "feeling," how do we know we are correct? And more importantly, *how do we communicate our intuition to other people?*

The role of mathematics, largely, is to help us turn these "feelings" into something precise. It gives us a language to communicate our technical intuitions to others in a way that's as unambiguous as possible. This allows us, and other people, to check if our intuition is right, and it makes it possible to share and spread ideas, for example, to solve those problems which require more than one person's work.

It also ensures our ideas live longer, so that new ideas can be built on top of them.

Category theory, in this light, can be seen as the *mathematics of mathematics*.[1] In mathematics, and in adjacent fields such as computer science, physics, statistics, and engineering, it is even *more common* to observe patterns than in everyday life. Sometimes, especially when these patterns involve several subjects at once, ordinary mathematics is not enough to express them. Category theory is then a way to make these higher-level ideas mathematically precise and to communicate them to others, even from different fields (provided that they know some category theory too). As with every mathematical theory, it comes with its own definitions, theorems, and *techniques*, albeit quite abstract. These abstract techniques, as we will see, often serve the purpose of *automating conceptual reasoning*. Using category theory, some ideas that could have otherwise taken great amounts of effort, intuition, or trial-and-error to fully understand become standard, almost routine, procedures to apply.

In this book, we focus on the basic notions and results of category theory, with particular emphasis on *why* things are defined the way they are and how they are used. We attempt to describe what intuitions, or "mathematical feelings," one is trying to make precise each time and how the corresponding abstract structure achieves that purpose. For example:

- the concept of *natural transformation* (see Section 1.4) formalizes the idea of a map which "makes sense regardless of how things are presented";
- the concept of *monad* (see Chapter 5) formalizes "generalized elements" or "formal operations" of a certain kind.

We will come across many more examples in the rest of the book.

---

[1] I learned this expression from Eugenia Cheng, see for example [Che23].

How to read this book: Categorical concepts, on their own, are quite abstract; however, as we have mentioned, they are often abstractions of something concrete, or formalizations of intuitions that one may have already had in the practice of mathematics or science. For some readers, it may be that a categorical definition or statement alone is not particularly suggestive or inspiring.[2] Because of that, alongside the abstract ideas, we always give many concrete examples and possible interpretations. There will be examples from different areas of mathematics (group theory, graph theory, probability, etc.), as well as some examples from adjacent fields, such as physics and computer science. Not every example will be helpful for every reader, but hopefully every reader can find at least one helpful example per concept. The reader is encouraged to read all of the examples — this way, they may learn something about a different field.

In order to always have something precise at hand, we begin each section by giving rigorous, abstract definitions and statements first. Immediately following this, we provide examples of what the abstraction aims to formalize. If the abstraction seems hard to understand at first glance, we recommend the reader turn to the examples and explanations and then return to the abstraction afterward. Look at as many examples as possible until you think you can "see the pattern," and only then compare that pattern to the categorical concept we are introducing.

In other words, *this book does not have to be read linearly*.

About this work: This book was originally developed as lecture notes for a course that I taught at the Max Planck Institute of Leipzig in the summer semester of 2019. It is rather different from other material on category theory, partly in content and largely in form:

- The audience of the course had very diverse backgrounds; there were algebraic geometers and topologists, as well as computer scientists, physicists, and chemists.

[2] Some people call category theory "abstract nonsense" — some use that term derogatively, some affectionately.

- The lecture notes were always written *after* the lecture had taken place to reflect what was discussed in class and include all the questions, remarks, and views of the participants.

Sources: This book is not a research or survey paper; it is an introduction to an entire field which was developed by others and which I too have learned from others. As such, it would be a nontrivial task to find out where each idea in this book, and each way to present it, originated. A lot easier is to say where I, the author, learned the concepts presented in this work:

- I learned the basics of category theory from Emily Riehl's book [Rie16] (in a course taught by Tobias Fritz in Leipzig in 2016). Its approach to the Yoneda lemma, universal properties, and the example for the case of matrices have been of great importance to me. For my 2019 course, particularly for the "pure math" part, I chose that book again as the recommended text. It has had great influence on this work as well as on all my category theory research.
- Another important source of understanding and examples has been (an early version of) Brendan Fong and David Spivak's book [FS19], which was the one I recommended for the "applied" part of my course. In particular, it is from there that I learned the interpretation of universal properties in terms of "probes" and "observations"and the adjoint functor theorem for preorders.
- Some of the material on monoidal categories emerged out of conversations with Noson Yanofsky, while reading an early version of his book [Yanng]. I learned about commutative monads from Martin Brandenburg's thesis [Bra14].
- Last but not least, many ideas in this book about how to present certain mathematical concepts are novel and are to be credited not only to me but also to the participants of my course. (Part of the treatment of monads already appeared in the introduction to my PhD thesis [Per18]).

Acknowledgments

Many people have helped me in writing this book, offering guidance in understanding the ideas behind the abstract concepts and providing suggestions for explaining these ideas to new learners:

- First of all, I want to thank all the participants in the course, without whom this work wouldn't exist. I want to thank in particular Emma Chollet, Wilmer Leal, Guillermo Restrepo, and Sharwin Rezagholi, who came up with many original ideas during the lectures, now recorded here.
- I also want to thank Carmen Constantin, Jules Hedges, Jürgen Jost, Slava Matveev, David Spivak, and more recently Walter Tholen and Noson Yanofsky for the interesting discussions, some of which were reflected in the way I taught this course and wrote this book.
- I want to thank all the people who, in the course of these years, pointed out typos, mistakes, or unclear reasoning in earlier versions or parts of this work. In particular, I want to thank Rob Cornish, Tobias Fritz, Mike Shulman, and Sam Staton and his group members, current and former.
- Finally, I want to express my heartfelt gratitude to Tobias Fritz for showing great patience in teaching me category theory.

About the Author

Paolo Perrone is a researcher at the University of Oxford (UK).

After completing his PhD at the University of Leipzig (Germany), he has worked at the Max Planck Institute for Mathematics in the Sciences (Leipzig, Germany), York University (Toronto, ON, Canada), and MIT (Cambridge, MA, USA).

His research focuses on category theory, probability, and information theory.

Contents

1

Basic Concepts

In this chapter, will look at the basic building blocks of category theory: categories, functors, and natural transformations.

As we mentioned in the introduction, whenever we introduce a new concept, we begin by giving a rigorous definition followed immediately by an intuitive explanation. So, in case the abstract definition is hard to understand at first, read on and come back to it after seeing the examples.

1.1 Categories

Definition 1.1.1. A **category** $\mathbf{C}$ consists of

- a collection $\mathbf{C}_0$, whose elements are called the **objects** of $\mathbf{C}$ and are usually denoted by uppercase letters, $X, Y, Z, \ldots$;
- a collection $\mathbf{C}_1$, whose elements are called the **morphisms**, or **arrows**, of $\mathbf{C}$ and are usually denoted by lowercase letters, f, g, $h, \ldots$;

such that

- each morphism is assigned two objects, called **source** and **target**, or **domain** and **codomain**. We denote the source and target of the morphism f by $s(f)$ and $t(f)$, respectively. If the morphism f has source X and target Y, we also write $f : X \to Y$, or, more graphically, $X \xrightarrow{f} Y$.

- each object X has a distinguished morphism $\mathrm{id}_X : X \to X$, called **identity morphism.**
- for each pair of morphisms f, g with $t(f) = s(g)$, there exists a specified morphism $g \circ f$, called the **composite morphism**, such that $s(g \circ f) = s(f)$ and $t(g \circ f) = t(g)$. More graphically,

$$X \xrightarrow{f} Y \xrightarrow{g} Z. \qquad (X \xrightarrow{g \circ f} Z)$$

These structures need to satisfy the following axioms:

- **Unitality**: For every morphism $f : X \to Y$, the compositions $f \circ \mathrm{id}_X$ and $\mathrm{id}_Y \circ f$ are both equal to f.
- **Associativity**: For $f : X \to Y$, $g : Y \to Z$, and $h : Z \to W$, the compositions $h \circ (g \circ f)$ and $(h \circ g) \circ f$ are equal.

A category is a very general structure: its objects and morphisms can be anything, provided that they satisfy the properties given above. In the following, we give some examples of categories arising from standard mathematical practice.

Warning. The composite $g \circ f$ means *first* applying f, then g:

$$X \xrightarrow{f} Y \xrightarrow{g} Z.$$

This choice of order, which is unfortunately the opposite of what one sees in diagrams, comes from traditional mathematics. In expressions such as

$$\sin(\cos(x)),$$

one must first take the cosine and then the sine. A way to avoid confusion is to read $g \circ f$ as "*g after f*."

Here is a depiction of associativity: Given three composable morphisms

$$X \xrightarrow{f} Y \xrightarrow{g} Z \xrightarrow{h} W,$$

we have *a priori* two ways to compose them. We could first compose f and g,

$$X \xrightarrow{f} Y \xrightarrow{g} Z \xrightarrow{h} W,$$
$$\underbrace{}_{g \circ f}$$

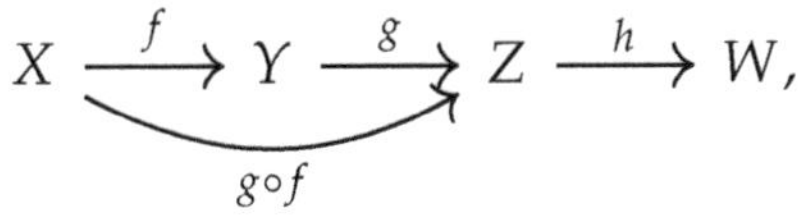

and then compose the resulting $g \circ f$ with h to obtain

$$X \xrightarrow{f} Y \xrightarrow{g} Z. \xrightarrow{h} W.$$
$$\underbrace{}_{g \circ f}$$
$$\underbrace{}_{h \circ (g \circ f)}$$

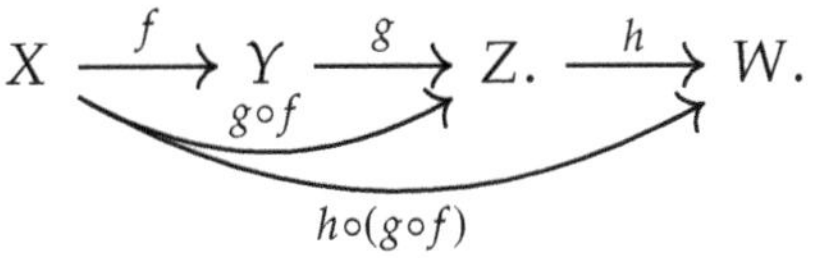

Alternatively, we could first compose g with h,

$$X \xrightarrow{f} Y \xrightarrow{g} Z \xrightarrow{h} W,$$
$$\underbrace{}_{h \circ g}$$

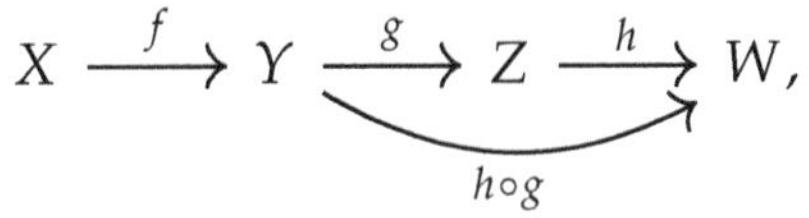

and then compose f with the resulting $h \circ g$ to obtain

$$X \xrightarrow{f} Y \xrightarrow{g} Z \xrightarrow{h} W.$$
$$\underbrace{}_{h \circ g}$$
$$\underbrace{}_{(h \circ g) \circ f}$$

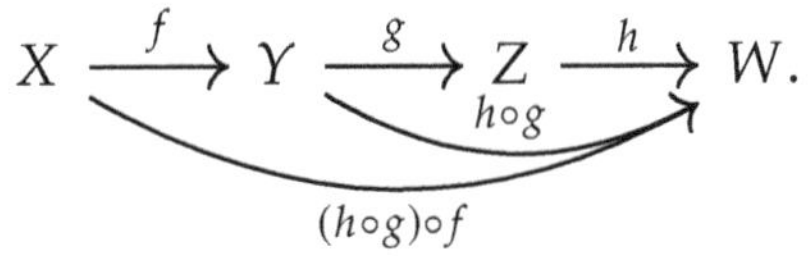

The associativity condition then states that the two methods give the same result, i.e. $h \circ (g \circ f) = (h \circ g) \circ f$. We can therefore write the triple composition without brackets, $h \circ g \circ f$, without any ambiguity.

An intuitive approach to understanding a category is as follows.

1.1.1 Categories defined by relations

Here is the first intuitive idea of what a category is.

Idea. *A category can look like a collection of objects which are related to each other in a consistent way.*

Let's see what this means by looking at examples.

Example 1.1.2 (sets and relations). An **equivalence relation** on a set X is a relation $\sim$ satisfying the following properties:

(a) **Reflexivity**: For each $x \in X$, we have $x \sim x$.

(b) **Transitivity**: For each $x, y, z \in X$, if $x \sim y$ and $y \sim z$, then $x \sim z$.

(c) **Symmetry**: For each $x, y \in X$, if $x \sim y$, then $y \sim x$ as well.

An equivalence relation is a mathematical way of formalizing the idea of "having something in common," such as "having the same shape." For example, here is what we obtain if, in a set of shapes, we draw an arrow whenever we want to say "has the same shape as" (regardless of color and size):

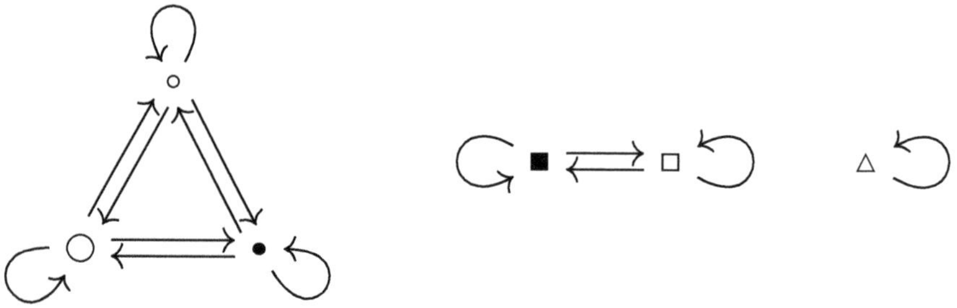

We note that:

(a) every x has the same shape as itself;

(b) if x has the same shape as y and y has the same shape as z, then x has the same shape as z;

(c) if x has the same shape as y, then y has the same shape as x.

An equivalence relation defines a category in the following way. It intuitively looks like the picture above:

- The objects are the elements of X.
- Given $x, y \in X$, there exists a unique morphism $x \to y$ if and only if $x \sim y$, and none otherwise.

The identity at an object x is the unique arrow $x \to x$ given by $x \sim x$. We know there is one due to reflexivity, and by definition, there is only one, so we don't have to worry about *which* arrow $x \longrightarrow x$ is the distinguished one. Similarly, the composition is given by transitivity: If we have arrows $x \longrightarrow y$ and $y \longrightarrow z$, it means that $x \sim y$ and $y \sim z$, and so by transitivity, $x \sim z$ as well, which means that there is an arrow $x \longrightarrow z$. Again, we don't have to worry about which $x \longrightarrow z$ arrow we want, since there is only one. The symmetry property states that if there is an arrow $x \longrightarrow y$, then there is an arrow $y \longrightarrow x$ as well. In other words, "we can always go back." This last property is not important to obtain a category; it is merely an extra property that this category has.

Example 1.1.3 (sets and relations). An **order relation** on a set X is a relation $\leq$ satisfying the following axioms:

(a) **Reflexivity**: For each $x \in X$ we have $x \leq x$.

(b) **Transitivity**: For each $x, y, z \in X$, if $x \leq y$ and $y \leq z$, then $x \leq z$.

(c) **Antisymmetry**: For each $x, y \in X$, if $x \leq y$ and $y \leq x$, then necessarily $x = y$.

We call a partially ordered set a **poset**. An order relation is a mathematical way of formalizing the idea of "comparing sizes," such as "being smaller or equal." Explicitly, the properties are as follows:

(a) Every x is smaller or equal to itself.

(b) If x is smaller or equal to y and y is smaller or equal to z, then x is smaller or equal to z.

(c) If x is smaller or equal to y and y is smaller or equal to x, then $x = y$.

An order relation defines a category analogously to what happens with equivalence relations: Objects are elements of X, and there is a unique arrow $x \longrightarrow y$ if and only if $x \leq y$. Again, reflexivity and transitivity are sufficient to have a category structure; antisymmetry is an additional property.

The most general relation which gives rise to a category is called a preorder, and it has no conditions on symmetry.

Example 1.1.4 (sets and relations, economics). A **preorder relation** on a set X is a relation $\lesssim$ satisfying only the following axioms:

(a) **Reflexivity**: For each $x \in X$, we have $x \lesssim x$.

(b) **Transitivity**: For each $x, y, z \in X$, if $x \lesssim y$ and $y \lesssim z$, then $x \lesssim z$.

Again, a preorder defines a category where there is a unique morphism between x and y if and only if $x \lesssim y$, with identities and composition given by reflexivity and transitivity. Equivalence and order relations are both special cases of preorder relations. A nontrivial example of a preorder is the one induced by *price* in economics, i.e. the relation of being "cheaper or equally priced." For example:

(a) x has lower or equal price to itself;

(b) if x has lower or equal price to y and y has lower or equal price to z, then x has lower or equal price to z.

However, differently from a partial order, if x has lower or equal price to y and y has lower or equal price to x, it does not necessarily imply that $x = y$: They will simply have the same price. They are not the same object but are equivalent. You can trade one for the other one, and vice versa.

Exercise 1.1.5 (sets and relations). Let $(X, \lesssim)$ be a preorder. Prove that the relation $\sim$ given by

$$x \sim y \quad \text{if and only if} \quad x \lesssim y \text{ and } y \lesssim x$$

is an equivalence relation.

For all these examples, given two objects, there is at most one morphism between them. This is not the case in general, and usually, the *choice* of the morphism matters.

1.1.2 Categories defined by operations

Here is another way to think about categories.

Idea. *A category can be viewed as a collection of operations which can be composed in a consistent manner.*

Here are some examples of categories where this interpretation is most helpful. We encourage the readers unfamiliar with group theory *not* to skip the following examples. Once one gets the basic idea of a group, many concepts in category theory are much easier to understand.

Example 1.1.6 (group theory). A **group** is a nonempty set G consisting of:

(a) a distinguished element $1 \in G$ called the **neutral element**, or **unit**;

(b) a binary operation $G \times G \to G$ called **multiplication**, which we denote by $(g, h) \mapsto g \cdot h$;

(c) for each element $g \in G$, an element g^{-1} called the **inverse**;

such that the following properties hold:

(a) **Unitality**: For each $g \in G$, the multiplications $g \cdot 1$ and $1 \cdot g$ are both equal to g.

(b) **Associativity**: For each $g, h, i \in G$, the multiplications $(g \cdot h) \cdot i$ and $g \cdot (h \cdot i)$ are equal.

(c) **Inverse law**: For each $g \in G$, the multiplications $g \cdot g^{-1}$ and $g^{-1} \cdot g$ are both equal to 1.

A group in mathematics is used to model the *symmetries* of some structure, namely the ways in which we can act on some object while keeping it "the same." For example, rotations of the plane form a group, as do permutations of a set. A group is similar to a category: There is the neutral element 1, which behaves like an identity, and multiplication, which behaves like composition (it is even associative). How can we then turn a group into a category? An important feature of a group, G, is that *we can compose any two elements of* G. There is no requirement of "matching source and target," as for a category. Therefore, in order to view a group as a particular category, we should make sure that the source and target of all morphisms always match. The only way to do this in general is to have *a unique object*. Here is the construction in detail.

Definition 1.1.7. Let G be a group. The **delooping** of G, denoted by $\mathbf{B}G$, is the following category:

- There is a single object $\bullet$.
- There is a morphism $\bullet \to \bullet$, for each element $g \in G$. We denote the morphism with the same letter, such as $g : \bullet \to \bullet$.
- The identity of the object $\bullet$ is the morphism given by $1 \in G$.
- The composition is given by the multiplication of G. That is, the composition $g \circ h$ of the morphisms given by $g, h \in G$ is the morphism given by the element $g \cdot h$ of G.

Since associativity and unitality hold for the elements of G, they also hold for the morphisms of $\mathbf{B}G$. Therefore, $\mathbf{B}G$ is a category.

Graphically, $\mathbf{B}G$ looks like a point with loops, with one special loop (the identity):

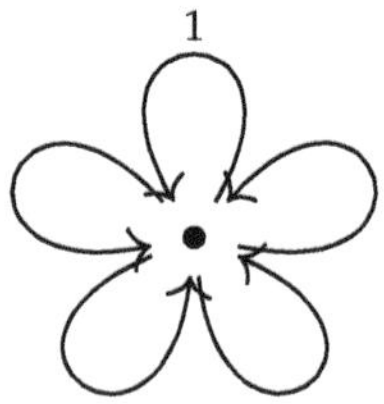

The term "delooping" comes from algebraic topology. There is a deep connection between algebraic topology and category theory, which is widely explored on the nLab,[1] and the idea of delooping is

[1] http://ncatlab.org.

part of this connection. For us, however, it will be merely a way of obtaining a category from a group (or monoid, see the following).

Remark 1.1.8. The category $\mathbf{B}G$ contains the same information as the group G and is simply a different way to express it. That is why some authors prefer omitting the symbol $\mathbf{B}$ and simply say that "a group is a certain category with a single object." In this text, we maintain a distinction between the two notions. Remember, however, that both encode the same data.

Remark 1.1.9. Since a group is usually modeling the "symmetries of something," sometimes we can make this delooping construction more concrete. For example, for the group of rotations of $\mathbb{R}^2$, we can take the object $\bullet$ to be $\mathbb{R}^2$ and the arrows to be rotations $\mathbb{R}^2 \to \mathbb{R}^2$ (as linear maps). This will be made precise later. However, for now, the delooping is purely formal.

In order to form the category $\mathbf{B}G$ from the group G, we never used the inverse elements and the inverse law. Similar to the symmetry property of equivalence relations, these are not needed to have a category and merely give to $\mathbf{B}G$ an additional structure, namely an "inverse" morphism g^{-1} to each morphism g (more on this in Section 1.1.5). Therefore, if we drop the inverse requirement, we can still get a category.

Example 1.1.10. A **monoid** is a nonempty set M consisting of:

(a) a distinguished element $1 \in M$ called the **neutral element** or **unit**;

(b) a binary operation $M \times M \to M$ called **multiplication**, which we denote by $(g, h) \mapsto g \cdot h$;

such that the following properties hold:

(a) **Unitality**: For each $g \in M$, the multiplications $g \cdot 1$ and $1 \cdot g$ are both equal to g.

(b) **Associativity**: For each $g, h, i \in M$, the multiplications $(g \cdot h) \cdot i$ and $g \cdot (h \cdot i)$ are equal.

The **delooping** of a monoid M is the category $\mathbf{B}M$ with, similar to groups, a single object $\bullet$, and it has as morphisms the elements of M with their identity and composition.

Just as a group is used to model the symmetries of some object, a monoid is used to model the transformations of an object which are not necessarily invertible. For example, consider instead of only the rotations of the plane $\mathbb{R}^2$, the set of *all linear maps* $\mathbb{R}^2 \to \mathbb{R}^2$, including the one that maps everything to $(0,0)$. These maps include the identity and can be composed, so they form a monoid. However, not all of them are invertible, so they don't form a group. There are many examples similar to this; a few of them are as follows:

- Given a vector space V, all linear maps $V \to V$ form a monoid.
- Given a topological space X, all continuous maps $X \to X$ form a monoid.
- Given a set X, all functions $X \to X$ form a monoid.

Here is another example from probability theory. Given a measurable space X, a **Markov kernel** on X is an assignment that takes points x of X and measurable subsets $B \subseteq X$ to numbers $k(B|x) \in [0,1]$ in such a way that k is measurable in x, for all B, and that it is a probability measure in B, for all x. We can view $k(B|x) \in [0,1]$ as a "probability of transitioning" into B if we start at point x. Two Markov kernels k and h can be composed as follows:

$$(h \circ k)(B|x) := \int_X h(B|x')\, k(dx'|x).$$

Now:

- Given a measurable space X and a Markov kernel $k : X \to X$, the set given by the identity id_X and the repeated applications of k, i.e. the set

$$\{\mathrm{id}_X,\ k,\ k^2,\ k^3,\ \dots\}$$

is a monoid. Sometimes, this is called the **Markov semigroup** generated by k.

- More generally, given a measurable space X, all Markov kernels $X \to X$ form a monoid. Sometimes, this is called the **full Markov semigroup on** X. (Can you write the identity as a Markov kernel?)

Remark 1.1.11. In fields such as probability theory, people sometimes prefer the notion of "semigroup" to the one of monoid. A semigroup is similar to a monoid, but without the unit (and the unitality property). In category theory, by convention, we always use monoids. The difference is minimal: One can almost always add the unit since it corresponds to "doing nothing," and it is largely only a matter of convention. (Requiring the identity is analogous to including the zero in the set of natural numbers.)

We conclude this part with an exercise.

> **Exercise 1.1.12 (Important!).** Prove that, in a group or in a monoid, the neutral element is unique. In other words, if both 1 and 1′ satisfy the properties of the neutral element, then $1 = 1'$.

1.1.3 Categories defined by spaces and maps with extra structure

In the case of categories defined by equivalence relations, orders, and preorders, we don't have to worry about the equality of composite arrows (for example to check associativity) since between any two objects, there is at most one arrow. In categories that arise as deloopings of groups or monoids, we don't have to worry about matching the source and target of arrows since there is only a single object. In the most general case, one needs to check both. Here is another idea, which is helpful to study the categories of this more general kind.

Idea. *A category can be viewed as a collection of sets or spaces equipped with extra structure and maps between them which are compatible with that structure.*

Categories of this form are usually named after their objects.

Example 1.1.13 (sets and relations). The category **Set** is the category whose objects are sets and whose morphisms are maps (functions) between them.

For each set X, the identity function is a function $X \to X$. Functions can be composed, and the composition is associative. Therefore, **Set** is a category.

Example 1.1.14 (several fields). Analogously to the category **Set**, the following are categories:

- **Top** has topological spaces as objects and continuous maps as morphisms.
- **Mfd** has smooth manifolds as objects and smooth maps as morphisms.
- **Vect** has real vector spaces as objects and linear maps as morphisms (one can choose any other field, such as $\mathbb{C}$, as long as it is kept fixed).
- **Meas** has measurable spaces as objects, and measurable maps as morphisms.
- **Grp** has groups as objects and group homomorphisms as morphisms, i.e. maps $f : G \to H$, such that for each $g, g' \in G$, we have $f(g \cdot g') = f(g) \cdot f(g')$.
- **Poset** has partially ordered sets as objects and monotone maps as morphisms, i.e. maps $f : X \to Y$, such that for each $x, x' \in X$ with $x \leq x'$ in X, we have $f(x) \leq f(x')$ in Y.

A category with sets as objects may have many choices of morphisms; for example, instead of functions, we may choose injective functions, bijections, or even relations. The choice of morphisms reflects the choice of *context* that we want to consider, or the choice of *structure* that we want to preserve (and study).

Example 1.1.15 (graph theory). For graphs, there are many choices of morphisms, depending on what one wants to do with them (graph theory is so general and versatile that many choices are meaningful). For example, for undirected, unweighted graphs, a possible choice of morphisms $f : G \to H$ is functions between the sets of vertices which preserve the adjacency relation: If the vertices x and y are connected by an edge in G, then $f(x)$ and $f(y)$ must be connected by an edge in H. With this choice of morphisms, the graphs and these morphisms form a category. This is, however, only one of many possible choices.

Exercise 1.1.16. In your field of mathematics (or physics, chemistry, computer science, economics, etc.), which structures do you work with the most? Can you construct a category of such structures? Are those structures better represented as objects or as morphisms of a category?

Here is also an example of what is *not* a category because composition fails.

Example 1.1.17 (calculus). Consider the following convex functions on $\mathbb{R}$:

- $x \mapsto f(x) := x^2$;
- $x \mapsto g(x) := x^2 - 1$.

Both are even strictly convex (their graphs are parabolas, see the following picture). However, the composite $f \circ g : \mathbb{R} \to \mathbb{R}$ is not convex: We have

$$f(g(x)) = (x^2 - 1)^2,$$

which is equal to 0 for $x = \pm 1$ and equal to 1 for $x = 0$.

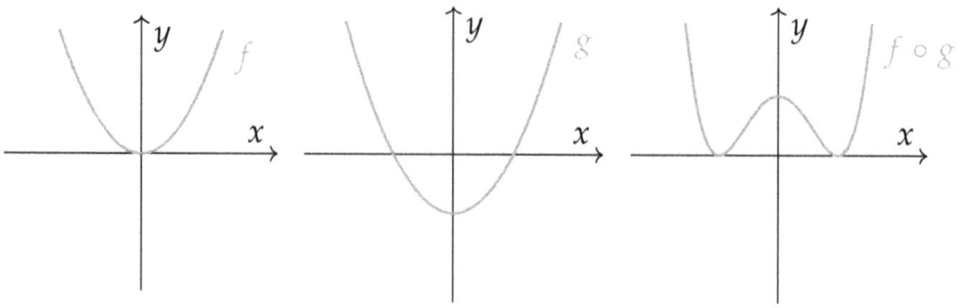

Therefore, convex functions on $\mathbb{R}$ do not form a category. The same is true for lower semicontinuous maps.

Exercise 1.1.18 (analysis). Show that there exists two lower semicontinuous maps $f, g : \mathbb{R} \to \mathbb{R}$ such that $f \circ g$ is not lower semicontinuous.

Therefore, whenever you define a category, check that all the axioms hold. Not everything is automatically a category.

Exercise 1.1.19 (analysis). Can we impose an additional requirement on convex maps in such a way as to make them closed under composition? Will they form a category that way? What about lower semicontinuous maps?

1.1.4 Set-theoretical considerations

The content of this section is a bit technical, and it will not be really needed to understand the rest of the text. It should nevertheless be mentioned in order to have a rigorous treatment and to avoid possible confusion. Moreover, the notation given in Definition 1.1.23 will be used later.

You may know from your set theory background that constructions such as a "set of all sets" often lead to problems. Therefore, if we want a "category of all sets," the objects of this category cannot form a set. That's why, in the definition of a category (Definition 1.1.1), we mentioned *collections* rather than sets: The objects and morphisms may in general be proper classes.

Definition 1.1.20. A category $\mathbf{C}$ is called **small** if $\mathbf{C}_0$ and $\mathbf{C}_1$ are sets.

A category $\mathbf{C}$ is called **locally small** if for every two objects X and Y of $\mathbf{C}$, the morphisms $X \to Y$ form a set.

Alternatively, one may refer to universes and **small and large sets** instead of sets and proper classes. In this formalism, a category is small if $\mathbf{C}_0$ and $\mathbf{C}_1$ are small sets, and locally small categories are defined analogously.

Most categories of interest are locally small. In particular, all the examples of categories analyzed so far are locally small. By construction, moreover, the categories obtained from relations on a set or as delooping of a monoid are small. You can prove converses of this last statement in the following exercise.

Exercise 1.1.21 (sets and relations). Prove that a small category with at most a single arrow between any two objects is a preorder. (*Hint*: There is not much to prove.)

Exercise 1.1.22. Similarly, prove that a locally small category with a single object is a monoid.

We conclude this section with a useful piece of notation, which we will use in the rest of this book.

Definition 1.1.23. Let X and Y be objects of a locally small category $\mathbf{C}$. The **hom-set**, or **hom-space**, of X and Y is the set of morphisms of $\mathbf{C}$ from X to Y. We denote it by $\mathrm{Hom}_{\mathbf{C}}(X, Y)$.

Hereinafter, we mostly work with locally small categories.

Warning. Sometimes, a mathematical structure can be an object of different categories. For example $\mathbb{R}$ and $\mathbb{R}^2$ are sets, topological spaces, vector spaces, and so on. That is why, it is important to keep track of the category when we write the hom-sets. For example:

- $\mathrm{Hom}_{\mathbf{Set}}(\mathbb{R}, \mathbb{R}^2)$ is the set of *all* functions $\mathbb{R} \to \mathbb{R}^2$, i.e. the morphisms of **Set**;

- $\mathrm{Hom}_{\mathbf{Top}}(\mathbb{R}, \mathbb{R}^2)$ is the set of all *continuous* functions $\mathbb{R} \to \mathbb{R}^2$, i.e. the morphisms of **Top**;
- $\mathrm{Hom}_{\mathbf{Vect}}(\mathbb{R}, \mathbb{R}^2)$ is the set of all *linear* functions $\mathbb{R} \to \mathbb{R}^2$.

If you want to learn more about the set-theoretical issues in category theory, you can read the notes by Mike Shulman [Shu08]. However, I recommend you read them after learning a bit more about category theory.

1.1.5 Isomorphisms and groupoids

For the case of equivalence relations and for delooping of groups, we had categories with a special property, which intuitively said that "we can always go back." Let's now try to make this intuition precise.

Definition 1.1.24. Let X and Y be objects in a category **C**. An **isomorphism** is a pair of morphisms

$$X \underset{g}{\overset{f}{\rightleftarrows}} Y$$

such that

(a) $g \circ f = \mathrm{id}_X$,

(b) $f \circ g = \mathrm{id}_Y$.

If there exists an isomorphism between X and Y, we say that X and Y are **isomorphic**. We also say that f is the **inverse** of g, and g is the **inverse** of f.

We see in this definition that g *not only* is in the opposite direction as f but also "undoes f," in the sense that applying g after f is like doing nothing at all. Similarly, f undoes g.

Warning. The two conditions do not imply one another, as the following example shows. Therefore, in order to have an isomorphism, both conditions need to be satisfied separately.

Example 1.1.25 (calculus, linear algebra). Let $X = \mathbb{R}^2$ and $Y = \mathbb{R}$. Let $f : \mathbb{R}^2 \to \mathbb{R}$ be the projection onto the x-axis $(x, y) \mapsto x$, and let $g : \mathbb{R} \to \mathbb{R}^2$ be the inclusion of the x-axis $x \mapsto (x, 0)$. Then, $f \circ g = \mathrm{id}_{\mathbb{R}}$, but $g \circ f \neq \mathrm{id}_{\mathbb{R}^2}$: In particular, $(g \circ f)(x, y) = g(x) = (x, 0)$, which is in general different from (x, y).

> **Exercise 1.1.26 (important!).** Let $f : X \to Y$ be a morphism in a category. Suppose that both g and $g' : Y \to X$ are inverse to f. Prove that $g = g'$. Conclude that, in particular, inverses in a group are unique.

Very often, when one speaks of an isomorphism, one does not mean a pair (f, g) as in Definition 1.1.24, but rather only one of the two maps. By the exercise above, if the other map exists, it is uniquely determined. Therefore, often, we only write one of the two maps and say that (for example) $f : X \to Y$ is an isomorphism. The other map is implied.

Here are some examples of isomorphisms: (If it is not clear why these are the isomorphisms of the respective categories, try to prove it yourself.)

Example 1.1.27 (several fields).

- In the category **Set**, the isomorphisms are the bijective functions.
- In **Top**, the isomorphisms are the homeomorphisms: invertible continuous maps whose inverse is continuous. Note that this is strictly stronger than saying continuous and bijective (can you give an example illustrating this?).
- In **Mfd**, the isomorphisms are the diffeomorphisms.
- In **Vect**, the morphisms are bijective linear maps.

> **Exercise 1.1.28 (important!).** Prove that the composition of invertible morphisms is invertible.

Let's now generalize equivalence relations and groups, i.e. categories where "we can always go back."

Definition 1.1.29. A **groupoid** is a category where all the morphisms are invertible.

Here are some examples of groupoids.

Example 1.1.30 (sets and relations). We have already seen equivalence relations. For an equivalence relation $\sim$ on a set X, we have that if $x \sim y$, i.e. an arrow $x \to y$, then $y \sim x$, i.e. an arrow $y \to x$. Checking that the two arrows invert each other is trivial and guaranteed by uniqueness. In detail, the composition $x \to y \to x$ has to be equal to the identity $x \to x$ because there is at most one arrow $x \to x$.

Example 1.1.31 (group theory). The delooping $\mathbf{B}G$ of a group G is a groupoid (this is why it's called a "groupoid": it generalizes a group). The inverses are given by the inverses in G.

Another example of a groupoid is the groupoid of sets and *bijections* since those are exactly the invertible morphisms. We can extend this more generally.

Definition 1.1.32. Let $\mathbf{C}$ be a category. The **core** of $\mathbf{C}$ is the groupoid whose objects are the objects of C and whose morphisms are the isomorphisms of C.

We have seen another example of a core already: In Exercise 1.1.5, we canonically obtained an equivalence relation from a preorder. Therefore, the core of a preorder is an equivalence relation.

We conclude this section with a "philosophical" remark. In category theory, one usually never talks about two objects being *equal*, but only *isomorphic*. In some sense, this reflects standard mathematical practice. For example, as vector spaces, every three-dimensional space is isomorphic to $\mathbb{R}^3$. But are they really *equal*? What would equality even mean in that context? What can be done in $\mathbb{R}^3$ can be done in any three-dimensional vector space, and that's what matters in the end. Therefore, isomorphism between objects is a more important notion than equality. When it comes to *morphisms*, on the other hand, two morphisms are often required to be *equal*. For example, in the definition of an isomorphism, $f^{-1} \circ f$ is *equal* to the identity id_X, not "isomorphic." Again, this reflects standard mathematical practice: Functions between spaces can be equal. For example, two functions $f, g : X \to Y$ are equal if and only if for every $x \in X$, we have $f(x) = g(x)$.

One may be tempted to ask whether it is possible to talk about *isomorphisms of maps* instead of equalities so that objects and morphisms are treated more on an equal footing. This is exactly the subject of *higher category theory*, in which one can talk about morphisms between morphisms, and so on. Higher category theory is very helpful, for example, in algebraic geometry and algebraic topology. The nLab[2] is the standard online wiki on this subject. In this book, for the most part, we consider *ordinary* category theory, where morphisms are allowed to be equal.

1.1.6 Diagrams: An informal definition

One of the most powerful methods of category theory is *reasoning in terms of diagrams*. We start with an informal (but consistent) definition of a diagram. A more formal definition will be given later on in terms of functors.[3]

Definition 1.1.33 (informal). A **diagram** in a category **C** is a directed (multi-)graph formed using objects and arrows of **C** such that:

- each object and morphism may appear more than once in the diagram;
- between any two objects, there may also be more than one morphism;
- for each object X in the diagram, the identity is implicitly present in the diagram (but generally not drawn);

[2] http://ncatlab.org.
[3] See Section 1.4.5.

- for each composable edges (arrows) $f : X \to Y$ and $g : Y \to Z$ which are appearing head-to-tail in the diagram, the composite $g \circ f : X \to Z$ is implicitly present in the diagram (but generally not drawn).

Since objects and morphisms may appear more than once in the diagram as different vertices and edges, we refer to vertices and edges to avoid ambiguity.

Exercise 1.1.34. Which morphisms are included in the following diagrams, implicitly or explicitly (without assuming any additional equalities besides those required by our axioms)?

$$X \xrightarrow{f} Y \qquad \begin{array}{ccc} X & \xrightarrow{f} & Y \\ & {\scriptstyle g\circ f}\searrow & \downarrow{\scriptstyle g} \\ & & Z \end{array} \qquad \begin{array}{ccc} X & \xrightarrow{f} & Y \\ & {\scriptstyle h}\searrow & \downarrow{\scriptstyle g} \\ & & Z \end{array} \qquad \begin{array}{ccc} X & \xrightarrow{f} & Y \\ & {\scriptstyle \mathrm{id}_X}\searrow & \downarrow{\scriptstyle g} \\ & & X \end{array}$$

$$X \underset{g}{\overset{f}{\rightleftarrows}} Y \qquad X \underset{f^{-1}}{\overset{f}{\rightleftarrows}} Y \qquad X \xrightarrow{f} X \qquad \begin{array}{ccc} X & \xrightarrow{f} & Y \\ \downarrow{\scriptstyle k} & & \downarrow{\scriptstyle h} \\ A & \xrightarrow{g} & B. \end{array}$$

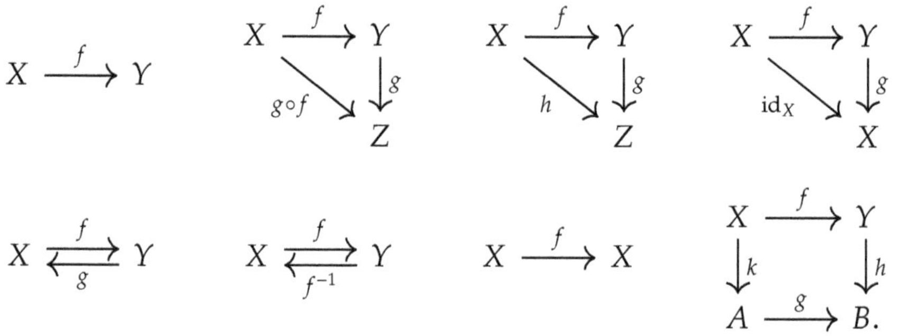

Definition 1.1.35. A diagram is **commutative** (or **commutes**) if for each two vertices X and Y in the diagram, all compositions along paths of composable arrows connecting X to Y are equal.

Warning. Not every diagram commutes!

Example 1.1.36. The diagram

$$\begin{array}{ccc} X & \xrightarrow{f} & Y \\ & {\scriptstyle h}\searrow & \downarrow{\scriptstyle g} \\ & & Z \end{array}$$

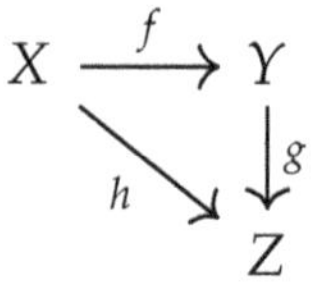

is commutative if and only if $g \circ f = h$. Similarly, the diagram

$$\begin{array}{ccc} X & \xrightarrow{f} & Y \\ \downarrow{\scriptstyle k} & & \downarrow{\scriptstyle h} \\ A & \xrightarrow{g} & B \end{array}$$

is commutative if and only if $h \circ f = g \circ k$.

Exercise 1.1.37. Which diagrams in Exercise 1.1.34 commute? For those which do not commute, which morphisms should we require to be equal in order for the diagrams to become commutative?

Exercise 1.1.38 (sets and relations). Prove that in a preorder (or in a poset, or in an equivalence relation), every diagram commutes.

Commutative diagrams can be pasted, as the following exercise shows.

Exercise 1.1.39. Consider the following diagram, which is drawn using two adjacent squares:

$$\begin{array}{ccccc} X & \xrightarrow{f} & Y & \xrightarrow{g} & Z \\ \downarrow{\scriptstyle k} & & \downarrow{\scriptstyle h} & & \downarrow{\scriptstyle \ell} \\ A & \xrightarrow{p} & B & \xrightarrow{q} & C. \end{array}$$

Suppose that the left and right squares separately commute. Prove that the whole diagram then commutes.

More generally, commutative diagrams of any shape can be pasted to give again commutative diagrams.

Recall the definition of an isomorphism (Definition 1.1.24). We can rewrite it purely in terms of commutative diagrams.

Definition 1.1.40 (alternative). Two morphisms $f : X \to Y$ and $g : Y \to X$ of a category $\mathbf{C}$ are inverse to each other, forming an isomorphism, if and only if the following two diagrams are commutative.

$$\begin{array}{ccc} X & \xrightarrow{f} & Y \\ & \overset{\mathrm{id}_X}{\searrow} & \downarrow{\scriptstyle g} \\ & & X \end{array} \qquad \begin{array}{ccc} Y & \xrightarrow{g} & X \\ & \overset{\mathrm{id}_Y}{\searrow} & \downarrow{\scriptstyle f} \\ & & Y. \end{array}$$

Again, keep in mind that both diagrams need to be checked separately, as the two conditions are independent.

How to draw diagrams in LaTeX: All the diagrams in this document have been drawn using the package `tikzcd`, which is based on PGF/TikZ. Today, there is also a more graphical and interactive web applet called *Quiver*, developed by Nathanael Arkor, which is compatible with `tikzcd` and freely available for use.[4]

1.1.7 The opposite category

Another useful technique in category theory is "reversing all the arrows."

Definition 1.1.41. Let C be a category. The **opposite category** of **C**, denoted as $\mathbf{C}^{\mathrm{op}}$, is the category obtained as follows:

- Objects are just the objects of **C**.
- A morphism, denoted as f^{op}, between two objects X and Y is a morphism f in **C** from Y to X (note the direction). Graphically,

$$\begin{array}{c} X \\ \downarrow{\scriptstyle f^{\mathrm{op}}} \\ Y \end{array} \quad \leftrightsquigarrow \quad \begin{array}{c} X \\ \uparrow{\scriptstyle f} \\ Y. \end{array}$$

- The identity at each object X is given by ${\mathrm{id}_X}^{\mathrm{op}}$.

[4] https://q.uiver.app/.

- Composition is the same as in **C**, but the order is reversed: $g^{\mathrm{op}} \circ f^{\mathrm{op}}$ is defined to be $(f \circ g)^{\mathrm{op}}$. In diagrams,

$$\begin{array}{ccc} X & & X \\ \big\downarrow{\scriptstyle f^{\mathrm{op}}} & & \big\uparrow{\scriptstyle f} \\ Y & \leftrightsquigarrow & Y \\ \big\downarrow{\scriptstyle g^{\mathrm{op}}} & & \big\uparrow{\scriptstyle g} \\ Z & & Z. \end{array}$$

Here are examples on how the opposite category looks like in practice.

Example 1.1.42 (sets and relations). Let $(X, \leq)$ be a poset, i.e. a set equipped with a partial order. As we have seen in Example 1.1.3, this is in particular a category with an arrow $x \to y$ if and only if $x \leq y$. The opposite category has then an arrow $x \to y$ if and only if $y \leq x$. We can consider it as the category obtained by the opposite relation: $\geq$ instead of $\leq$. For example, for $X = \mathbb{R}$ with its usual order, $(\mathbb{R}, \leq)^{\mathrm{op}}$ is $(\mathbb{R}, \geq)$, ordered downward.

Example 1.1.43 (group theory). Let G be a group. We have seen in Section 1.1.2 that the delooping $\mathbf{B}G$ is a category. The opposite category $(\mathbf{B}G)^{\mathrm{op}}$ will again have a single object, and morphisms will be again elements of G; however, the composition is reversed: $(g \circ h)^{\mathrm{op}} = h^{\mathrm{op}} \circ g^{\mathrm{op}}$.

Since the axioms of categories are symmetric with respect to reversing all arrows, the following is true: *A statement in a category* **C** *is true if and only if the dual statement is true in* $\mathbf{C}^{\mathrm{op}}$, i.e. the statement that we obtain by switching the source and target of all maps, and the order of composition.

Warning. A statement is true in **C** if and only if the dual statement is true in $\mathbf{C}^{\mathrm{op}}$; however, in general, the two statements do not appear the same.

This is why this principle is so powerful: For each result we prove in category theory, there is always a dual result that we obtain for free. Here is an example.

Exercise 1.1.44. Let $f, g : X \to Y$ be morphisms in $\mathbf{C}$. Prove that $f = g$ if and only if $f^{\mathrm{op}} = g^{\mathrm{op}}$ in $\mathbf{C}^{\mathrm{op}}$. (*Hint*: There is not much to prove.)

Given the result of the exercise, the following statements simply follow as corollaries.

Corollary 1.1.45. *A diagram in* $\mathbf{C}$ *commutes if and only if the dual diagram in* $\mathbf{C}^{\mathrm{op}}$ *commutes.*

As an example, the diagram

$$\begin{array}{ccc} X & \xrightarrow{f} & Y \\ \downarrow{\scriptstyle k} & & \downarrow{\scriptstyle h} \\ A & \xrightarrow{g} & B \end{array}$$

commutes if and only if $h \circ f = g \circ k$. This happens if and only if $f^{\mathrm{op}} \circ h^{\mathrm{op}} = k^{\mathrm{op}} \circ g^{\mathrm{op}}$, i.e. if the dual diagram commutes:

$$\begin{array}{ccc} X & \xleftarrow{f^{\mathrm{op}}} & Y \\ \uparrow{\scriptstyle k^{\mathrm{op}}} & & \uparrow{\scriptstyle h^{\mathrm{op}}} \\ A & \xleftarrow{g^{\mathrm{op}}} & B. \end{array}$$

Moreover, since isomorphisms can be defined purely in terms of commutative diagrams (Definition 1.1.40), we get the following.

Corollary 1.1.46. *A morphisms* $f : X \to Y$ *is invertible if and only if* $f^{\mathrm{op}} : Y \to X$ *is invertible. (What will the inverse be?)*

In particular, if X *and* Y *are isomorphic in* $\mathbf{C}$*, then they are also isomorphic in* $\mathbf{C}^{\mathrm{op}}$.

In the following section, we present more examples of how reversing the arrows can give meaningful statements.

1.2 Mono and Epi

So far, we have expressed in terms of diagrams the notion of isomorphism, which gives the idea of when two objects are, in some sense, "the same." This generalizes the notion of "bijective map" between sets and gives a proper notion of "sameness" in other categories (such as homeomorphisms between topological spaces). We would now like to present a diagrammatic approach to describing subspaces and quotients, generalizing injective and surjective maps. In category theory, these maps are called *monomorphisms* (from the Greek word meaning "one," as in "one-to-one") and *epimorphisms* (from the Greek word meaning "onto").

1.2.1 Monomorphisms

Definition 1.2.1. A morphism $m : X \to Y$ in a category $\mathbf{C}$ is called a **monomorphism**, or **mono**, or is said to be **monic**, if the following holds. For every object A of $\mathbf{C}$ and for every pair of maps $f, g : A \to X$, i.e. fitting into the diagram

$$A \underset{g}{\overset{f}{\rightrightarrows}} X \xrightarrow{m} Y,$$

and such that $m \circ f = m \circ g$, we have $f = g$ (i.e. the whole diagram commutes).

Intuitively, a monomorphism is a map which "does not map different things of X to the same thing in Y." Let's try to see what this means using the interpretation of categories in terms of spaces and maps. If A is a space and a is a point of A, then the image $f(a)$ of a through f will be a point of X which, in general, may be different from the image $g(a)$ along g. We now want to say that if the points $f(a)$ and $g(a)$ are different in X, then applying m to both will *keep them different* in Y. In other words, if $f(a) \neq g(a)$, then $m \circ f(a) \neq m \circ g(a)$. Equivalently, if the points are equal in Y, i.e. $m \circ f(a) = m \circ g(a)$, then they must have been equal already in X, i.e. $f(a) = g(a)$. This is what the definition implies. If this property holds for every a of A and every A, then the map m "behaves as if it did not identify different things of X," and we call it a monomorphism.

Here are some examples of monomorphism: (Can you prove that these morphisms are indeed the monomorphisms of their respective categories?)

Example 1.2.2 (several fields).

- In **Set**, the monomorphisms are the injective maps.
- In **Top**, the monomorphisms are the injective continuous maps.
- In the category **FVect** of finite-dimensional real vector spaces and linear maps, the monomorphisms are the injective linear maps.
- In the category **Grp**, the monomorphisms are the injective group homomorphisms.

Exercise 1.2.3 (several fields). Prove the above claims (or at least the ones related to a field of math which is familiar to you). *Hint*: A convenient choice of A may help.

In general, not every monomorphism looks like an injective map, as observed in the following examples.

Exercise 1.2.4 (sets and relations). Prove that in a preorder, every morphism is a monomorphism.

Exercise 1.2.5 (sets and relations; difficult!). What are the monomorphisms of $\mathbf{Set}^{\mathrm{op}}$?

(The answer to this question will hopefully become clear at the end of this section.)

From the examples given above, you may have noted something: Isomorphisms seem to be monomorphisms in all the categories seen so far. For example, in **Set**, every invertible function is in particular injective. This is always the case.

Proposition 1.2.6. *Every isomorphism of* **C** *is in particular a monomorphism of* **C**.

Before looking at the proof, try to do it yourself.

Proof. Let $m : X \to Y$ be an isomorphism with inverse m^{-1}. Let now A be any object of **C**, and let $f, g : A \to X$ be such that $m \circ f = m \circ g$. We can apply m^{-1} to both terms, and since $m^{-1} \circ m = \mathrm{id}_X$, we get

$$\begin{aligned} m^{-1} \circ m \circ f &= m^{-1} \circ m \circ g \\ f &= g. \end{aligned}$$

□

1.2.2 Split monomorphisms

You may have noted that, in the proof of Proposition 1.2.6, we only applied m^{-1} on the left of m, never on the right. In other words, we only used the *left* diagram of Definition 1.1.40, not the right one. Therefore, the proof above also works for maps which admit only a "left inverse." Here is what this means rigorously.

Definition 1.2.7. Let $m : X \to Y$ be a morphism of **C**. A **left inverse**, or **retraction**, of m is a map $r : Y \to X$ such that $r \circ m = \mathrm{id}_X$. Equivalently, r is such that the following diagram commutes:

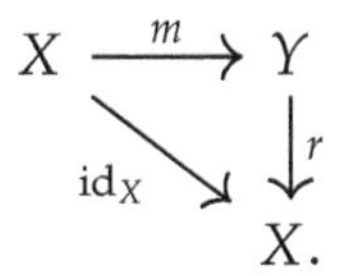

If m admits a left inverse, we call m a **split monomorphism** and r its **splitting**.

A split monomorphism is less than an isomorphism since an isomorphisms is required to have a *two-sided* inverse (there is an additional diagram that has to commute). In particular, we have the following.

Proposition 1.2.8. *Every isomorphism is a split monomorphism.*

Moreover, since the proof of Proposition 1.2.6 only requires left inverses, we have the following.

Proposition 1.2.9. *Every split monomorphism is a monomorphism.*

In general, the converse is not true. You may know from topology, or from graph theory, that there are injective maps which do not have a retraction. Let's see some examples in detail.

Example 1.2.10 (sets and relations, linear algebra).

- In **Set**, every injective map from a nonempty set has a left inverse. Therefore, every monomorphism of sets with nonempty domain is split.
- In **FVect**, every injective linear map has a left inverse. Therefore, every monomorphism is split.

Let's look now at the case of topological spaces.

Example 1.2.11 (geometry, topology). Let S^1 be a circle and D^2 be a two-dimensional disc in $\mathbb{R}^2$. Let $m : S^1 \to D^2$ be the embedding of the circle as the boundary of D^2. This map is injective and continuous; therefore, it is a monomorphism of **Top**. However, there is no retraction of m: Any such retraction $r : D^2 \to S^1$ would have to be such that $r \circ m = \mathrm{id}_{S_1}$; that is, it has to map the boundary of D^2 to the corresponding point of the circle S^1. This cannot be done in a continuous way: Where would r map the center of the disc? Any such assignment would have to "pierce a hole" somewhere in the disc. (We will see a formal proof of this fact in Example 1.3.38.)

Whenever a continuous map $m : X \to Y$ admits a continuous retraction $Y \to X$, we say that X is a **retract** of Y. As the previous example shows, this is a stronger condition than simply requiring m to be injective (but of course it is *necessary* that m is injective, by Proposition 1.2.9). Therefore, almost by definition, the split monomorphisms of **Top** are precisely the embeddings of retracts.

A similar phenomenon happens in the category of groups.

Exercise 1.2.12 (group theory). Prove that the inclusion $\mathbb{Z} \hookrightarrow \mathbb{R}$ is a monomorphism in **Grp**, but it is not split mono.

In a generic category, monomorphisms and split monomorphisms are not the same. This is something that may seem counterintuitive because they are almost the same in **Set**. But keep in mind that the generic case is what happens in **Top**.

Exercise 1.2.13 (graph theory). In the category of graphs, are all monomorphisms split? What about the category **Poset** of partial orders and monotone maps?

1.2.3 Epimorphisms

Let's now present the dual notion.

Definition 1.2.14. A morphism $e : X \to Y$ is called an **epimorphism**, or **epi**, if the following holds. For every object A of $\mathbf{C}$ and every $f, g : Y \to A$, sitting in the diagram

$$X \xrightarrow{e} Y \underset{g}{\overset{f}{\rightrightarrows}} A$$

and such that $f \circ e = g \circ e$, we have $f = g$ (i.e. the full diagram commutes).

Let's try to interpret this definition. An epimorphism, intuitively, is "something having full image." Again, if we consider objects and morphisms as spaces and maps, the definition means the following. Suppose that the image of e in Y is the whole of Y. Then, if any two maps f and g on Y agree on the image of e, they *must agree on the whole of* Y. That is, if $f \circ e = g \circ e$, then $f = g$. This is what the definition is saying. If this holds for each f, g into A and for each A, then e "behaves as if it had full image," and we call it an epimorphism.

This notion of epimorphism is exactly the dual notion to the notion of monomorphism.

Exercise 1.2.15. Prove that $f : X \to Y$ is epi in $\mathbf{C}$ if and only if $f^{\mathrm{op}} : Y \to X$ is mono in $\mathbf{C}^{\mathrm{op}}$.

Therefore, we get for free the dual statement to Proposition 1.2.6.

Proposition 1.2.16. *Every isomorphism is an epimorphism.*

This makes intuitive sense: For example, every bijection is surjective.

Exercise 1.2.17. You may not be comfortable yet with believing that Proposition 1.2.16 is true just because the dual statement is true. If that is the case, try to prove the proposition yourself from scratch, and note that the diagrams appearing in the proof are dual to those appearing in the proof of Proposition 1.2.6.

Here are some examples of epimorphisms.

Example 1.2.18 (sets and relations, linear algebra, topology).

- In **Set**, the epimorphisms are the surjective maps.
- In **FVect**, the epimorphisms are the surjective linear maps.
- In **Top**, the epimorphisms are the surjective continuous maps.

(Can you prove these statements?)

Again, not in every category epimorphisms look like surjective maps. For example, in $\mathbf{Set}^{\mathrm{op}}$, an epimorphism is the opposite of an injective map. Again, in a preorder, every arrow is an epimorphism. More interestingly, try to prove the following.

Exercise 1.2.19 (algebra). In the category **Mon** of monoids and monoid morphisms, the inclusion $\mathbb{N} \hookrightarrow \mathbb{Z}$ is an *epimorphism*.

1.2.4 Split epimorphisms

Definition 1.2.20. Let $e : X \to Y$ be a morphism of **C**. A **right inverse**, or **section**, of e is a map $s : Y \to X$ such that $e \circ s = \mathrm{id}_Y$. Equivalently, s is such that the following diagrams commutes:

$$\begin{array}{ccc} Y & \xrightarrow{s} & X \\ & {\scriptstyle \mathrm{id}_Y} \searrow & \downarrow{\scriptstyle e} \\ & & Y. \end{array}$$

If e admits a right inverse, we call e a **split epimorphism** and call s its **splitting.**

Again, this is the dual notion to the one of split monomorphism. Therefore, we get the following dual results.

Proposition 1.2.21. *Every isomorphism is a split epimorphism.*

Proposition 1.2.22. *Every split epimorphism is an epimorphism.*

Moreover, note the following: If s is a right inverse of e, then e is a left inverse of s. So, in particular, if e is split epi, then its section s is split mono, and vice versa. The pair of maps

$$X \underset{s}{\overset{e}{\rightleftarrows}} Y$$

is sometimes called a **section–retraction pair**, or it is said (generalizing the case of topological spaces) that Y is a **retract** of X.

This gives then an intuitive way to think about split monos and epis: *A retract is simultaneously a subspace and a quotient in a compatible way*. For example, $\mathbb{R}$ is simultaneously a subspace and a quotient of $\mathbb{R}^2$, both in **Vect** and **Top**. The notion is not symmetric: $\mathbb{R}^2$ is not a retract of $\mathbb{R}$ (intuitively, the retract is often "smaller" in some sense). In Example 1.2.11, we saw that the circle is a subspace of the disc, but not a quotient; therefore, it is not a retract. This should give us an idea of what to expect as examples of split epimorphisms.

Example 1.2.23 (sets and relations, linear algebra).

- In **Set**, every surjective map has a right inverse. Therefore, every epimorphism is split.[5]
- In **FVect**, every surjective linear map has a right inverse. Therefore, again, every epimorphism is split.

Example 1.2.24 (topology). Consider the half-open interval $[0,1)$ and the map $e : [0,1) \to S^1$, which closes the interval to a circle (mapping t to the pair $(\cos(2\pi\, t), \sin(2\pi\, t))$ if we consider the circle as a subset of $\mathbb{R}^2$). This map is surjective since it reaches the whole circle, so it is an epimorphism of **Top**. However, it is not split: Any splitting would have to map the circle back to the interval, and to do so, it would have to "break the circle open." This cannot be done in a continuous way.

In **Top**, the split epimorphisms are the retractions (or the quotients to a retract), and being a retraction is a strictly stronger requirement than just being a surjective map.

> **Exercise 1.2.25 (algebra).** Consider the map $\mathbb{R} \to S^1$ of **Grp** mapping all integers to the neutral element of S^1 (as time is mapped to a clock, or as $t \mapsto e^{2\pi i t}$ if we consider the circle as a subset of $\mathbb{C}$). This is surjective, so it is an epimorphism of **Grp**. Show that it does not admit a section.

Let's observe again the map e from Example 1.2.24. We have that $e : [0,1) \to S^1$ is even *bijective*. That is, it is mono and epi. But it is not an isomorphism; in particular, it is not split epi (or, more intuitively, the circle and the interval are different spaces). Therefore, *a map which is both epi and mono is not necessarily an isomorphism.*

[5] If you are familiar with set theory, try to show that the axiom of choice is equivalent to the statement that in **Set**, every epimorphism is split. If you are not familiar with set theory, you can safely ignore this comment.

However, the following is true.

Proposition 1.2.26. *If $f : X \to Y$ is epi and split mono, it is an isomorphism. Dually, if it is mono and split epi, it is an isomorphism.*

This is why, in **Set**, a map is invertible when it is injective and surjective: Those are not just mono and epi, they are also *split epi*. But this is not the general case.

Exercise 1.2.27 (important!). Prove Proposition 1.2.26. (*Hint*: You only need to prove one of the two statements, as the other one is dual.)

1.3 Functors

We now look at functors, which can be seen as a convenient choice of "arrows between categories."

Definition 1.3.1. Let **C** and **D** be categories. A **functor** $F : \mathbf{C} \to \mathbf{D}$ consists of the following data:

- for each object X of **C**, an object FX of **D**;
- for each morphism $f : X \to Y$ of **C**, a morphism $Ff : FX \to FY$ of **D** (note that the domain and codomain of Ff must be exactly FX and FY);

such that the following **functoriality axioms** hold:

- **Unitality** (or sometimes **normalization**): For every object X of **C**, $F(\mathrm{id}_X) = \mathrm{id}_{FX}$. That is, F maps identities into identities.
- **Compositionality** (or, sometimes, **cocycle condition**): For every pair of composable morphisms

$$X \xrightarrow{f} Y \xrightarrow{g} Z$$

in **C**, we have $F(g \circ f) = Fg \circ Ff$. That is, the following diagram must commute:

$$
\begin{array}{ccc}
 & FY & \\
\overset{Ff}{\nearrow} & & \overset{Fg}{\searrow} \\
FX & \xrightarrow[\;F(g \circ f)\;]{} & FZ.
\end{array}
$$

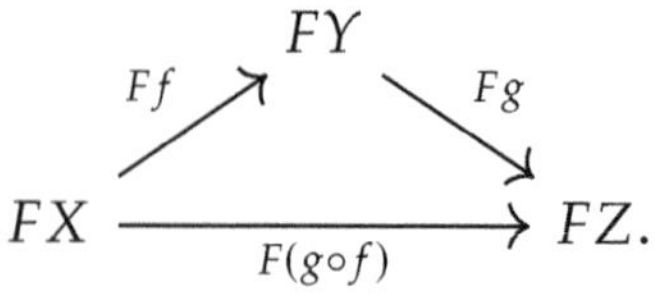

In other words, F respects the composition of arrows.

Note that in the expression $F(\mathrm{id}_X) = \mathrm{id}_{FX}$, we have on the left the identity of X in **C** and on the right the identity of FX in **D**. Similarly, in the expression $F(g \circ f) = Fg \circ Ff$, we have on the left the composition in **C** and on the right the composition in **D**.

In the following, we look at some ways to interpret functors between different types of categories.

1.3.1 Functors as mappings preserving relations

Idea. *A functor can look like a mapping preserving or respecting the relations between objects.*

Example 1.3.2 (sets and relations). Let $(X, \leq)$ and $(Y, \leq)$ be partial orders. As we have seen in Example 1.1.3, these are in particular categories. A functor $F : (X, \leq) \to (Y, \leq)$ consists first of all of a mapping between the objects, i.e. a function $F : X \to Y$. Moreover, to each arrow in X, there has to be a corresponding arrow of Y. That is, if $x \leq x'$ in X, then $Fx \leq Fx'$ in Y. The functoriality axioms are immediately satisfied since between any two objects, there is at most one morphism: The identity at $x \in X$ has to be mapped to the identity at $Fx \in Y$ since there is no other arrow $Fx \to Fx$, and the same is true for composition. Therefore, *a functor between partial orders is the same thing as a monotone map.* If x and x' are related (x is less or equal than x'), then their images under F, the elements Fx and Fx', must be related too (Fx is less than or equal to Fx').

Example 1.3.3 (sets and relations). Let $(X, \sim)$ and $(Y, \sim)$ be equivalence relations. As we have seen in Example 1.1.2, these are in particular categories. A functor $F : (X, \sim) \to (Y, \sim)$ is first of all a function $F : X \to Y$. Again, arrows of X have to be mapped into arrows of Y; that is, if $x \sim x'$ in X, then $Fx \sim Fx'$ in Y. Again, the functoriality axioms are guaranteed by uniqueness. Therefore, *a functor between equivalence relations is the same thing as a map respecting the equivalence.* If x and x' are equivalent in X, then they have to be mapped into equivalent elements of Y. This is sometimes called an **equivariant map** for the equivalence relations.

Warning. In the target category, there may be *more relations* than in the source category. In the partial order case, $F : (X, \leq) \to (Y, \leq)$, it could be that $x \nleq x'$, but still $Fx \leq Fx'$. Relations can be "created," with the important point being that they are not "destructed." Similarly, for $F : (X, \sim) \to (Y, \sim)$, it could be that $x \nsim x'$, but still $Fx \sim Fx'$. The important thing is that if $x \sim x'$, then $Fx \sim Fx'$. In the terminology of category theory, relations have to be *preserved*, but not necessarily *reflected*.

Exercise 1.3.4 (sets and relations). Let $(X, \sim)$ be an equivalence relation. Let $X/\sim$ be the **quotient space** of the relation, i.e. the set of equivalence classes. We can consider $X/\sim$ as a category where the only arrows are identities. Prove that the quotient map $(X, \sim) \to X/\sim$, assigning to each $x \in X$ its equivalence class $[x] \in X/\sim$, is a functor.

Exercise 1.3.5 (sets and relations). More generally, let $(X, \lesssim)$ be a preorder. Consider the equivalence relation $\sim$ given by Exercise 1.1.5. Define a partial order $\leq$ on the quotient space $X/\sim$ in such a way that the quotient map $(X, \lesssim) \to (X/\sim, \leq)$ is a functor.

For all these examples, we didn't have to worry about checking the functoriality axioms due to uniqueness. However, in general, this check is necessary, as the following examples show.

1.3.2 Functors as mappings preserving operations

Idea. *A functor can be viewed as a mapping that preserves or respects operations and their composition.*

Example 1.3.6 (group theory). Let G and H be groups. *A function $f : G \to H$ is a group homomorphism if and only if the induced mapping $\mathbf{B}G \to \mathbf{B}H$ is a functor.* Let's see why. First of all, f is a group homomorphism if and only if the following properties are satisfied:

- **Unit condition**: $f(1) = 1$, where on the left we have the unit of G and on the right we have the unit of H.
- **Multiplication condition**: For each $g, g' \in G, f(g \cdot g') = f(g) \cdot f(g')$, where on the left we have multiplication in G and on the right we have multiplication in H.

Instead, a functor $\mathbf{B}G \to \mathbf{B}H$ consists of the following data:

- There is a mapping between the objects of $\mathbf{B}G$ to the objects of $\mathbf{B}H$. However, since both categories have only a single object, this is trivial: We just map the single object of $\mathbf{B}G$ to the single object of $\mathbf{B}H$.
- Now, we need a map from the morphisms of $\mathbf{B}G$ to the morphisms of $\mathbf{B}H$. Their source and targets will always match since there is just one object; therefore, this amounts simply to a function $f : G \to H$ (remember that the morphisms of $\mathbf{B}G$ are just the elements of the group G, and the same is true for H).
- The function $f : G \to H$ has to map the identity to the identity and the multiplication to the multiplication by functoriality. This says precisely that f has to be a group homomorphism.

Example 1.3.7. Similarly, let M and N be monoids. *A function $f : M \to N$ is a monoid homomorphism if and only if it induces a functor* $\mathbf{B}M \to \mathbf{B}N$. The proof is the same since in Example 1.3.6, we never used inverses.

Here are two very important examples.

Example 1.3.8 (group theory). Let G be a group. A **linear representation** of G is a functor $R : \mathbf{B}G \to \mathbf{Vect}$. (If you don't know what a linear representation is, you can take this as a definition.) Let's see what this means in practice:

- First of all, we need to map the single object $\bullet$ of $\mathbf{B}G$ to an object of **Vect**, i.e. a vector space. In other words, we need to select a particular vector space, which will be the space where our representation acts. Let's call this space V, so $R(\bullet) = V$. In representation theory, one says that G *is acting on the space* V.
- For each morphism of $\mathbf{B}G$, i.e. for each element $g \in G$, we need a linear map $Rg : V \to V$. This map is an isomorphism (see the following exercise).
- The assignment $g \mapsto Rg$ has to be such that the neutral element 1 is mapped to the identity id_V and that a multiplication of two elements is mapped to their composition.

Exercise 1.3.9 (group theory, important!). Show that for each $g \in G$, the morphism Rg is an isomorphism.

Therefore, $R : \mathbf{B}G \to \mathbf{Vect}$ induces a group homomorphism $G \to \mathrm{Aut}(V)$. ($\mathrm{Aut}(V)$, called the **automorphism group** of V, is the set of isomorphisms from V to itself. If you are not familiar with it, prove that it has a natural group structure.)

A possible interpretation of a group representation is the following. A group is a rather abstract object, but it is usually an abstraction of the idea of *symmetry* of more concrete objects. A representation is then a way to assign to the group G a *concrete* space where G acts. For example, you may view the group S^1 as "rotations" in a plane. A way to make this intuition rigorous is then to take a vector space modeling our plane, for example $\mathbb{R}^2$, and to view S^1 as acting on $\mathbb{R}^2$ via linear maps $\mathbb{R}^2 \to \mathbb{R}^2$, the rotations, one for each element of S^1.

This allows us to have a much more concrete idea of what S^1 looks like "in practice."

Of course, similarly to how not every function is a bijection, in general, the representation of a group may lose information. For example, one could represent $S^1 \times S^1$ again on the plane $\mathbb{R}^2$ by letting the first factor S^1 act on the plane while forgetting the second factor. This is a well defined representation, but it is not a *faithful* representation of the group $S^1 \times S^1$.

A similar version of representation of a group involves letting it act on a set, simply by permuting its elements, rather than on a vector space.

Example 1.3.10 (group theory). Let G be a group. A **permutation representation** or **action** of G is a functor $R : \mathbf{B}G \to \mathbf{Set}$.

As in the case of vector spaces, this picks out a particular set X on which the group G acts. A set equipped with such a group action is sometimes called a G-**set**.

> **Exercise 1.3.11 (group theory).** Show that a permutation representation of G contains the same data as a set X together with a group homomorphism $G \to \mathrm{Aut}(X)$. (A concrete way to prove this is to establish a bijection between the two structures, given G and X.)

Again, in general, a permutation representation may lose information. However, for each group, there is always a *faithful* permutation representation, i.e. a representation which does not lose any information.

Theorem 1.3.12 (Cayley). *Let G be any group. Then, there exists a set X such that G is isomorphic to a subgroup of* $\mathrm{Aut}(X)$.

> **Exercise 1.3.13 (group theory, linear algebra).** Using Cayley's theorem, derive an analogous statement for linear representations: Show that every group embeds into the linear automorphisms of a (possibly infinite-dimensional) vector space.

Cayley's theorem can be considered a special case of a more general result in category theory, known as the *Yoneda lemma*. We explore this in more detail (and prove it) in Chapter 2 (see in particular Example 2.2.3).

Let us now look at the most general case of functors, where neither objects nor morphisms between two given objects are unique.

1.3.3 Functors defining induced maps

Idea. *A functor can be viewed as a consistent way of defining new* induced maps *from existing maps.*

Example 1.3.14 (sets and relations). The **power set functor** is a functor $P : \mathbf{Set} \to \mathbf{Set}$ defined in the following way:

- On objects, it maps a set X to its **power set** PX, i.e. the set of subsets of X. An element $S \in PX$ is a subset $S \subseteq X$.
- On morphisms, it maps a function $f : X \to Y$ to the induced function $Pf : PX \to PY$ given by the **image of subsets**: Given a subset $S \subseteq X$, we obtain a subset of Y as follows. We apply f to all the elements in S, and we get several elements of Y. These form a subset of Y, i.e. an element of PY. Equivalently,

$$(Pf)(S) \;:=\; \{y \in Y \mid y = f(x) \text{ for some } x \in S\}.$$

In set theory, usually, one denotes the image of subsets with the same letter as the original function, namely $f(S)$. This makes sense since the map $Pf : PX \to PY$ is *induced* by $f : X \to Y$. However, technically, it is not the same map as f (it has different domain and codomain). This intuitive idea of "having a map which is still defined by f, but with different domain and codomain" is made precise exactly by functoriality. So, in category-theoretical terminology, the map $f : X \to Y$ induces, via the *functor* P, a map $Pf : PX \to PY$.

Exercise 1.3.15 (sets and relation). Check that P satisfies the functoriality axioms.

Just as we can induce maps between subsets, we can do so between probability measures too. We first look at finitely supported probability measures for readers who are not familiar with measure theory. More sophisticated constructions can be done as well, using measure theory, and will be given in the subsequent exercises.

Example 1.3.16 (basic probability). The **probability functor,** also called **distribution functor,** is a functor $\mathcal{P} : \textbf{Set} \to \textbf{Set}$ defined as follows:

- On objects, it maps a set X to the set of **finitely supported probability measures** on X. Those are functions $p : X \to [0,1]$ with only finitely many nonzero entries such that their sum is 1, that is,

$$\sum_{x \in X} p(x) = 1.$$

  These can be considered as "finite normalized histograms over the elements of X."

- On morphisms, to a function $f : X \to Y$, we assign the function $\mathcal{P}f : \mathcal{P}X \to \mathcal{P}(Y)$ given by the **pushforward of measures** along f. In particular, given $p \in PX$, the probability measure $(\mathcal{P}f)(p) \in \mathcal{P}Y$ is given by

$$(\mathcal{P}f)(p)\,(y) := \sum_{x \in f^{-1}(y)} p(x).$$

  Intuitively, this "moves the columns of the histogram on X along f to give a histogram on Y, stacking columns on top of each other whenever they end up over the same element."

In the probability and measure theory literature, usually the map $\mathcal{P}f$ is written as f_*. In terms of random variables, the map giving **image random variables** (which have the pushforward of the measure as law) is instead denoted again by f. Similar to the power set case, these notations reflect the fact that the map $\mathcal{P}f$ is induced by f functorially.

Here are more general constructions for readers with a background in probability (or who want to learn more about probability).

Exercise 1.3.17 (measure theory, probability). Consider the category **Meas** of measurable spaces and measurable maps. Recall that the pushforward of measures, in general, is defined as follows. Let $f : X \to Y$ be measurable. Let p be a measure on X, and let A be a measurable subset of Y. Then, since f is measurable, by definition, $f^{-1}(A)$ is a measurable subset of X, and the pushforward of p along f is defined to be the assignment

$$A \longmapsto (f_* p)(A) := p(f^{-1}(A)).$$

Now, define a probability functor $\mathcal{P} : \mathbf{Meas} \to \mathbf{Meas}$ in such a way that:

- to each measurable space X, we assign a space $\mathcal{P}X$ of probability measures on X;
- to each measurable map $f : X \to Y$, we get a map $\mathcal{P}f : \mathcal{P}X \to \mathcal{P}Y$ via the pushforward of measures.

Keep in mind that, in order for $\mathcal{P}$ to be a functor **Meas** $\to$ **Meas**, we need $\mathcal{P}X$ to be an object of **Meas**, i.e. a measurable set, and $\mathcal{P}f$ to be a morphism of **Meas**, i.e. a measurable function. Therefore, when you define the spaces $\mathcal{P}X$, you have to equip them with a σ-algebra, and you have to do it in such a way that the map $\mathcal{P}f$ will be measurable.

One convenient way to do it is the following. Given $A \subseteq X$ to be measurable, consider the mapping $i_A : \mathcal{P}X \to [0,1]$ given by evaluating

$$i_A(p) := p(A).$$

Equip now $\mathcal{P}X$ with the initial σ-algebra of all the maps i_A in the form above, i.e. the coarsest σ-algebra which makes all the maps i_A measurable. Prove that, using this approach, for every measurable map $f : X \to Y$, the map $\mathcal{P}f : \mathcal{P}X \to \mathcal{P}Y$ is measurable, and check the functoriality axioms so that $\mathcal{P} : \mathbf{Meas} \to \mathbf{Meas}$ is a functor.

This functor is called the **Giry functor**.

Exercise 1.3.18 (measure theory, probability). Consider the category **CHaus** of compact Hausdorff spaces and continuous maps. Assign to each object X (which is a compact Hausdorff space) the space of Radon probability measures on X, equipped with the weak topology (which is again a compact Hausdorff space). Show that this assignment is part of a functor, with the assignment on morphisms again given by the pushforward of measures.

This functor is called the **Radon functor**.

Here is another functor giving an "induced map," this time coming from computer science.

Example 1.3.19 (basic computer science, combinatorics). Consider again the category **Set** of sets and functions. (Readers familiar with computer science may want to use types instead of sets.) Given a set X, we can form a new set LX whose elements are *lists of elements of* X. A list is an expression in the form $[x_1, \dots, x_n]$, where x_i are elements of X and the length n is finite but can be arbitrarily large (it can also be zero, giving the empty list [ ]).

The list construction is functorial: Given $f : X \to Y$, we can induce a function $Lf : LX \to LY$ by applying f elementwise to the list. The empty list of LX is mapped to the empty list of LY, and $[x_1, \dots, x_n]$ is mapped to $[f(x_1), \dots, f(x_n)]$. Extending a function f to *lists of possible inputs*, giving as output the list of the results, is then the action of the functor on the morphisms. This functionality is for example given (roughly) by `map` in Python, `map` (and `fmap`) in Haskell, and it corresponds to the attribute `Listable` in Mathematica.

Another class of examples which are widely used in the applications of category theory is forgetful functors. These are functors which intuitively "forget the extra structures." Let's see what this means through examples.

Example 1.3.20 (topology). Consider the functor U : **Top** $\to$ **Set**:

- On objects, it maps a topological space X to the **underlying set** X. That is, it considers X only as a set now, "forgetting" the topology. (The letter U comes from "underlying.")

- On morphisms, it maps a continuous function $f : X \to Y$ to the **underlying function** $f : X \to Y$, which is now considered only as a function between sets, "forgetting" the continuity.

Here are some things to keep in mind:

- A continuous function is in particular a function, so we can always do this. However, not every function is continuous; therefore, the functions between the *sets* X and Y may in general be more than those obtained by forgetting the continuity.
- One can define many possible topologies on the same set. So, in general, the "underlying set" assignment may map different spaces to the same set; it is many-to-one.

This idea of "forgetting the structure" can be done for many structures, not just topologies.

Exercise 1.3.21 (several fields). Define the following forgetful functors:

- **Grp** $\to$ **Set**, forgetting the group structure;
- **Ring** $\to$ **Grp**, forgetting the multiplication (but keeping the addition);
- let **TopGrp** be the category of topological groups and continuous group homomorphisms, and define forgetful functors **TopGrp** $\to$ **Top** and **TopGrp** $\to$ **Grp** that forget the group structures and the topology, respectively.

In category theory, many ideas are intuitively related to *structures, preserving structures*, and *forgetting structures*. Functors can often make this intuition precise: Every functor can be considered "forgetful" in some way, but we can keep track of exactly what they forget (see Section 1.5.3).

For people with a more pure math background, here is a classic example of a functor: the fundamental group. The construction is quite long, and the details will be given as an exercise for interested readers. (This is difficult for people without a background in topology. If you cannot solve this exercise, but you would still like to know about the fundamental group, an excellent reference is [Hat02].)

Example 1.3.22 (algebraic topology). Let $\mathbf{Top}_*$ be a category of **pointed topological spaces**, defined as follows:

- Objects are topological spaces with a distinguished point, i.e. pairs (X, x), where X is a topological space and x is a point of X, usually called the **base point.**
- A morphism $f : (X, x) \to (Y, y)$ is a continuous map $f : X \to Y$ which preserves the base point, i.e. $f(x) = y$.

Let now (X, x) be a pointed topological space. A **loop in** X **based at** x is a continuous function $\ell : [0, 1] \to X$ such that $\ell(0) = \ell(1) = x$. Intuitively, this looks indeed like a loop inside X from the point x to itself.

Now, consider the loops $\ell, m : [0, 1] \to X$ at x. The **concatenated loop** ℓm is the loop at x given as follows. First, we walk along the loop ℓ and then along the loop m. This takes twice the time; that is, this gives a function $[0, 2] \to X$ rather than $[0, 1] \to X$. To obtain a function $[0, 1] \to X$, we have to walk twice as fast. Formally, $\ell m : [0, 1] \to X$ is the loop at x given by

$$\ell m(t) := \begin{cases} \ell(2t), & 0 \leq t \leq 1/2; \\ m(2t - 1), & 1/2 < t \leq 1. \end{cases}$$

A **homotopy** between the loops $\ell, m : [0, 1] \to X$ at x is a continuous map $h : [0, 1] \times [0, 1] \to X$ such that:

- for each $s \in [0, 1]$, the map $t \mapsto h(s, t)$ is a loop in X based at x;
- the map h is equal to ℓ at $s = 0$ and equal to m at $s = 1$. That is, for each $t \in [0, 1]$, $h(0, t) = \ell(t)$ and $h(1, t) = m(t)$.

Intuitively, h is a way to deform the loop ℓ continuously into the loop m, while keeping the base points fixed. If there exists a homotopy h between ℓ and m, we say that ℓ and m are **homotopic**. This is an equivalence relation (see the following exercise), and the equivalence classes are called **homotopy classes**. We denote the space of homotopy classes by $\pi_1(X, x)$

The concatenation of loops induces a concatenation between the homotopy classes which equips $\pi_1(X, x)$ with a group structure. The unit is given by the constant loop at x, and the inverses are given by "walking the loop backwards" (again, see the following exercise). Therefore, we call $\pi_1(X, x)$ the **fundamental group of** X **at** x.

Let now $f : (X, x) \to (Y, y)$ be a base point-preserving continuous function. We can map a loop at x to a loop of y by simply applying f to it; that is, given a loop ℓ at $x \in X$, we form the loop

$$[0, 1] \xrightarrow{\ell} X \xrightarrow{f} Y.$$

Intuitively, this applies f to each point of the loop ℓ, and since f is continuous, the resulting points in Y will form again a loop based at y. We then get a function from loops at x in X to loops at y in Y. This function respects homotopy equivalence (see the following exercise); therefore, it induces a map between the equivalence classes, $\pi_1(X, x) \to \pi_1(Y, y)$. We denote this resulting map $\pi_1(f)$.

The assignment given by $(X, x) \mapsto \pi_1(X, x)$ and $f \mapsto \pi_1(f)$ is a functor $\mathbf{Top}_* \to \mathbf{Grp}$.

Exercise 1.3.23 (algebraic topology). In the notation used in the previous example:

- Show that homotopy of loops is an equivalence relation.
- Show that the concatenation of loops up to homotopy makes $\pi_1(X, x)$ a group. (*Hint*: To form inverses, walk the loop backward. Why is it an inverse up to homotopy?)

- Show that if ℓ and m are homotopic, then applying f to both gives homotopic loops. Conclude that f will induce a well-defined mapping $\pi_1(f) : \pi_1(X, x) \to \pi_1(Y, y)$.
- Show, moreover, that the induced map $\pi_1(f) : \pi_1(X, x) \to \pi_1(Y, y)$ is a group homomorphism.
- Verify the functoriality axioms, and conclude that $\pi_1 : \mathbf{Top}_* \to \mathbf{Grp}$ is a functor.

Exercise 1.3.24 (graph theory, algebraic topology). Can one construct a similar functor on a category of graphs?

In different fields of mathematics, the functoriality condition $F(g \circ f) = Ff \circ Fg$ appears under different names. One name under which this condition is known is **chain rule**. This is the case, for example, of the chain rule of derivatives.

The usual *derivative* of calculus (or *differential*, or *gradient*, or *Jacobian matrix*) can be considered a *linear approximation* of a differentiable function $f : X \to Y$ in the neighborhood of a point x. Linear maps live between vector spaces; therefore, in order to have a linear map, we need to first replace our original spaces X and Y with vector spaces and then replace f with a linear map. This is precisely what we can do with a functor: We assign to each space X a vector space and to each map f a linear map, the derivative. There are many ways of constructing this in practice. Here is one.

Example 1.3.25 (calculus). Define the category of **pointed Euclidean spaces** $\mathbf{Euc}_*$ as follows:

- As objects, we take Euclidean spaces ($\mathbb{R}^n$ for different $n \in \mathbb{N}$) with a distinguished point $x \in \mathbb{R}^n$. Let's denote these objects by $(\mathbb{R}^n, x)$.
- As morphisms $f : (\mathbb{R}^n, x) \to (\mathbb{R}^m, y)$, we take smooth (i.e. differentiable infinitely many times) functions $\mathbb{R}^n \to \mathbb{R}^m$ such that $f(x) = y$.

These correspond to the functions that we want to differentiate at the point at which we want to take the differential.

The derivative is now a functor $D : \mathbf{Euc}_* \to \mathbf{Vect}$ defined in the following way:

- On objects, it maps $(\mathbb{R}^n, x)$ to $\mathbb{R}^n$ (now seen as a vector space).
- On morphisms, it maps $f : (\mathbb{R}^n, x) \to (\mathbb{R}^m, y)$ to the derivative $Df|_x$ of the function f at the point x, which is a linear function $\mathbb{R}^n \to \mathbb{R}^m$ (usually represented by the Jacobian matrix).

The intuition is that on objects, it maps $(\mathbb{R}^n, x)$ to the space of "vectors starting at x," which is canonically isomorphic to $\mathbb{R}^n$. On morphisms, the differential gives the map between vectors at x and vectors at $f(x) \in \mathbb{R}^m$. This is functorial because of the following:

- The derivative of the identity map $(\mathbb{R}^n, x) \to (\mathbb{R}^n, x)$ is simply the identity of $\mathbb{R}^n$ (the identity matrix).
- Considering the composable maps

  $$(\mathbb{R}^n, x) \xrightarrow{f} (\mathbb{R}^m, y) \xrightarrow{g} (R^p, z),$$

  we have that, *by the chain rule of derivatives*,

  $$D(g \circ f)|_x = Dg|_y \circ Df|_x,$$

  where the composition on the right is the composition of linear maps. Equivalently, in components,

  $$\frac{\partial (g \circ f)^k}{\partial x^i} = \sum_{j=1}^{m} \frac{\partial g^k}{\partial y^j} \frac{\partial f^j}{\partial x^i},$$

  for each $i = 1, \ldots, n$ and $k = 1, \ldots, p$.

Therefore, the derivative is a functor, with functoriality guaranteed by the chain rule.

Readers with a background in differential geometry can also try to do the same with manifolds. The intuition is quite similar, replacing the spaces and the maps with *linear approximations*.

Exercise 1.3.26 (differential geometry). Define the category of **pointed manifolds $\mathbf{Mfd}_*$** as follows:

- As objects, we take smooth manifolds M with a distinguished point $x \in M$. Let's denote these objects by (M, x).
- As morphisms $f : (M, x) \to (N, y)$, we take smooth maps $M \to N$ such that $f(x) = y$.

Consider the following assignment:

- On objects, it maps (M, x) to the tangent space $T_x M$, which is a vector space.
- On morphisms, it maps $f : (M, x) \to (N, y)$ to the derivative $Df|_x$ of the function f at the point x, which is a linear function $T_x M \to T_y N$.

Show that this is a functor $\mathbf{Mfd}_* \to \mathbf{Vect}$.

Exercise 1.3.27 (differential geometry; difficult!). Can you construct a functor which takes the tangent *bundle* of a manifold, rather than the tangent space at a single point?

For readers with a background on information theory or dynamical systems, here is another exercise.

Exercise 1.3.28 (information theory; difficult!). It is well known that the Shannon entropy also satisfies a sort of chain rule: For random variables X, Y on finite state spaces,

$$H(X, Y) = H(X) + H(Y|X).$$

Let now **FinProb** be the category of *finite probability spaces and measure-preserving maps*. That is:

- an object (X, p) consists of a finite set X equipped with a probability measure p;
- a morphism $f : (X, p) \to (Y, q)$ is a function $f : X \to Y$ such that $f_* p = q$ (the latter denotes the pushforward of measures, see Example 1.3.16).

Now, show that, due to the chain rule, entropy is functorial on **FinProb**. The main question is: *to which category?*

1.3.4 Functors and cocycles

In some fields of mathematics, functoriality appears under the name of *cocycle condition*, or, more specifically, 1-*cocycle condition*. The reason for this terminology comes from algebraic topology. The intuitive idea is that a 1-cocycle represents a map on the paths of a given space which only depends on the endpoints. For example, given a 2-simplex (triangle)

$$
\begin{array}{ccc}
A & \overset{f}{\rule{1cm}{0.4pt}} & B \\
 & {}_{h}\diagdown & \downarrow{}^{g} \\
 & & C
\end{array}
$$

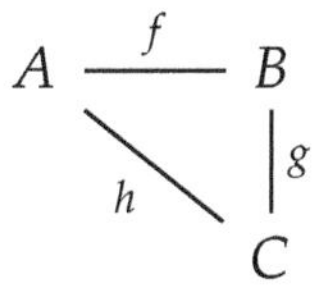

in a simplicial complex, a 1-cocycle F with coefficients in an abelian group (say, $\mathbb{Z}$) satisfies the identity

$$ Ff + Fg = Fh. $$

This is analogous to a functor in the following sense: Suppose that we have a *commutative* triangle in a category $\mathbf{C}$

$$
\begin{array}{ccc}
A & \overset{f}{\longrightarrow} & B \\
 & {}_{h}\searrow & \downarrow{}^{g} \\
 & & C
\end{array}
$$

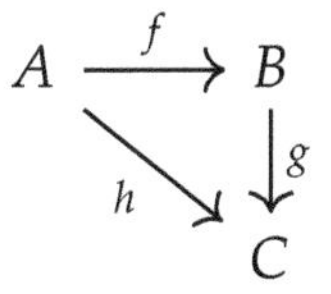

and a functor $F : \mathbf{C} \to \mathbf{B}\mathbb{Z}$. Then, we have

$$ Ff \circ Fg = F(g \circ f) = Fh. $$

Both conditions can be intuitively interpreted as "F only depends on the endpoints." Note that, for the analogy to work, it is important that we start with a *commutative* triangle. A mapping that depends only on the endpoints regardless of whether the triangle commutes would have the stricter property of being a *coboundary*, and not every cocycle is a coboundary. In some sense, a commutative triangle, or, more generally, a commutative diagram, can be interpreted geometrically as a "filled region," such as a filled triangle, not just its perimeter.

This intuition can be made precise and concrete, and the resulting correspondence between functors and cocycles is actually very deep. This is, however, quite advanced in its full generality and well outside the scope of this book. If you are interested, you can look at the nLab page about cocycles,[6] perhaps beginning with the motivation page on cohomology.[7] What is important is to keep in mind that *a functor is similar to a* 1*-cocycle*. That's why, in different fields of math, people call "cocycle conditions" certain conditions that make some mappings analogous to cocycles. More than often, those mappings are actually *functors*.

The term "cocycle condition" is also used outside geometry and topology, for example, in stochastic processes. In particular, the cocycle condition for a Markov process can be seen as an instance of functoriality.

Example 1.3.29 (probability). Let (X,p) be a probability space, i.e. a measurable space equipped with a probability measure. Given probability spaces (X,p) and (Y,q), we say that a Markov kernel $K:(X,p)\to(Y,q)$ is **measure-preserving** if for every measurable subset B of Y,

$$q(B)=\int_X K(B|x)\,p(dx).$$

[6]http://ncatlab.org/nlab/show/cocycle.

[7]https://ncatlab.org/nlab/show/motivation+for+sheaves%2C+cohomology+and+higher+stacks.

This generalizes the idea of a measure-preserving function (defined in Exercise 1.3.28 for the finite case) to the nondeterministic case.

Let now X be a measurable space. A **continuous-time Markov process** on X is (equivalently) a collection of measures p_t, for each $t \in \mathbb{R}$, and a collection of measure-preserving Markov kernels $K_{s,t} : (X, p_s) \to (X, p_t)$, for each $s \leq t \in \mathbb{R}$, satisfying the following conditions:

- **Unitality**: For each $t \in \mathbb{R}$, $K_{t,t} = \mathrm{id}_X$.
- **Cocycle condition**: For each $r \leq s \leq t \in \mathbb{R}$,

$$K_{r,s} \circ K_{s,t} = K_{r,t}.$$

These conditions precisely state that the assignment $(r,s) \mapsto K_{r,s}$ is a functor from $(\mathbb{R}, \leq)$ (which is a partial order, hence a category) into the category of probability spaces and measure-preserving Markov kernels. Analogously, this can be said for discrete-time processes, processes defined only for positive time, etc.

In strict terms, we don't really have a cohomology theory for Markov processes (at least not in an obvious way). The term "cocycle" here is rather used because it's *analogous* to a cocycle. But actually, it is a *functor*, and functors are analogous to 1-cocycles.

People with a background in geometry and topology may find the following example helpful: The (Čech) cocycle condition for vector and principal bundles is also a form of functoriality.

Example 1.3.30 (algebraic topology, differential geometry). Let X be a topological space, and let $\mathcal{U} = (U_i)_{i \in I}$ be an open cover of X. Suppose, for simplicity, that it is a **good cover**, i.e. all U_i and their finitary intersections are contractible. The **Čech groupoid $\mathbf{C}(\mathcal{U})$** of the cover $\mathcal{U}$ is the following groupoid:

- The objects of $\mathbf{C}(\mathcal{U})$ are the open sets U_i of the cover $\mathcal{U}$.
- There is an isomorphism $U_i \to U_j$ for each finite sequence $(i = i_0, i_1, \dots, i_n = j)$ such that for each $k = 1, \dots, n$, the sets U_{i_k} and $U_{i_{k-1}}$ have nonempty intersection.

- Given three sets U_i, U_j, and U_k with nonempty pairwise intersection, the diagram of isomorphisms

$$\begin{array}{ccc} U_i & \xrightarrow{\cong} & U_j \\ & {\scriptstyle\cong}\nwarrow \qquad \swarrow{\scriptstyle\cong} & \\ & U_k & \end{array} \tag{1.3.1}$$

  commutes whenever the triple intersection $U_i \cap U_j \cap U_k$ is nonempty.

The definition of a vector bundle in terms of local trivializations and transition functions can be seen as a functor $F : \mathbf{C}(\mathcal{U}) \to \mathbf{Vect}$. All the objects FU_i in the image of F are isomorphic whenever X is connected and correspond to the *fiber* of the bundle. To each morphism of $\mathbf{C}(\mathcal{U})$, i.e. to each intersection $U_i \cap U_j$, there corresponds an invertible linear map $F_{ij} : FU_i \to FU_j$, or **transition function**, with inverse F_{ji}. By functoriality, the transition functions have to satisfy the following:

- **Identity**: For $i = j$, the transition function F_{ii} has to be the identity.
- Composition: For every commutative diagram in the form (1.3.1), the F-image of that diagram has to commute as well. In other words, whenever a triple intersection $U_i \cap U_j \cap U_k$ is nonempty, then $F_{jk} \circ F_{ij} = F_{ik}$. This is usually called the **cocycle condition for transition functions.**

In this case, there actually *is* a cohomology theory in which this construction is a cocycle, namely the *Čech cohomology*. However, to make the analogy precise, one has to take a non-abelian version of this theory, and so, on the surface, it does not look like the usual cohomology theories that one comes across in traditional algebraic topology (if you are interested, see the nLab entry on non-abelian cohomology[8]).

[8]http://ncatlab.org/nlab/show/nonabelian+cohomology.

1.3.5 Functors, mono and epi

We have said that functors preserve relations between objects. Let's now see what they preserve more concretely, as well as what they do *not* preserve.

Let's start with a simple example. Let $F : \mathbf{C} \to \mathbf{D}$ be a functor, and consider the following diagram in $\mathbf{C}$:

$$X \underset{g}{\overset{f}{\rightrightarrows}} Y.$$

If we apply F, we get a diagram in $\mathbf{D}$:

$$FX \underset{Fg}{\overset{Ff}{\rightrightarrows}} FY.$$

Remark 1.3.31. If the first diagram commutes (in $\mathbf{C}$), then the second diagram commutes (in $\mathbf{D}$).

The reason is actually trivial: The first diagram commutes if and only if $f = g$. However, that implies $Ff = Fg$, which means exactly that the second diagram commutes. This can be done for each pair of objects in a diagram and for each pair of arrows between them. Therefore, we have the following.

Corollary 1.3.32. *Functors preserve commutative diagrams. If a diagram commutes in* $\mathbf{C}$*, its image under a functor* $F : \mathbf{C} \to \mathbf{D}$ *commutes in* $\mathbf{D}$*.*

This is trivial, but very important. Here is another very important fact.

Proposition 1.3.33. *Let* $F : \mathbf{C} \to \mathbf{D}$ *be a functor, and suppose that we have an isomorphism*

$$X \underset{g}{\overset{f}{\rightleftarrows}} Y$$

in $\mathbf{C}$*. Then, the diagram*

$$FX \underset{Fg}{\overset{Ff}{\rightleftarrows}} FY$$

is an isomorphism in **D**. *In particular, F maps isomorphic objects of* **C** *to isomorphic objects of* **D**.

Proof. By the definition of isomorphism, we have

$$g \circ f = \mathrm{id}_X \quad \text{and} \quad f \circ g = \mathrm{id}_Y.$$

Applying F, we get

$$F(g \circ f) = F(\mathrm{id}_X) \quad \text{and} \quad F(f \circ g) = F(\mathrm{id}_Y),$$

which by functoriality of F gives

$$Fg \circ Ff = \mathrm{id}_{FX} \quad \text{and} \quad Ff \circ Fg = \mathrm{id}_{FY}.$$

This means precisely that the pair (Ff, Fg) is an isomorphism between FX and FY. □

Recall that split monomorphisms and split epimorphisms satisfy by definition only one of the two conditions of isomorphism (Definitions 1.2.7 and 1.2.20), namely, $g \circ f = \mathrm{id}_X$ means that f is split mono and g is split epi. Therefore, the same proof also gives us the following result.

Proposition 1.3.34. *Let* $F : \mathbf{C} \longrightarrow \mathbf{D}$ *be a functor, and let* $m : X \longrightarrow Y$ *be a split monomorphism of* **C**, *with retraction* $r : Y \longrightarrow X$. *Then,* Fm *is a split monomorphism of* **D**, *with retraction* Fr.

Dually, let $e : X \longrightarrow Y$ *be a split epimorphism of* **C** *with section* $s : Y \longrightarrow X$. *Then,* Fe *is a split epimorphism of* **D** *with section* Fs.

In particular, functors map split monomorphisms to split monomorphisms and split epimorphisms to split epimorphisms.

Warning. If a monomorphism f is not split, its image Ff under a functor f may not be a monomorphism. The same is true for epimorphisms.

Here is a simple example of that fact.

Example 1.3.35 (algebra). The inclusion map $i : (\mathbb{N}, +) \hookrightarrow (\mathbb{Z}, +)$ is an epimorphism of the category **Mon** of monoids and monoid homomorphisms. Take now the forgetful functor U : **Mon** $\to$ **Set** that maps a monoid to the underlying set and a monoid homomorphism to the underlying function. The function $Ui : \mathbb{N} \to \mathbb{Z}$ is not surjective, so it is not an epimorphism of **Set**.

Another example, which is quite helpful in practice, is the following. We encourage readers unfamiliar with graph theory to try to understand this example and the ones that follow on the same topic: Graphs and categories are very intimately related.

Example 1.3.36 (graph theory). Let **Graph** be the category of undirected graphs and functions between the vertices preserving adjacency. There is a functor π_0 : **Graph** $\to$ **Set** which:

- to each graph G, it assigns the set of connected components of G, denoted by $\pi_0(G)$.
- to each morphism $f : G \to H$, it assigns the induced maps between the connected components, in the following way. For each connected component $[x] \in \pi_0(G)$, take a representative vertex $x \in [x]$. Map it with f to $f(x) \in H$, and take its connected component $[f(x)] \in \pi_0(H)$. This is well defined since if we take another $x' \in [x]$, then there is a path of edges $x \to x'$, and since f preserves adjacency, there is a path of edges $f(x) \to f(x')$. Therefore, $[f(x)] = [f(x')]$.

Consider now the morphism f between the following two graphs:

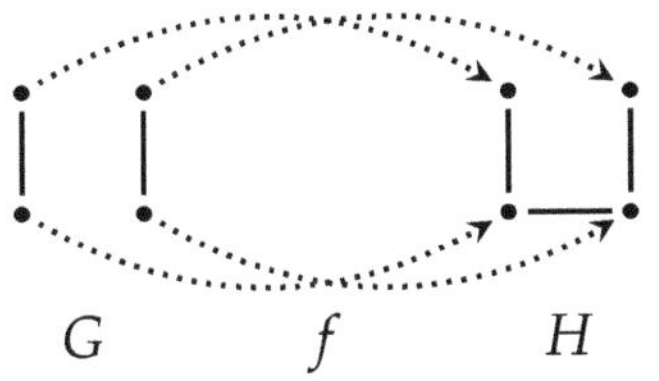

The map $f : G \to H$ is injective, and injective maps that preserve adjacency are monomorphisms of **Graph**. However, the induced map $\pi_0(f) : \pi_0(G) \to \pi_0(H)$ is the following function:

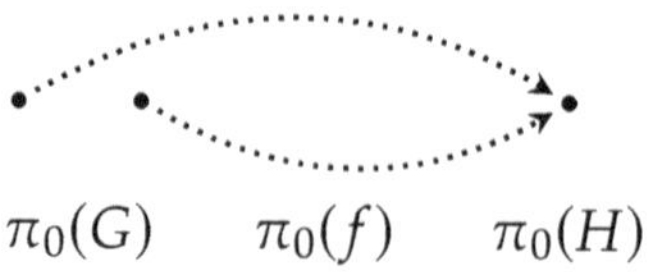

which is clearly not injective. Therefore, even though f is mono, $\pi_0(f)$ is not.

We can use this fact to our own advantage. Suppose that f were *split* mono. Then, $\pi_0(f)$ would have been split mono too. The fact that $\pi_0(f)$ is not mono (and so in particular not split mono) necessarily means that f cannot be split. In other words, we have the following corollary.

Corollary 1.3.37. *Let $F : \mathbf{C} \to \mathbf{D}$ be a functor. Suppose that Ff is* not *mono (resp., epi). Then, f is not split mono (resp., split epi), even when it happens to be mono (resp., epi). In other words, f does not admit a retraction (resp., section).*

This is quite useful in practice: *Functors can be used to prove nonexistence of retractions and sections*. Proving that something does not exist is often quite hard, and functors give us a tool to do exactly that. Sometimes, functors are said to *detect obstructions to the existence of retractions and sections*. This is the case, for example, for many functors in homological algebra. Let's show how this is used in algebraic topology using the following example, which is a category-theoretical solution to Exercise 1.2.25.

Example 1.3.38 (algebraic topology). Consider the inclusion map $i : S^1 \to D^2$. This is injective and continuous, so it is a monomorphism of **Top**. Let's now show that it has no retraction. For convenience, fix a point $x \in S^1$, and call $y := i(x) \in D^2$ so that $i : (S^1, x) \to (D^2, y)$ is a morphism of the category $\mathbf{Top}_*$ of pointed topological spaces.

Consider now the fundamental group functor $\pi_1 : \mathbf{Top}_* \to \mathbf{Grp}$ (Example 1.3.22). The fundamental group of S^1 is (isomorphic to) $(\mathbb{Z}, +)$, as loops are generated by going once around the perimeter, while the fundamental group of D^2 is the trivial group 0 since the disk is contractible. Therefore, the map $\pi_1(i) : \pi_1(S^1) \to \pi_1(D^2)$ is the zero map $\mathbb{Z} \to 0$, sending everything to zero. This map is not injective, so it is not a monomorphism of **Grp**. Therefore, i is not split mono, i.e. it has no retraction.

The same can be done for spheres and discs of higher dimensions using higher homotopy groups. The nonexistence of this retraction can be used, for example, to give a very short proof of Brouwer's fixed-point theorem (see for example [Hat02, Theorem 1.9]).

1.3.6 What is not a functor?

Of course, *many* things are not functors. Here are, however two examples of a construction which *may seem functorial at first look, but is not*. Given a set X, we can assign to it its automorphism group $\mathrm{Aut}(X)$, the group of all invertible functions from X to itself. (If X is finite, this is usually called the **permutation group**, or **symmetric group**, of X.)

One may be tempted to think of Aut as a functor $\mathbf{Set} \to \mathbf{Grp}$, but there is no nontrivial way to make this construction functorial. The reason is what it would do on *morphisms*: A function $f : X \to Y$ does not induce a well-defined mapping $\mathrm{Aut}(X) \to \mathrm{Aut}(Y)$.[9]

Similarly, the set of endomorphisms $\mathrm{End}(X)$, i.e. the functions from X to itself, is a monoid. Again, this does not give a functor $\mathbf{Set} \to \mathbf{Mon}$.

This is true in most categories (where Aut and End are the isomorphisms and endomorphisms in the category). There are categories where this construction can be made functorial in a convenient way;

[9] Of course, one could always pick the map sending everything to the identity of $\mathrm{Aut}(Y)$, but there is no *nontrivial* way.

however, in the generic case, this does not work, so Aut and End are not functors. (Aut and End do have convenient categorical descriptions, just not as functors — but they are beyond the scope of this book.)

Exercise 1.3.39 (group theory). Take the category of finite sets and *injective* maps. Can you make Aut functorial in a nontrivial way on this category?

1.3.7 Hom-functors, contravariant functors, and presheaves

Among the most important functors in category theory are the *hom-functors*, which associate to an object the arrows in or out of it. Let's see this in detail.

Definition 1.3.40. Let $\mathbf{C}$ be a (locally small) category, and let A be an object of $\mathbf{C}$. The **hom-functor** $\mathrm{Hom}_{\mathbf{C}}(A,-) : \mathbf{C} \to \mathbf{Set}$ is defined as follows:

- It maps an object X of $\mathbf{C}$ to the set $\mathrm{Hom}_{\mathbf{C}}(A,X)$ of morphisms $A \to X$ of $\mathbf{C}$.
- It maps a morphism $f : X \to Y$ of $\mathbf{C}$ to the function between hom-sets given by postcomposition with f:

$$\begin{aligned} \mathrm{Hom}_{\mathbf{C}}(A,X) &\xrightarrow{f\circ -} \mathrm{Hom}_{\mathbf{C}}(A,Y) \\ (A \xrightarrow{p} X) &\longmapsto (A \xrightarrow{p} X \xrightarrow{f} Y). \end{aligned}$$

Let's check that this is indeed a functor. First of all, it maps the identity of X to the following function:

$$\begin{aligned} \mathrm{Hom}_{\mathbf{C}}(A,X) &\xrightarrow{\mathrm{id}_X\circ -} \mathrm{Hom}_{\mathbf{C}}(A,X) \\ p &\longmapsto \mathrm{id}_X \circ p = p, \end{aligned}$$

which by left unitality is the identity function between hom-sets. Similarly, given $f : X \to Y$ and $g : Y \to Z$, it maps the composition $g \circ f$ to the following function:

$$\begin{aligned} \mathrm{Hom}_{\mathbf{C}}(A, X) &\xrightarrow{(g\circ f)\circ -} \mathrm{Hom}_{\mathbf{C}}(A, Z) \\ p &\longmapsto (g \circ f) \circ p = g \circ (f \circ p), \end{aligned}$$

which by associativity is the composition of the procedures of post-composing with f and with g.

The hom-functor $\mathrm{Hom}_{\mathbf{C}}(A, -)$ gives us arrows *out of* A. Let's now try to construct a functor $\mathbf{C} \to \mathbf{Set}$ that gives us arrows *to* A. We run immediately into the problem of what it does on morphisms. Given a morphism $f : X \to Y$, we want to construct a function $\mathrm{Hom}_{\mathbf{C}}(X, A) \to \mathrm{Hom}_{\mathbf{C}}(Y, A)$ induced by f. However, there is no natural choice of such a function. What we *can* do, instead, is induce a function the other way, namely $\mathrm{Hom}_{\mathbf{C}}(Y, A) \to \mathbf{C}_{\mathbf{Set}}(X, A)$, in the following way. Given $q \in \mathrm{Hom}_{\mathbf{C}}(Y, A)$, i.e. $q : Y \to A$, we can *precompose it* with f to obtain the morphism $q \circ f : X \to A$:

$$X \xrightarrow{f} Y \xrightarrow{q} A.$$

This assignment is like a functor $\mathbf{Set} \to \mathbf{Set}$, except that *it reverses all the arrows*. Therefore, it is not quite a functor $\mathbf{Set} \to \mathbf{Set}$ but rather a functor $\mathbf{Set}^{\mathrm{op}} \to \mathbf{Set}$.

Definition 1.3.41. Let $\mathbf{C}$ be a (locally small) category, and let A be an object of $\mathbf{C}$. The (other) **hom-functor** $\mathrm{Hom}_{\mathbf{C}}(-, A) : \mathbf{C}^{\mathrm{op}} \to \mathbf{Set}$ is defined as follows:

- It maps an object X of $\mathbf{C}$ to the set $\mathrm{Hom}_{\mathbf{C}}(X, A)$ of morphisms $X \to A$ of $\mathbf{C}$.
- It maps a morphism $f : X \to Y$ of $\mathbf{C}$ to the function between hom-sets given by *pre*composition with f:

$$\begin{aligned} \mathrm{Hom}_{\mathbf{C}}(A, X) &\xleftarrow[f\circ -]{} \mathrm{Hom}_{\mathbf{C}}(A, Y) \\ (X \xrightarrow{f} Y \xrightarrow{q} A) &\longleftarrow\!\shortmid (Y \xrightarrow{q} A). \end{aligned}$$

Functoriality is similar to that of the functor $\mathrm{Hom}_{\mathbf{C}}(A, -)$, except for the direction of the arrows.

In general, given categories **C** and **D**, a functor $\mathbf{C}^{\mathrm{op}} \to \mathbf{D}$ behaves like a usual functor, except that the direction of the arrows and their composition are reversed. (You may be wondering why $\mathbf{C}^{\mathrm{op}} \to \mathbf{D}$ and not $\mathbf{C} \to \mathbf{D}^{\mathrm{op}}$? The two concepts are actually the same. Can you see why?)

Remark 1.3.42. In some areas of mathematics, instead of a functor $\mathbf{C}^{\mathrm{op}} \to \mathbf{D}$, one refers to a **contravariant functor** $\mathbf{C} \to \mathbf{D}$ (and the ordinary functors are called **covariant functors**). This is analogous to what happens in linear algebra, where one could alternatively consider covariant and contravariant vectors, or vectors belonging to the dual space. In category theory, it is customary not to use the word "contravariant" except informally and prefer the use of functors $\mathbf{C}^{\mathrm{op}} \to \mathbf{D}$.

Whichever convention you choose, be consistent: Avoid saying things like "a contravariant functor $\mathbf{C}^{\mathrm{op}} \to \mathbf{D}$" — this could lead to confusion.

Example 1.3.43 (linear algebra). Consider the category **Vect** of vector spaces and linear maps. The **dual space** is a functor $\mathbf{Vect}^{\mathrm{op}} \to \mathbf{Vect}$ (or, a contravariant functor) constructed in the following way, similar to the function spaces above:

- It assigns to each vector space V its dual space V^*, i.e. the vector space of linear functionals (i.e. linear maps) $V \to \mathbb{R}$.
- It assigns to each linear map $f : V \to W$ the map $f^* : W^* \to V^*$. The map f^* takes an element ω of W^*, which is a linear functional $\omega : W \to \mathbb{R}$, and gives the linear functional $V \to \mathbb{R}$ (i.e. element of V^*) given by the composition

$$V \xrightarrow{f} W \xrightarrow{\omega} \mathbb{R}.$$

This functor reverses all the arrows. We now want to discuss the **double dual**, so we apply the functor again, and this time the arrows are back to the normal direction. That is, the double dual is a functor **Vect** $\to$ **Vect**, which explicitly works as follows:

- It assigns to each vector space V its **double dual space** V^{**}, i.e. the vector space of linear functionals $V^* \to \mathbb{R}$.
- It assigns to each linear map $f : V \to W$ the map $f^{**} : V^{**} \to W^{**}$. This map takes an element k of V^{**}, which is a linear functional $k : V^* \to \mathbb{R}$, and gives the linear functional $W^* \to \mathbb{R}$ (i.e. an element of W^{**}) given by the composition

$$W^* \xrightarrow{f^*} V^* \xrightarrow{k} \mathbb{R},$$

  where f^* is the dual map to f defined above.

Now, what follows is a very important definition.

Definition 1.3.44. Let **C** be a category. A **presheaf** on **C** is a functor $\mathbf{C}^{\mathrm{op}} \to$ **Set**.

The reason why we pick **Set** as the target category is that, in our setting, *the hom-spaces of* **C** *are sets*. That is, given any two objects X and Y of **C**, then $\mathrm{Hom}_{\mathbf{C}}(X, Y)$ is a *set*. (Of course, for this to be technically correct, we need to be in a locally small category. Let's assume we are, from now on.)

Remark 1.3.45. There are settings, such as algebraic geometry, where one works in categories whose hom-sets admit additional structure. For example, in the category of abelian groups, the hom-sets can themselves be considered abelian groups. In that case, one may be interested in functors to **AbGrp** instead of **Set**. In other applications, one may be interested in different structures that one can put on the hom-sets. These additional structures are called *enrichments* and are the topic of *enriched category theory*. In this book, we only study ordinary category theory, where hom-sets are just sets. (If you are interested in enriched category theory, this book can still help you: Most concepts in ordinary category theory have analogues in the enriched case.)

The "function space functor" $\mathrm{Hom}_{\mathbf{Set}}(-, \mathbb{R})$ is, for example, a presheaf on **Set**. More generally, given a category **C** and an object X of **C**, one can analogously form a presheaf $\mathrm{Hom}_{\mathbf{C}}(-, X) : \mathbf{C}^{\mathrm{op}} \to \mathbf{Set}$.

Not all presheaves are in this form. Here is a more general example from basic graph theory.

Example 1.3.46 (graph theory). Let **Par** be a category consisting only of the objects and morphisms specified by this diagram of "parallel arrows,"

$$V \underset{t}{\overset{s}{\rightrightarrows}} E,$$

plus the implicit identities at the two objects. A presheaf on **Par**, i.e. a functor $F : \mathbf{Par}^{\mathrm{op}} \to \mathbf{Set}$, consists explicitly of the following:

- Two distinguished sets FV and FE;
- Two functions $Fs, Ft : FE \to FV$ (note the direction).

These can be interpreted as *directed multigraphs*. The set FV is the set of vertices, the set FE is the set of edges, and the maps Fs and Ft associate to each edge its source and target, respectively. For example, the graph

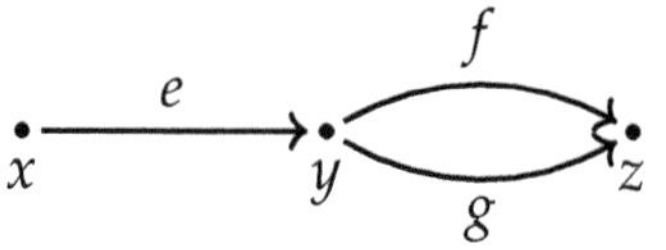

can be constructed as follows by "attaching" the edges to their source and target:

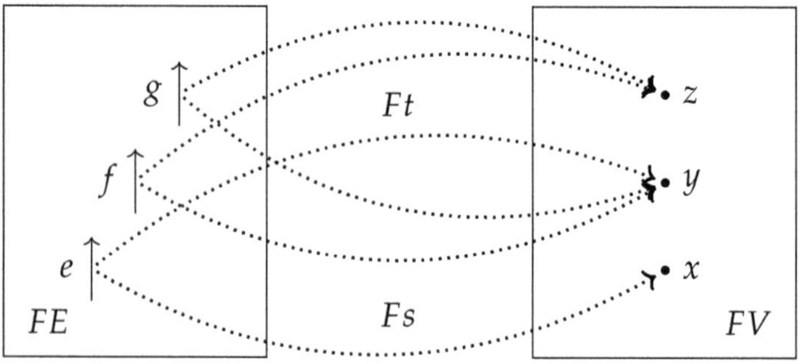

Note that, with constructions of this form, between any two vertices there may be multiple edges, and there may be loops, hence the term *multi*graph. (Although it may slightly resemble a category, note that identities and compositions here are missing.)

Presheaves, though originating in algebraic geometry, find applications beyond it. As the example above shows, they can also be used to understand graphs. This will be of use later on.

Here is another example of a presheaf, which is historically the first and remains a very important one.

Example 1.3.47 (topology). Let X be a topological space. Let $O(X)$ be the topology of X, i.e. the poset of open subsets of X, ordered by inclusion. As a poset, $O(X)$ can be considered a category. We can form a presheaf $O(X)^{\mathrm{op}} \to \mathbf{Set}$ in the following way:

- To each open subset $U \in O(X)$, we associate the set of continuous functions $U \to \mathbb{R}$ (where U inherits the topology from X).
- To each inclusion $U \subseteq V$, which is a morphism of $O(X)$, we need a mapping from continuous functions $V \to \mathbb{R}$ to continuous functions $U \to \mathbb{R}$. As such mapping, we apply the following restriction: Given $g : V \to \mathbb{R}$, we can restrict g to the subset $U \subseteq V$ to get the function $g|_U : U \to \mathbb{R}$. This is again continuous.

This preserves the identities and composition; therefore, it is a functor $O(X)^{\mathrm{op}} \to \mathbf{Set}$, i.e. a presheaf on $O(X)$. Note that, by construction, this reverses the arrows.

Remark 1.3.48. The term *presheaf* is one of the few words in category theory whose translation to other languages is not obvious. In German, it translates to *Prägarbe*, in French *prefaisceau*, in Spanish *prehaz*, and in Italian *prefascio*. It derives from the word *sheaf*, which is itself inspired by agriculture (such as in *a sheaf of grain*). We will not cover sheaves in this book, no pun intended. (For stalks, see Example 3.2.73.)

1.3.8 Further particular functors

The hom-functor $\mathrm{Hom}_{\mathbf{C}}(-,-) : \mathbf{C}^{\mathrm{op}} \times \mathbf{C} \to \mathbf{C}$ is a "functor of two variables." We would like now to show that it is not only functorial in both arguments but also *jointly* functorial. In order to make this precise, let's define rigorously what we mean by the expression $\mathbf{C}^{\mathrm{op}} \times \mathbf{C}$.

Definition 1.3.49. Let $\mathbf{C}$ and $\mathbf{D}$ be categories. Their **product**, or **cartesian product**, is the category $\mathbf{C} \times \mathbf{D}$ whose:

- objects are pairs (C, D), where C is an object of $\mathbf{C}$ and D is an object of $\mathbf{D}$;
- morphisms $(C, D) \to (C', D')$ are pairs (f, g), where $f : C \to C'$ is a morphism of $\mathbf{C}$ and $g : D \to D'$ is a morphism of $\mathbf{D}$;
- identity at (C, D) is given by $(\mathrm{id}_C, \mathrm{id}_D)$, and composition is defined entrywise.

Consider now a functor from the product category $F : \mathbf{C} \times \mathbf{D} \to \mathbf{E}$. By fixing one of its arguments, we see that it is functorial in the other one. For example, given an object C of $\mathbf{C}$, the mapping $F(C, -) : \mathbf{D} \to \mathbf{E}$ is a functor defined as follows:

- On objects, it maps an object D of $\mathbf{D}$ to the object $F(C, D)$ of $\mathbf{E}$.
- On morphisms, it maps a morphism $g : D \to D'$ of $\mathbf{D}$ to the morphism $F(\mathrm{id}_C, g)$ of $\mathbf{E}$, the image under the functor F of the morphism $(\mathrm{id}_C, g) : (C, D) \to (C, D')$ of $\mathbf{C} \times \mathbf{D}$.

Because of this, we can equivalently write $F(C, g)$ or $F(\mathrm{id}_C, g)$. The other argument works similarly, and we can set $F(f, D) = F(f, \mathrm{id}_D)$.

The converse is not true: A mapping which is separately functorial in both arguments is not necessarily *jointly functorial*, i.e. a functor on the product category. Here is a necessary and sufficient condition for that.

Proposition 1.3.50. *Let $F : \mathbf{C} \times \mathbf{D} \to \mathbf{E}$ be an assignment from objects and arrows of $\mathbf{C} \times \mathbf{D}$ to objects and arrows of $\mathbf{E}$, which is functorial in both* *arguments separately, i.e. such that for every object C of* **D** *and D of* **D**, *the mappings $F(C,-) : \mathbf{D} \to \mathbf{E}$ and $F(-,D) : \mathbf{C} \to \mathbf{E}$ are functors.*

We have that F is jointly functorial if and only if for each morphism $f : C \to C'$ of **C** *and $g : D \to D'$ of* **D**, *the following diagram of* **E** *commutes:*

$$\begin{array}{ccc} F(C,D) & \xrightarrow{F(f,D)} & F(C',D) \\ {\scriptstyle F(C,g)}\downarrow & & \downarrow{\scriptstyle F(C',g)} \\ F(C,D') & \xrightarrow[F(f,D')]{} & F(C',D'), \end{array} \tag{1.3.2}$$

which is sometimes called the ***interchange law***.

Proof. First of all, note that the following diagram of $\mathbf{C} \times \mathbf{D}$ commutes:

$$\begin{array}{ccc} (C,D) & \xrightarrow{(f,\mathrm{id}_D)} & (C',D) \\ {\scriptstyle (\mathrm{id}_C,g)}\downarrow & & \downarrow{\scriptstyle (\mathrm{id}_{C'},g)} \\ (C,D') & \xrightarrow[(f,\mathrm{id}_{D'})]{} & (C',D'). \end{array}$$

Therefore, if F is jointly functorial on $\mathbf{C} \times \mathbf{D}$, then diagram (1.3.2) commutes.

Conversely, suppose that F is separately functorial and that diagram (1.3.2) commutes. Then, given a morphism $(f,g) : (C,D) \to (C',D')$ of $\mathbf{C} \times \mathbf{D}$, we can set $F(f,g)$ to be exactly the composite given by (either path of) diagram (1.3.2). This makes F jointly functorial. (Why does this assignment preserve composition?) □

Proposition 1.3.51. *The hom-functor* $\mathrm{Hom}_{\mathbf{C}}(-,-) : \mathbf{C}^{\mathrm{op}} \times \mathbf{C} \to \mathbf{C}$ *is jointly functorial.*

Proof. Using Proposition 1.3.50, it suffices to prove that for each morphism $f : C \to C'$ of **C** and $g : D \to D'$ of **C**, the following diagram commutes (note the direction of the arrows):

$$\begin{array}{ccc} \mathrm{Hom}_{\mathbf{C}}(C',D) & \xrightarrow{-\circ f} & \mathrm{Hom}_{\mathbf{C}}(C,D) \\ {\scriptstyle g\circ -}\downarrow & & \downarrow{\scriptstyle g\circ -} \\ \mathrm{Hom}_{\mathbf{C}}(C',D') & \xrightarrow[-\circ f]{} & \mathrm{Hom}_{\mathbf{C}}(C,D'). \end{array}$$

Now, let $h : C' \to D$. The commutativity of the diagram above exactly corresponds to the fact that $(g \circ h) \circ f = g \circ (h \circ f)$, i.e. showing associativity. □

To conclude this section, let's look at two special types of functors: the *identity functor* and the *composition* of two functors.

Definition 1.3.52. Let $\mathbf{C}$ be a category. The **identity functor** $\mathrm{id}_{\mathbf{C}} : \mathbf{C} \to \mathbf{C}$ maps each object and each morphism of $\mathbf{C}$ to itself. That is,

$$\mathrm{id}_{\mathbf{C}}(X) := X; \quad \mathrm{id}_{\mathbf{C}}(f) := f.$$

When there is no ambiguity, we denote $\mathrm{id}_{\mathbf{C}}$ simply by id.

The functoriality of this construction is immediate (can you see why?). Similar to identity morphisms, identity functors are not very useful in their applications, but it's useful to *keep track* of when a functor does nothing. (Seen as a "forgetful functor," for example, the identity functor forgets exactly nothing.)

Definition 1.3.53. Let $\mathbf{C}$, $\mathbf{D}$ and $\mathbf{E}$ be categories and $F : \mathbf{C} \to \mathbf{D}$ and $G : \mathbf{D} \to \mathbf{E}$ be functors. The **composite functor** $G \circ F : \mathbf{C} \to \mathbf{E}$ (or more briefly GF) maps:

- each object X of $\mathbf{C}$ to the object $G(FX)$ of $\mathbf{E}$ (or more briefly, GFX);
- each morphism f of $\mathbf{C}$ to the morphism $G(Ff)$ of $\mathbf{E}$ (or more briefly, GFf).

Let's check functoriality. On identity morphisms,

$$GF(\mathrm{id}_X) = G(\mathrm{id}_{FX}) = \mathrm{id}_{GFX}$$

by the functoriality of F and G.

On composite morphisms,

$$GF(g \circ f) = G(Fg \circ Ff) = GFg \circ GFf,$$

again by the functoriality of F and G.

Note that, similar to morphisms, two functors F and G can be composed if and only if the codomain of F is the domain of G.

The composition of functors is associative and unital. Thus, categories and functors form themselves a category (see Section 1.5).

1.4 Natural Transformations

We now turn to natural transformations, which can be considered a convenient choice of "arrows between functors."

Naturality is a concept in category theory that often doesn't immediately "click" to newcomers (*why does that diagram have to commute?*). In this section, we present many points of view and examples to illustrate the idea and make it as concrete as possible. If you need even more examples, just read on: In the rest of the book, we see so many instances of naturality that, through usage, you will likely develop an intuitive "feeling" for it.

Definition 1.4.1. Let $\mathbf{C}$ and $\mathbf{D}$ be categories, and let F and G be functors $\mathbf{C} \to \mathbf{D}$. A **natural transformation** α **from** F **to** G consists of the following data:

- for each object C of $\mathbf{C}$, a morphism $\alpha_C : FC \to GC$ in $\mathbf{D}$, called the **component of** α **at** C;
- for each morphism $f : C \to C'$ of $\mathbf{C}$, the following diagram has to commute (**naturality condition**):

$$\begin{array}{ccc} FC & \xrightarrow{Ff} & FC' \\ \big\downarrow{\scriptstyle \alpha_C} & & \big\downarrow{\scriptstyle \alpha_{C'}} \\ GC & \xrightarrow{Gf} & GC'. \end{array}$$

Notation 1.4.2. We write natural transformations as double arrows, $\alpha : F \Rightarrow G$, to distinguish them in diagrams from functors (which are denoted by single arrows):

$$\mathbf{C} \underset{G}{\overset{F}{\rightrightarrows}} \mathbf{D}. \quad (\Downarrow \alpha)$$

The notation suggests that these are "arrows between functors." We make this precise shortly.

Sometimes, when we say that morphisms $\alpha_C : FC \to GC$ in $\mathbf{D}$ are part of a natural transformation $\alpha : F \Rightarrow G$, we say that *the maps* $\alpha_C : FC \to GC$ *are natural in* C. We use this especially for the case where F and G are functors to **Set**, and in that case, we say that the *functions* (for example, bijections) $\alpha_C : FC \to GC$ are natural in C.

Let's now give some examples for an intuitive understanding.

1.4.1 Natural transformations as systems of arrows

Idea. *A natural transformation can be viewed as a consistent system of arrows between the images of two functors.*

Example 1.4.3 (sets and relations). Let $(X, \leq)$ and $(Y, \leq)$ be partial orders, and consider the monotone maps $f, g : (X, \leq) \to (Y, \leq)$. We can view f and g as functors between two categories, where the arrows of Y (and X) are simply the inequalities $y \leq y'$. A natural transformation between f and g is, for each $x \in X$, an arrow $f(x) \to g(x)$; that is, for each $x \in X$, we must have $f(x) \leq g(x)$. Traditionally, one says that $f \leq g$ pointwise. The naturality condition is trivial here since in a partial order, all diagrams commute. So, there is a unique natural transformation $f \Rightarrow g$ if and only if $f \leq g$ pointwise.

Note that in the example above, the order on X and the monotonicity of f and g are not really used to define the pointwise order; only the order on Y is used. In general, however, the arrows of the domain category and functoriality will matter (see the following section).

Exercise 1.4.4 (sets and relations). Suppose that Y is equipped with an equivalence relation instead of a partial order. What do natural transformations between functors into Y look like?

1.4.2 Natural transformations as structure-compatible mappings

Idea. *A natural transformation can be viewed as a mapping between functors that preserves specified actions, symmetries, or other structures.*

Example 1.4.5 (group theory). Let G be a group. We know (Example 1.3.8) that a linear representation of G is a functor $\mathbf{B}G \to \mathbf{Vect}$, mapping the single object of $\mathbf{B}G$ to a specified vector space V, on which the group G "acts." Let now $R, S : \mathbf{B}G \to \mathbf{Vect}$ be linear representations acting on V and W, respectively. A natural transformation $\alpha : R \to S$ consists of the following:

- For each object of $\mathbf{B}G$, we need a morphism of $\mathbf{Vect}$ between the images of this object under R and S. As $\mathbf{B}G$ has a single object and its images under R and S are V and W, this amounts to a linear map $\alpha : V \to W$.
- For each morphism of $\mathbf{B}G$, i.e. for each $g \in G$, the following diagram has to commute:

$$\begin{array}{ccc} V & \xrightarrow{Rg} & V \\ \downarrow{\scriptstyle \alpha} & & \downarrow{\scriptstyle \alpha} \\ W & \xrightarrow{Sg} & W. \end{array}$$

  That is, for each vector $v \in V$, acting with g commutes with the map α. Denoting the two actions of G by just $g\cdot$, this reads

$$\alpha(g \cdot v) = g \cdot \alpha(v).$$

A map commuting with a group action is called an **equivariant map**. The natural transformations between two representations of G are then the linear G-equivariant maps. In representation theory, these are also known as **morphisms of representations**.

Example 1.4.6 (group theory). A group G acting on a space encodes an idea of *symmetry* of the given space. Analogously to the linear case, a G-**space** is a functor $\mathbf{B}G \to \mathbf{Top}$. For example, the group D_3, the group of symmetries of an equilateral triangle, can also act on the following spaces via rotations and reflections:

The natural transformations between G-spaces are exactly the G-equivariant continuous maps. These can be interpreted as the maps which *do not break the symmetry*: For example, denote by ϕ the map from the figure on the left to that on the right, which collapses the petals into segments and then includes them in the figure on the right:

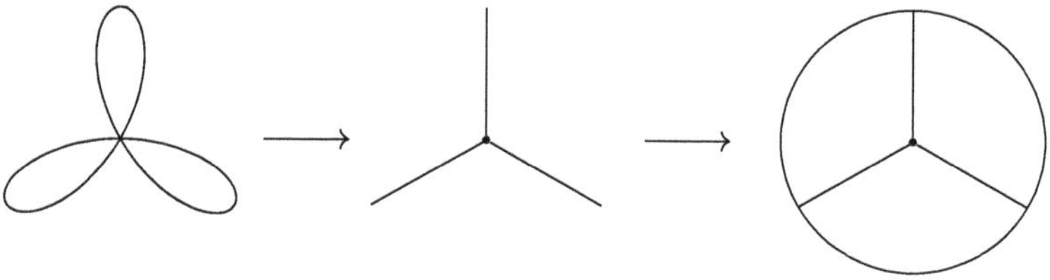

Rotating 120° counterclockwise before and after applying the map gives the same result, as the following illustration shows:

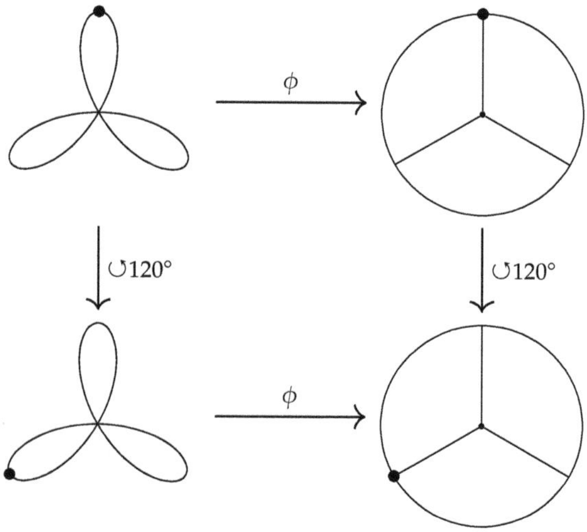

where we marked one of the petals with a dot to make the rotation visible.

To better understand why this map preserves the symmetry, let's also look at what would *break* the symmetry. The easiest example would be collapsing only one of the petals instead of all three. That would break the symmetry. The resulting figure wouldn't even be symmetric under rotations, so we wouldn't even have a map between G-spaces. There is however a subtler way to break the symmetry, while keeping the codomain symmetric: proceed as in the example above, but collapse one (and only one) of the petals completely to the center point, then embed the resulting shape again in the right-most rounded figure. This breaks the symmetry even if the codomain is symmetric, because *it does not treat all the petals equally*. One can come up with even more subtle ways to break the symmetry, for example permuting two petals. The notion of equivariance is a way to exclude any kind of "symmetry breaking," in a very strong sense.

Warning. Note the difference between an *invariant map* and an *equivariant map*. An invariant map satisfies $\phi(g \cdot x) = \phi(x)$, while an equivariant map satisfies $\phi(g \cdot x) = g \cdot \phi(x)$. Although both notions are related to "respecting the symmetry," they are not the same in general. Equivariance does not mean that the map *is independent of the action of* g, but rather it means that *it is compatible with the action of* g. In particular, for a map $\phi : X \to Y$ to be equivariant, we have to specify an action on Y as well, and different choices of actions give us different instances of equivariance. (Invariance is a special case of equivariance for the case where the action of g on Y is trivial.)

Example 1.4.7 (dynamical systems). Let M be a monoid, for example $(\mathbb{N}, +)$. A **topological dynamical system with monoid** M is a functor $\mathbf{B}M \to \mathbf{Top}$. This picks a space X and lets the dynamics indexed by the monoid M act on the given space X. Given dynamical systems $F, G : \mathbf{B}M \to \mathbf{Top}$ acting on the spaces X and Y, respectively, a **morphism of dynamical systems** is a natural transformation $\alpha : F \to G$. This amounts to the following: First of all, we need a map $\alpha : X \to Y$. Moreover, for each $m \in M$, which we can interpret as a time interval, the following diagram has to commute:

$$\begin{array}{ccc} X & \xrightarrow{Fm} & X \\ \downarrow{\scriptstyle\alpha} & & \downarrow{\scriptstyle\alpha} \\ Y & \xrightarrow{Gm} & Y. \end{array}$$

In a way, this means that α is compatible with the dynamics.

Keep in mind that those working in the field of dynamical systems have certain preferences, as follows:

- preference for semigroups rather than monoids (see Remark 1.1.11);
- focus on *compact* spaces instead of all topological spaces;
- greater interest in *surjective* maps $\alpha : X \to Y$ (making the square above commute).

Example 1.4.8 (graph theory). We have seen in Example 1.3.46 that presheaves on the category **Par** (defined there) can be seen as directed multigraphs. Now, let $F, G : \mathbf{Par}^{\mathrm{op}} \to \mathbf{Set}$ be presheaves (i.e. multigraphs). A natural transformation $\alpha : F \to G$ consists of two functions $\alpha_E : FE \to GE$ and $\alpha_V : FV \to GV$ such that the following diagram commutes:

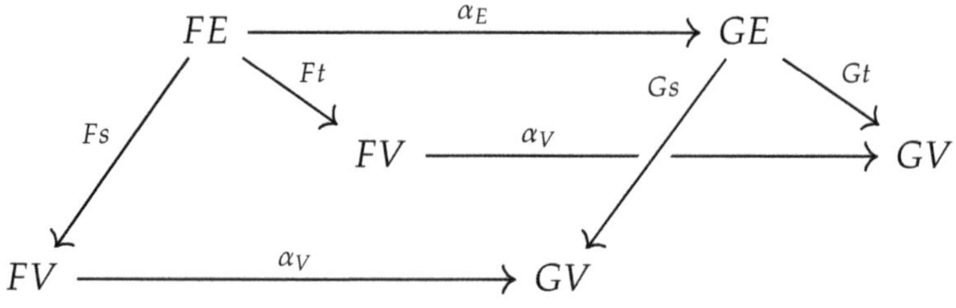

This means that, for an edge $e \in FE$, the source of $\alpha_E(e)$ is exactly the result of applying α_V to the source of e. The same is true for targets. In other words, α is a mapping between edges and vertices that *preserves the incidence relations*, i.e. the relations pertaining to which edges are between which vertices. Such a mapping is sometimes known as a **graph homomorphism**.

1.4.3 Natural transformations as compatible systems of maps

Idea. *A natural transformation can look like a mapping or assignment which is canonical, systematic, or "natural" in the intuitive sense (hence the name).*

Example 1.4.9 (sets and relations). Consider the power set functor $P : \mathbf{Set} \to \mathbf{Set}$ of Example 1.3.14. Given a set X, there is a canonical embedding of X into its power set PX, namely the map $\sigma : X \to PX$, which assign to each $x \in X$ the singleton $\{x\} \in PX$. This map is natural in the following sense. Let $f : X \to Y$ be any function. Then, the following diagram commutes:

$$\begin{array}{ccc} X & \xrightarrow{f} & Y \\ \downarrow{\scriptstyle \sigma} & & \downarrow{\scriptstyle \sigma} \\ PX & \xrightarrow{Pf} & PY. \end{array}$$

Indeed, let $x \in X$. Then, on one side of the diagram,

$$\sigma(f(x)) = \{f(x)\},$$

and on the other side,

$$Pf(\sigma(x)) = \{f(x)\},$$

recalling that the action of Pf on a set involves simply applying f to all the elements of the set (in this case, the set $\{x\}$ has only one element).

Note that the top of the diagram has $f : X \to Y$ without any functor applied to it, or with *the identity functor* applied to it. Therefore, σ is a natural transformation from the identity functor on **Set** to P, i.e. $\sigma : \mathrm{id}_{\mathbf{Set}} \Rightarrow P$. (Note that we have denoted σ_X and σ_Y simply by σ. We do this, for brevity, when this does not cause ambiguity.)

What do we mean when we say that the map above is *canonical*? We don't mean that "there is only one such map" in general. Rather, we mean the following. Let $p : X \to X$ be a permutation, i.e. a bijective map, which we can view as a "relabeling" of the elements of X.

By the naturality of σ, the following diagram commutes:

$$\begin{array}{ccc} X & \xrightarrow{p} & X \\ \downarrow{\sigma} & & \downarrow{\sigma} \\ PX & \xrightarrow{Pp} & PX. \end{array}$$

This can be interpreted as follows: If we relabel the elements of X and then those of PX (i.e. the subsets of X) accordingly, then the map σ will appear the same after the relabeling. In other words, the assignment $\sigma : X \to PX$ *is compatible with relabeling of elements*. It is in this sense that the map σ is called canonical. (The same condition holds if p is not a bijection, and "different elements might get the same label." The map σ is also "equivariant" under these "bad" relabelings.)

We should remark that σ is not *invariant* under relabeling but is *equivariant*, similar to the maps in the previous section. Just as equivariance requires an action on the codomain (and on the domain), naturality requires the codomain (and the domain) to come from a *functor*. Without these two functors, one cannot discuss naturality, and different functors give different instances of naturality.

Here is an analogous map for the probability functor of Example 1.3.16. Again, we first give it for the finitely supported case without using measure theory — the equivalents for more general probability measures are given in the following exercises.

Exercise 1.4.10 (basic probability). Consider the probability functor $\mathcal{P}$ on **Set**, given in Example 1.3.16. Given $x \in X$, define $\delta_x \in PX$ to be the function

$$\delta_x(y) := \begin{cases} 1, & x = y; \\ 0, & x \neq y. \end{cases}$$

Show that the map $\delta : X \to \mathcal{P}X$ assigning to each $x \in X$ the measure δ_x is a natural transformation from the identity functor to $\mathcal{P}$, analogous to the singleton map for the power set.

Let's now see the analogue for general probability measures, not necessarily finitely supported.

Exercise 1.4.11 (measure theory, probability). Let X be a measurable space, and let $x \in X$. Recall that the **Dirac measure at** x is defined to be the one that maps each measurable set $A \subseteq X$ to

$$\delta_x(A) := \begin{cases} 1, & x \in A; \\ 0, & x \notin A. \end{cases}$$

Consider now either the Giry functor $\mathcal{P}$ on **Meas**, given in Exercise 1.3.17, or the Radon functor $\mathcal{P}$ on **CHaus**, given in Exercise 1.3.18 (and make the topological spaces measurable using the Borel σ-algebra). Show that in both cases, the map $\delta : X \to \mathcal{P}X$ assigning to each $x \in X$ the measure δ_x is again a natural transformation from the identity functor to $\mathcal{P}$.

Exercise 1.4.12 (basic computer science, combinatorics). Construct a natural transformation from the identity functor to the list functor of Example 1.3.19, analogous to the ones given in the exercises above.

A similar phenomenon to the "natural maps" above can also be seen in vector spaces. In particular, you probably know from linear algebra that a finite-dimensional vector space V is always isomorphic to its dual V^*, as well as to the dual of the dual, i.e. the double dual V^{**}. However, in some way, V^{**} is "a lot more similar to V than V^* is." That's why sometimes (such as in differential geometry), V^{**} is directly identified with V. (Hence the term "dual": it's as if there were only two of them, V and V^*.) In terms of natural transformations, we can make this intuition precise: There is a *natural* isomorphism $V \to V^{**}$ (but not $V \to V^*$).

Example 1.4.13 (linear algebra). Consider the double dual functor $(-)^{**}$: **Vect** $\to$ **Vect** of Example 1.3.43, mapping a vector space V to its double dual V^{**}, i.e. the space of linear functionals $V^* \to \mathbb{R}$. Every element v of V defines canonically an element of the double dual V^{**} by **evaluation**. Let's see what this means. An element of V^{**} is a linear functional $V^* \to \mathbb{R}$. We can map a linear functional $\omega \in V^*$

to $\mathbb{R}$ by just feeding it v, i.e. $\omega \mapsto \omega(v) \in \mathbb{R}$. Thus, from $v \in V$, we get a linear map $V^* \to \mathbb{R}$, i.e. an element of V^{**}. Let's call this map $\eta : V \to V^{**}$. Let's now prove that this map is natural. This means that for every linear map $f : V \to W$, the following diagram must commute:

$$\begin{array}{ccc} V & \xrightarrow{f} & W \\ \downarrow{\scriptstyle \eta} & & \downarrow{\scriptstyle \eta} \\ V^{**} & \xrightarrow{f^{**}} & W^{**}. \end{array}$$

That is, for each $v \in V$, we need to prove that $\eta(f(v))$ and $f^{**}(\eta(v))$ are equal (as elements of W^{**}). Now, the elements of W^{**} are linear maps $W^* \to \mathbb{R}$, so let $\omega \in W^*$. We have to prove that

$$\eta(f(v))(\omega) = f^{**}(\eta(v))(\omega).$$

We have, by the definition of η,

$$\eta(f(v))(\omega) = \omega(f(v)).$$

On the other hand, by the definition of f^{**} (Example 1.3.43), as well as that of η and f^*,

$$f^{**}(\eta(v))(\omega) = \eta(v)(f^*\omega) = (f^*\omega)(v) = \omega(f(v)).$$

Therefore, η is natural; that is, it is a natural transformation from the identity of **Vect** to the double dual functor.

You probably know from linear algebra that for finite-dimensional vector spaces, η is even an isomorphism. Therefore, a finite-dimensional vector space V is *naturally* isomorphic to its double dual V. If the space is not finite-dimensional, the map η is not an isomorphism, but it is still natural.

Again, let's see why η is "canonical." Let $c : V \to V$ be a linear isomorphism, which we can interpret as a "change of basis." Then, the following diagram commutes.

$$\begin{array}{ccc} V & \xrightarrow{c} & V \\ \downarrow{\scriptstyle \eta} & & \downarrow{\scriptstyle \eta} \\ V^{**} & \xrightarrow{c^{**}} & V^{**}. \end{array}$$

We can interpret this as follows: If we change the basis of V through the map c and also change the basis of V^{**} accordingly, then the map η will look the same after the change of basis. That is, η *is compatible with changes of basis* (again, in an *equivariant* way, not invariant). In particular, if V is finite-dimensional, then η is an isomorphism that identifies the elements of V with those of V^{**}, and this identification is stable under consistent basis changes (as specified by the functors id and $(-)^{**}$). Again, it is in this sense that we say that the map η is canonical, or "natural" in the intuitive sense. This idea is made precise by the naturality condition — actually, the naturality condition is precisely this idea, except that the map c is allowed to be an arbitrary morphism, not necessarily an isomorphism.

For finite-dimensional vector spaces, we have seen that the map η is an isomorphism. Such natural transformations are called natural isomorphisms:

Definition 1.4.14. Let $\mathbf{C}$ and $\mathbf{D}$ be categories, $F, G : \mathbf{C} \to \mathbf{D}$ be functors, and $\alpha : F \Rightarrow G$ be a natural transformation. We call α a **natural isomorphism** if for each object C of $\mathbf{C}$, the component $\alpha_C : FC \to GC$ is an isomorphism. In that case, we also say that the functors F and G are **(naturally) isomorphic**.

Some authors use the term *natural equivalence* instead of *natural isomorphism*. Equivalence is however a different concept from isomorphism in category theory (we study it in Section 1.5.4), so we avoid using the term *equivalence* in this context.

1.4.4 What is not natural?

In order to understand better what naturality really means, here are examples of constructions which may seem natural but are not.

Example 1.4.15 (linear algebra). The *single* dual V^* is not naturally isomorphic to V, not even in finite dimension. A reason is that, on the category of vector spaces, the single dual functor is contravariant; it reverses all the arrows (see Example 1.3.43). Therefore, there can be no natural transformation from the identity to the single dual: One is a functor **Vect** $\to$ **Vect**, and the other is a functor $\mathbf{Vect}^{\mathrm{op}} \to \mathbf{Vect}$. But there is a deeper reason: Even if we try to generalize the notion of naturality appropriately, all isomorphisms $V \to V^*$ that one can construct (in finite dimension) are not stable under a change of basis (such as the dual basis), and this is not natural.

For example, let V be an n-dimensional vector space, and let $e_1, \dots, e_n$ be a basis of V. We can construct the dual basis $e^1, \dots, e^n$ of V^* as the unique one such that $e^i(e_j) = \delta^i_j$. Given these bases, one can construct an isomorphism $d : V \to V^*$ by setting $d(e_i) := e^i$ for all i. This corresponds to "copying the coordinates," in the sense that we map a vector v of V to the vector of V^* that *has the same coordinates as* V (where the coordinates in V are with respect to the basis $e_1, \dots, e_n$, and the coordinates in V^* are with respect to the dual basis). That is, almost by definition,

$$d\left(\sum_{i=1}^{n} v^i e_i\right) = \sum_{i=1}^{n} v^i e^i.$$

However, this procedure uses the basis in such a way that a different choice of basis gives us a different map, even if we change the basis of V^* accordingly (show this!). Therefore, the map d is not natural.

Exercise 1.4.16 (linear algebra). You may know from linear algebra that a finite-dimensional vector space equipped with a positive-definite scalar product is "canonically" isomorphic to its dual space. Can you make this statement precise in terms of natural transformations? (The question primarily asks: in which category?)

Here is another well-known case of a map that fails to be natural in linear algebra (and quantum mechanics).

Example 1.4.17 (linear algebra, quantum physics). Consider the category **Vect**. The tensor product of vector spaces[10] gives us a functor

[10] If you are not familiar with the tensor product, see the beginning of Section 2.3.2.

$F : \mathbf{Vect} \to \mathbf{Vect}$ that sends:

- a vector space V to the vector space $FV := V \otimes V$ (the tensor product, not the cartesian one);
- a linear map $f : V \to W$ to the linear map $Ff : V \otimes V \to W \otimes W$, defined on a basis as

$$Ff(e_i \otimes e_j) := f(e_i) \otimes f(e_j).$$

The expression of the map $f \otimes f$ does not depend on the choice of the coordinates used (prove it!), so that we have a well-defined functor.

However, suppose we want a natural transformation $\alpha : \mathrm{id} \Rightarrow F$, with components $\alpha_V : V \to V \otimes V$. It turns out that *there is no such natural map except the zero map*. Results of this kind, especially for the case of Hilbert spaces, are known in the quantum information literature as *no-broadcasting theorems* (and are related to the *no-cloning theorem*, see [CK17, Section 5.2.8]) since they can be interpreted in quantum physics as the fact that we cannot duplicate a generic quantum state. One may be tempted to define such a map, for example, in the following way using coordinates:

$$\alpha\left(\sum_{i=1}^{n} v^i e_i\right) := \sum_{i=1}^{n} v^i e_i \otimes e_i.$$

However, just as in Example 1.4.15, this map is not compatible with changes of basis, and so it is not natural. (In physics, this means that this map can only duplicate the state of basis vectors but not generic states.) Further context will be given in Section 6.2.3.

1.4.5 Functor categories and diagrams

Definition 1.4.18. Let **C** and **D** be categories. The **functor category** $[\mathbf{C}, \mathbf{D}]$ is constructed as follows:

- Objects are functors $F : \mathbf{C} \to \mathbf{D}$.
- Morphisms are natural transformations $\alpha : F \Rightarrow G$.

Note that the functors are the *objects* of the category, not the morphisms.

Exercise 1.4.19 (important!). Prove that $[\mathbf{C}, \mathbf{D}]$ is indeed a category, where the identity of a functor F is the natural transformation with components $\mathrm{id}_{FC} : FC \to FC$, and the composition of the two natural transformations $\alpha : F \Rightarrow G$ and $\beta : G \Rightarrow H$ is the natural transformation $\beta \circ \alpha : F \Rightarrow H$ with components

$$FC \xrightarrow{\alpha_C} GC \xrightarrow{\beta_C} HC.$$

(Sometimes, this is called **vertical composition**, see Section 1.4.6.) Why are the identity and the composite natural transformations natural again?

Example 1.4.20 (several fields). The functor category generalizes the following known constructions:

- If $\mathbf{C}$ and $\mathbf{D}$ are partial orders, $[\mathbf{C}, \mathbf{D}]$ is the partial order of monotone functions $\mathbf{C} \to \mathbf{D}$, ordered by the pointwise order (see Example 1.4.3).
- If $\mathbf{C} = \mathbf{B}G$ for a group G and $\mathbf{D} = \mathbf{Vect}$, the category $[\mathbf{C}, \mathbf{D}]$ is the category of linear representations of G and G-equivariant (linear) maps (see Example 1.4.5).
- Analogously, if $\mathbf{C} = \mathbf{B}G$ and $\mathbf{D} = \mathbf{Top}$, the category $[\mathbf{C}, \mathbf{D}]$ is the category of G-spaces and G-equivariant (continuous) maps (see Example 1.4.6).
- Again in the same spirit, if instead $\mathbf{C} = \mathbf{B}M$ for M a monoid, the category $[\mathbf{C}, \mathbf{D}]$ is the category of dynamical systems with dynamics indexed by M and their morphisms (see Example 1.4.7).
- If $\mathbf{C}$ is the category $\mathbf{Par}$ with only two parallel arrows of Example 1.3.46, then $[\mathbf{C}, \mathbf{D}]$ is the category of directed multigraphs and incidence-preserving mappings (see Example 1.4.8). We denote this category by $\mathbf{MGraph}$.

Warning. The functor category $[\mathbf{C}, \mathbf{D}]$ may not be locally small, even if $\mathbf{C}$ and $\mathbf{D}$ are (why?).

Now that we know about functors, we are also ready for a rigorous definition of a diagram. We had defined diagrams explicitly, but informally, in Section 1.1.6. Here is the real, but more abstract, definition.

Definition 1.4.21 (This time for real!). Let $\mathbf{C}$ be a category and $\mathbf{I}$ be a small category.[11] A **diagram in C of shape I** is a functor $\mathbf{I} \to \mathbf{C}$.

The **category of I-shaped diagrams in C** is the functor category $[\mathbf{I}, \mathbf{C}]$.

This is analogous to how one defines a loop in a topological space X as a continuous map $S^1 \to X$, or a curve as a continuous map $\mathbb{R} \to X$. We have a category $\mathbf{I}$ of a certain shape, for example

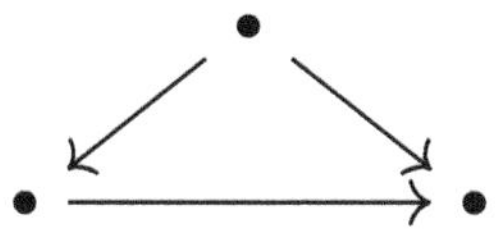

(Identity and compositions are not drawn, but are present — in particular, there are two distinct morphisms from the top object to the bottom-right one.) Therefore, a functor $D : \mathbf{I} \to \mathbf{C}$ will pick a diagram in $\mathbf{C}$ as its "image," such as

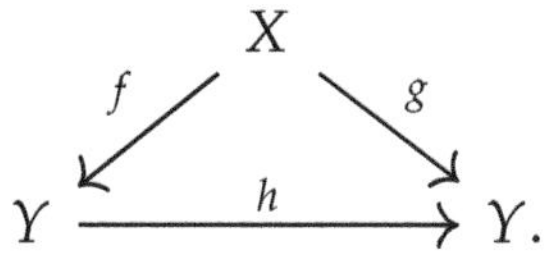

Note that D doesn't need to be injective in any sense; therefore, objects and morphisms can appear more than once in the diagram. Also, note that this triangle doesn't need to be commutative. However, if the original triangle in $\mathbf{I}$ commutes (in the sense given in Section 1.1.6),

[11] Some authors allow diagrams to be indexed by any category, not necessarily small.

then this has to commute in **C** too (and we have a commutative diagram).

A morphism of diagrams $\alpha : D \Rightarrow D'$ is a natural transformation. This means that we have a diagram

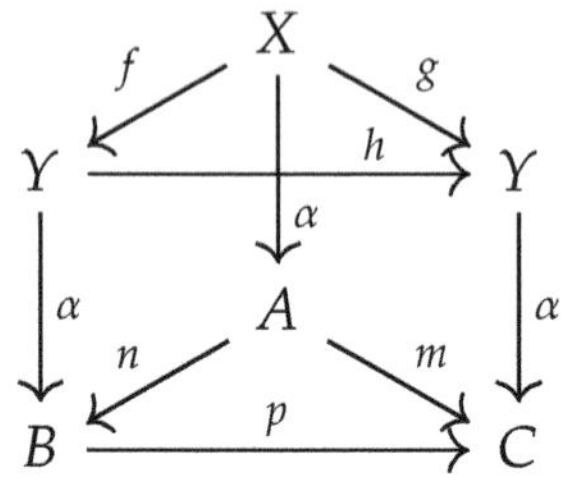

such that all the vertical squares (the ones involving α) are commutative. Readers familiar with algebraic topology may find it helpful to think of this diagram as something analogous to a *homotopy* between two loops $S^1 \to X$.

1.4.6 Whiskering and horizontal composition

Consider three categories $\mathbf{C}, \mathbf{D}, \mathbf{E}$, a functor $F : \mathbf{C} \to \mathbf{D}$, functors $H, I : \mathbf{D} \to \mathbf{E}$, and a natural transformation $\beta : H \Rightarrow I$. The situation is better represented as the following diagram:

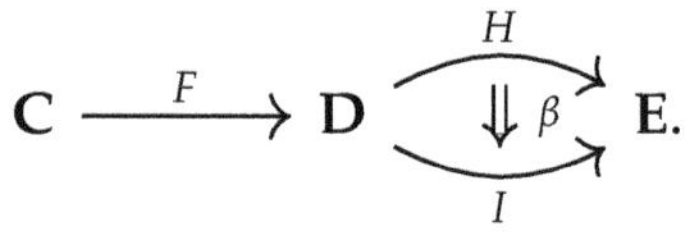

Then, we also have a natural transformation $H \circ F \Rightarrow I \circ F$, i.e. between the top composition and the bottom composition in the diagram. This is obtained as follows. The natural transformation β has as components arrows $\beta_D : HD \to ID$ for each object D of **D**. In particular, if we take a D in the image of F, i.e. $D = FC$, for some object C of **C**, there is an arrow $\beta_{FC} : HFC \to IFC$ of **E**. We can therefore define a natural transformation, which we denote by $\beta F : H \circ F \Rightarrow I \circ F$, where:

- its component $(\beta F)_C$, at each object C of $\mathbf{C}$, is the arrow given by the component $\beta_{FC} : HFC \to IFC$ of β, applied to FC;
- for each arrow $f : C \to C'$ of $\mathbf{C}$, the naturality diagram

$$\begin{array}{ccc} HFC & \xrightarrow{HFf} & HFC' \\ \big\downarrow{\scriptstyle \beta_{FC}} & & \big\downarrow{\scriptstyle \beta_{FC'}} \\ IFC & \xrightarrow{IFf} & IFC' \end{array}$$

  commutes because of the naturality of β.

We call the natural transformation βF the **whiskering** of the natural transformation β on the left by the functor F, since the functor F is added as a sort of "whisker" on the left of β.

Analogously, now suppose that we have the situation depicted in the following diagram:

$$\mathbf{C} \underset{G}{\overset{F}{\rightrightarrows}} \mathbf{D} \xrightarrow{\;H\;} \mathbf{E}. \qquad (\Downarrow \alpha : F \Rightarrow G)$$

Again, we get a natural transformation $HF \Rightarrow HG$, which we denote as $H\alpha$, as follows. For each object C of $\mathbf{C}$, the natural transformation α gives an arrow $\alpha_C : FC \to GC$ of $\mathbf{D}$. We can now apply H to this arrow to get the arrow $H(\alpha_C) : HFC \to HGC$ of $\mathbf{E}$. This is how the natural transformation $H\alpha : HF \Rightarrow HG$ is constructed:

- Its component at each object C of $\mathbf{C}$ is given by the arrow $H\alpha_C : HFC \to HGC$ of $\mathbf{E}$;
- For each arrow $f : C \to C'$ of $\mathbf{C}$, consider the following diagrams:

$$\begin{array}{ccc} FC & \xrightarrow{Ff} & FC' \\ \big\downarrow{\scriptstyle \alpha_C} & & \big\downarrow{\scriptstyle \alpha_{C'}} \\ GC & \xrightarrow{Gf} & GC' \end{array} \qquad \begin{array}{ccc} HFC & \xrightarrow{HFf} & HFC' \\ \big\downarrow{\scriptstyle H\alpha_C} & & \big\downarrow{\scriptstyle H\alpha_{C'}} \\ HGC & \xrightarrow{HGf} & HGC'. \end{array}$$

  The diagram on the left commutes by the naturality of α. The diagram on the right can be obtained from the diagram on the left by applying the functor H, and since functors preserve commutative diagrams, the diagram on the right commutes too. The right-hand side diagram is precisely the naturality condition for $H\alpha_C$.

We call the natural transformation $H\alpha_C$ the **whiskering** of α on the right by H. Note that for each object C of $\mathbf{C}$, the map $H\alpha_C$ may be interpreted either as $H(\alpha_C)$, i.e. the map obtained by applying H to the component α_C of α, or as $(H\alpha)_C$, i.e. the component at C of $H\alpha$. These two coincide by definition, so writing $H\alpha_C$ does not lead to ambiguity.

Exercise 1.4.22 (important!). Consider the situation depicted in the following diagram:

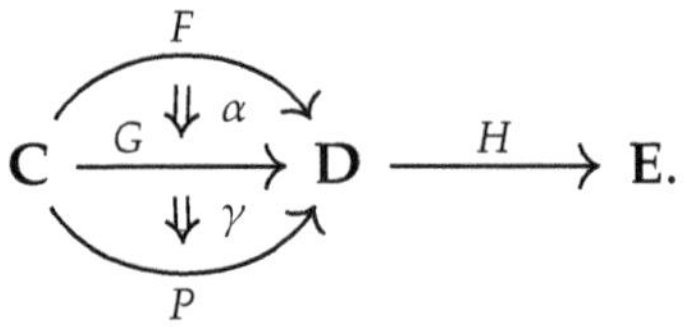

Prove, using the functoriality of H, that $H(\gamma \circ \alpha) = (H\gamma) \circ (H\alpha)$. (How about the dual situation?)

Suppose now that we have the following situation:

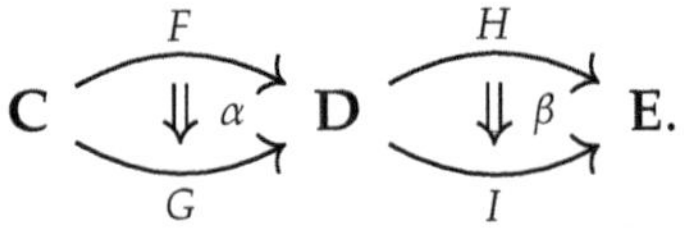

One could view this situation as follows. First, we whisker α on the right with H,

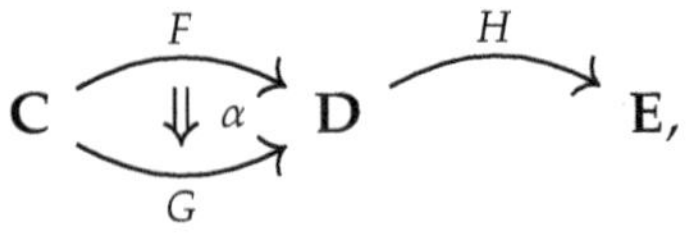

then we whisker β on the left with G,

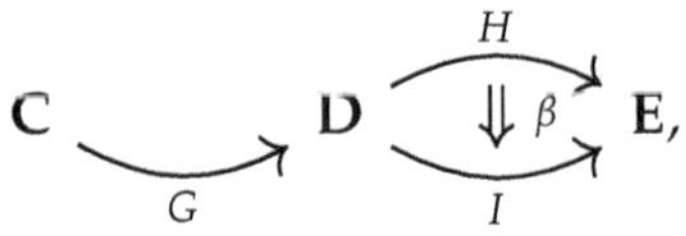

and finally, we compose the two, "gluing" the diagrams,[12]

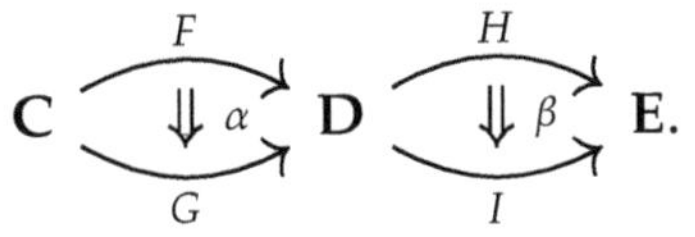

Alternatively, we could first whisker β with F on the left,

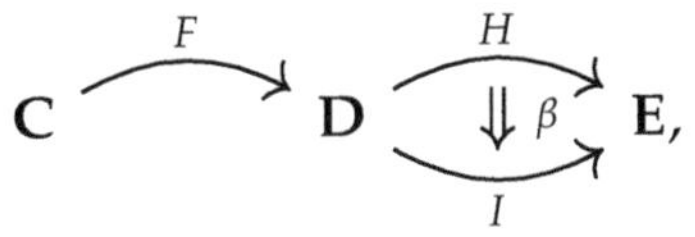

whisker α with I on the right,

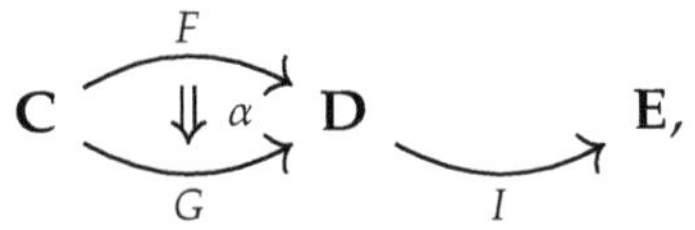

and again compose the two,

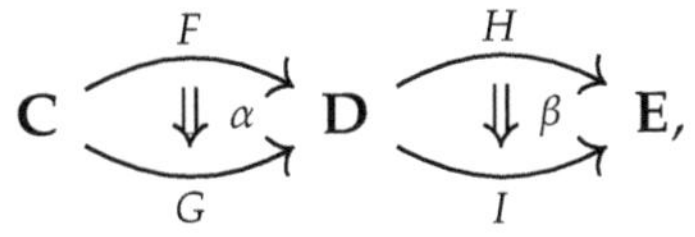

and the diagram would look the same. This turns out to yield exactly the same result so that the diagram is not ambiguous. In other words, we have the following.

Proposition 1.4.23. *Let* $\mathbf{C}$, $\mathbf{D}$, *and* $\mathbf{E}$ *be categories. Let* $F, G : \mathbf{C} \to \mathbf{D}$ *and* $H, I : \mathbf{D} \to \mathbf{E}$ *be functors, and let* $\alpha : F \Rightarrow G$ *and* $\beta : H \Rightarrow I$ *be natural transformations. Then,*

$$(\beta G) \circ (H\alpha) = (I\alpha) \circ (\beta F).$$

[12]Note that these are more general than the diagrams defined in either Section 1.1.6 or Section 1.4.5 since they contain double arrows (2-cells). We will not give their full definition here, but it should be clear what we mean.

Proof. It suffices to check that these two composite natural transformations have the same components. So, let C be an object of $\mathbf{C}$. Plugging in the definition of whiskering, the component of $(\beta G)\circ(H\alpha)$ at C is the arrow of $\mathbf{E}$ given by the composition

$$HFC \xrightarrow{H\alpha_C} HGC \xrightarrow{\beta_{GC}} IGC.$$

Analogously, the component of $(I\alpha)\circ(\beta F)$ at C is the arrow of $\mathbf{E}$ given by the composition

$$HFC \xrightarrow{\beta_{FC}} IFC \xrightarrow{I\alpha_C} IGC.$$

These two compositions are equal if and only if the following diagram commutes:

$$\begin{array}{ccc} HFC & \xrightarrow{H\alpha_C} & HGC \\ \downarrow{\scriptstyle \beta_{FC}} & & \downarrow{\scriptstyle \beta_{GC}} \\ IFC & \xrightarrow{I\alpha_C} & IGC; \end{array}$$

indeed, this diagram commutes: This is granted by the naturality of β. □

Naturality, in other words, means that writing these diagrams does not lead to ambiguity.

Definition 1.4.24. Consider the hypothesis of Proposition 1.4.23, fitting in the diagram

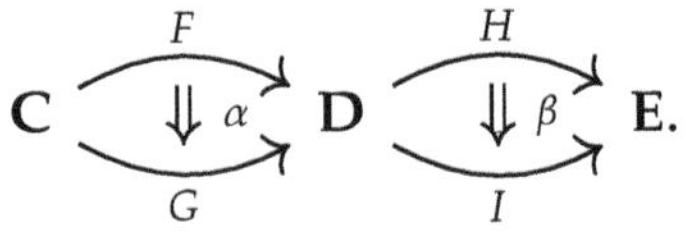

We call the resulting natural transformation $HF \Rightarrow IG$, which can be written equivalently (by Proposition 1.4.23) as $(\beta G)\circ(H\alpha)$, or $(I\alpha)\circ(\beta F)$, the **horizontal composition of α and β.** We denote it simply by juxtaposition: $\beta\alpha$.

Warning. The horizontal composition $\beta\alpha$ is *different* from the ordinary composition $\beta \circ \alpha$, which we can call the **vertical composition.** In this case, the vertical composition $\beta \circ \alpha$ is not even defined since the target of α (the functor G) is not necessarily equal to the source of β (the functor H).

However, the vertical and horizontal compositions are related by the so-called **interchange law** of natural transformations, as follows.

Proposition 1.4.25 (interchange law). *Consider the situation depicted in the following diagram:*

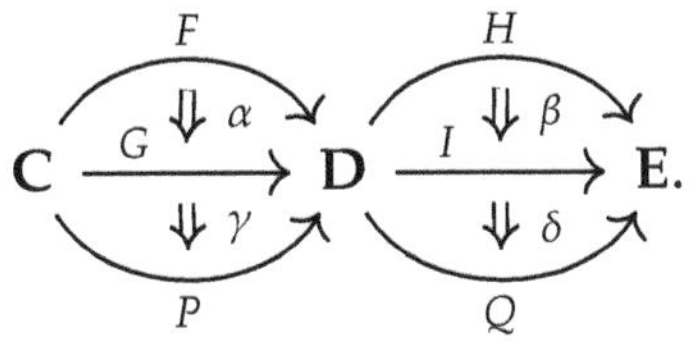

The diagram is not ambiguous in the sense that the vertical composition of the horizontal compositions

$$(\delta\gamma) \circ (\beta\alpha)$$

is equal to the horizontal composition of the vertical compositions

$$(\delta \circ \beta)(\gamma \circ \alpha).$$

Proof. By definition, $\beta\alpha$ can be written as $(I\alpha) \circ (\beta F)$. Similarly, $\delta\gamma$ can be written as $(\delta P) \circ (I\gamma)$. Now, using Exercise 1.4.22 and Proposition 1.4.23,

$$\begin{aligned}(\delta\gamma) \circ (\beta\alpha) &= (\delta P) \circ (I\gamma) \circ (I\alpha) \circ (\beta F)\\ &= (\delta P) \circ I(\gamma \circ \alpha) \circ (\beta F)\\ &= (\delta P) \circ (\beta P) \circ H(\gamma \circ \alpha)\\ &= (\delta \circ \beta)P \circ H(\gamma \circ \alpha),\end{aligned}$$

which is a way to write $(\delta \circ \beta)(\gamma \circ \alpha)$. □

Exercise 1.4.26. Try to understand this proof graphically by drawing the whiskerings explicitly.

Before we conclude, here are two exercises that may help in gaining a better intuition for naturality.

Exercise 1.4.27. Let $F : \mathbf{C} \times \mathbf{D} \to \mathbf{E}$ be an assignment which is separately functorial in both arguments. Show that it is jointly functorial if and only if the assignment

$$\begin{aligned} \mathbf{C} &\longrightarrow [\mathbf{D}, \mathbf{E}] \\ C &\longmapsto F(C, -) \end{aligned}$$

is functorial in $\mathbf{C}$. (*Hint*: For each $f : C \to C'$ of $\mathbf{C}$, *something* needs to be natural in D.)

We interpret this in Section 6.5 as the fact that **Cat** is *cartesian-closed* (see that section for more).

Exercise 1.4.28. Denote by $\mathbf{2}$ the category with two objects 0 and 1 and a single non-identity morphism between them. Given functors $F, G : \mathbf{C} \to \mathbf{D}$, show that a natural transformation is equivalently given by a (joint) functor $H : \mathbf{2} \times \mathbf{C} \to \mathbf{D}$ such that $H(0, -) = F$ and $H(1, -) = G$. Compare this with homotopies of maps between topological spaces.

1.5 Studying Categories by Means of Functors

The main philosophy of category theory involves studying objects not by looking "inside" them, but rather in terms of the morphisms between them. This can be done for categories too: They can be studied in terms of functors.

1.5.1 The category of categories

Categories themselves form a category, with functors as the morphisms. In order to avoid size issues (such as when considering the category of *all* categories), we have to restrict ourselves to small categories.

Definition 1.5.1. The category **Cat** has:

- as objects, small categories;
- as morphisms, the functors between them.

The identities are given by the identity functors and the composition by the composition of functors.

Given small categories $\mathbf{C}$ and $\mathbf{D}$, the functors between them are the elements of the set $\mathrm{Hom}_{\mathbf{Cat}}(\mathbf{C}, \mathbf{D})$. However, as we saw in Section 1.4.5, they are also the *objects of a category*, namely the functor category $[\mathbf{C}, \mathbf{D}]$. As we said at the end of Section 1.1.5, in a set, elements can be equal or not, while in a category, objects can be either *isomorphic* or not — equality is not really a helpful concept inside a category. Along the same line, it is much more helpful to discuss *isomorphisms* between functors than equality. It is much more convenient, in mathematical practice, to think of functors $\mathbf{C} \to \mathbf{D}$ as objects of a category than as elements of a set. In other words, the category $[\mathbf{C}, \mathbf{D}]$ is mathematically a more useful construction than the *set* $\mathrm{Hom}_{\mathbf{Cat}}(\mathbf{C}, \mathbf{D})$.

Therefore, the category **Cat**, as defined above, is not very much used in category theory. There is a more refined notion of **Cat** in which the hom-spaces $\mathrm{Hom}_{\mathbf{Cat}}(\mathbf{C}, \mathbf{D})$ are themselves *categories* instead of just sets. This is called a **2-category**, and it is one of the core concepts of higher category theory, which is beyond the scope of this basic book.[13]

[13]For the interested reader, the nLab (http://ncatlab.org) has extensive material on this. See also Refs. [JY21; Lac09], but in order to understand them, I recommend to finish this book first.

In particular, the notions of mono and epi of **Cat** are not a convenient way to study categories since these concepts are based on the concept of equality of arrows and not isomorphism of functors. Much more helpful notions are given in Section 1.5.3.

1.5.2 Subcategories

Definition 1.5.2. Let **C** be a category. A **subcategory S** of **C** consists of a subcollection of the objects of **C** and a subcollection of the morphisms of **C** such that:

- for each object S of **S**, the identity of S is also in **S**;
- for each pair of composable arrows $f : X \to Y$ and $g : Y \to Z$ of **S**, their composite $g \circ f$ is also in **S**.

Definition 1.5.3. A subcategory **S** of **C** is called **wide** if all objects of **C** are also objects of **S** (but **S** could have fewer morphisms).

A subcategory **S** of **C** is called **full** if given any two objects X and Y of **S** (and **C**), all their morphisms in **C** are also morphisms in **S**. In other words,

$$\mathrm{Hom}_{\mathbf{S}}(X, Y) = \mathrm{Hom}_{\mathbf{C}}(X, Y).$$

(But not all objects of **C** are necessarily objects of **S** too.)

Example 1.5.4 (several fields). The following are wide subcategories:

- The category **Lip** of metric spaces and *Lipschitz* maps is a wide subcategory of the category **Met** of metric spaces and continuous maps.
- The category of sets and *injective* maps is a wide subcategory of **Set**.
- For every category, its core (Definition 1.1.32) is a wide subcategory.
- For S a subgroup of G, the category $\mathbf{B}S$ is a wide subcategory of $\mathbf{B}G$.

The following are full subcategories:

- The category **AbGrp** of abelian groups and group homomorphisms is a full subcategory of **Grp**.
- The category **FVect** of finite-dimensional vector spaces and linear maps is a full subcategory of **Vect**.
- The category **CHaus** of compact Hausdorff spaces and continuous maps is a full subcategory of **Top**.
- A subposet $(S, \leq)$ of $(X, \leq)$ with the order induced from $(X, \leq)$ is a full subcategory of $(X, \leq)$ seen as a category.

1.5.3 Full, faithful, and essentially surjective

A function $f : X \to Y$ between sets can be:

- **injective**: for $x, x' \in X$, we have that if $f(x) = f(x')$ in Y, then $x = x'$ in X;
- **surjective**: for each $y \in Y$, there exists $x \in X$ such that $f(x) = y$.

For functors, we have instead three possible properties. These roughly translate to "injective on arrows," "surjective on arrows," and "surjective on objects up to isomorphism."

Definition 1.5.5. A functor $F : \mathbf{C} \to \mathbf{D}$ can be

- **faithful**: for every objects C, C' of $\mathbf{C}$ and every pair of arrows $f, f' : C \to C'$ of $\mathbf{C}$, we have that if $Ff = Ff'$ in $\mathbf{D}$, then $f = f'$ in $\mathbf{C}$.
- **full**: for every objects C, C' of $\mathbf{C}$ and every arrow $g : FC \to FC'$ of $\mathbf{D}$, there exists an arrow $f : C \to C'$ of $\mathbf{C}$ such that $Ff = g$;
- **essentially surjective**: for every object D of $\mathbf{D}$, there exist an object C of $\mathbf{C}$ and an isomorphism $FC \to D$.

We also call F **fully faithful** if it is full and faithful.

Remark 1.5.6. Given objects C, C' of $\mathbf{C}$, a functor $F : \mathbf{C} \to \mathbf{D}$ induces a function

$$\mathrm{Hom}_{\mathbf{C}}(C, C') \to \mathrm{Hom}_{\mathbf{D}}(FC, FC').$$

This function is injective for all C, C' precisely if F is faithful, surjective for all C, C' precisely if F is full, and bijective for all C, C' precisely if F is fully faithful.

Compare the definition of essential surjectivity with that of surjectivity of functions. An essentially surjective functor is "surjective up to isomorphism." This is the only version of surjectivity that we really need between categories since we look at *isomorphic objects* rather than equal ones. Fully faithful functors are "injective up to isomorphism," as the following proposition shows.

Proposition 1.5.7. *Let $F : \mathbf{C} \to \mathbf{D}$ be a fully faithful functor. Let C, C' be objects in $\mathbf{C}$, and let $\phi : FC \to FC'$ be an isomorphism of $\mathbf{D}$. Then, there is a unique isomorphism $\tilde{\phi} : C \to C'$ such that $F\tilde{\phi} = \phi$.*

In particular, if $F(C)$ and $F(C')$ are isomorphic (in $\mathbf{D}$), then C and C' are isomorphic (in $\mathbf{C}$).

Proof. Let $\psi : F(C') \to F(C)$ be the inverse of the isomorphism ϕ. Since F is fully faithful, necessarily $\phi = F\tilde{\phi}$ for a unique $\tilde{\phi} : C \to C'$, and $\psi = F\tilde{\psi}$ for a unique $\tilde{\psi} : C' \to C$. It now suffices to prove that $\tilde{\phi}$ and $\tilde{\psi}$ are inverse to each other. Now, as ϕ and ψ are inverse to each other, $\psi \circ \phi = \mathrm{id}_{FC}$, or, in other words, $F\tilde{\psi} \circ F\tilde{\phi} = \mathrm{id}_{FC}$. By functoriality, the last equation is equivalent to $F(\tilde{\psi} \circ \tilde{\phi}) = F(\mathrm{id}_C)$. Since F is faithful, this implies $\tilde{\psi} \circ \tilde{\phi} = \mathrm{id}_C$. In the same way, from $\phi \circ \psi = \mathrm{id}_{FC'}$, one can conclude that $\tilde{\phi} \circ \tilde{\psi} = \mathrm{id}_{C'}$. Therefore, $\tilde{\phi}$ and $\tilde{\psi}$ are inverse to each other. □

Example 1.5.8 (algebra, analysis).

- The forgetful functor $\mathbf{AbGrp} \to \mathbf{Grp}$ is fully faithful. It is not essentially surjective: Not every group is isomorphic to an abelian group. More generally, a full subcategory always defines a fully faithful "inclusion" functor, and vice versa.

- The forgetful functor **Lip** $\to$ **Met** is faithful and essentially surjective, but not full: Not every continuous map is Lipschitz.

Let's see what happens for posets, equivalence relations, and preorders.

Example 1.5.9 (sets and relations). Let $(X, \leq)$ and $(Y, \leq)$ be posets, and let $f : (X, \leq) \to (Y, \leq)$ be monotone:

- f is always trivially faithful since for any two objects $x, x' \in X$, there can be up to one unique arrow between them.
- f is full precisely if it is **order-reflecting**; that is, $f(x) \leq f(x')$ implies $x \leq x'$. Note the direction of the implication: This is a stronger condition than monotonicity (monotonicity is just functoriality).
- Since two elements in a poset are isomorphic if and only if they are equal, essential surjectivity is just surjectivity.

For equivalence relations and preorders, again, every functor is automatically faithful and full precisely if it *reflects* (and not just preserves) the relation. Essential surjectivity is more subtle: Given $f : (X, \sim) \to (Y, \sim)$, the function f is essentially surjective if for every $y \in Y$, there exists $x \in X$ such that $f(x) \sim y$. In other words, the image of f is not necessarily the whole of Y, but *it hits every equivalence class of* Y.

Let's now see what happens for groups (and analogously, for monoids).

Example 1.5.10 (group theory). Let G and H be groups. A group homomorphism $f : G \to H$ corresponds to a functor $F : \mathbf{B}G \to \mathbf{B}H$, and vice versa. We have that:

- F is always trivially essentially surjective since $\mathbf{B}H$ has a single object, which is the image of the single object of $\mathbf{B}G$.
- F is faithful exactly if f is injective by Remark 1.5.6.

- F is full exactly if f is surjective by Remark 1.5.6.
- Therefore, F is fully faithful if and only if f is a group isomorphism.

The term *faithful* originally comes from representation theory, and here is why.

Example 1.5.11 (group theory). Let G be a group. A linear representation of G is a functor $R : \mathbf{B}G \to \mathbf{Vect}$. Let V be a unique object in the image of R,i.e. the space on which G acts. The functor R is faithful if and only if different elements, $g \neq h \in G$, give different linear maps, $V \to V$. In other words, for every $g \neq h \in G$, there exists $v \in V$ such that $g \cdot v \neq h \cdot v$. In representation theory, a representation with this property is called a **faithful representation of** G. The intuitive idea is that this representation respects all the structure of G without forgetting anything. As the terminology suggests, it "represents G faithfully."

The same can be said about group actions in other categories, such as **Set** and **Top**.

Remark 1.5.12. One may ask, *why do we have three properties for functors*? Why not two, as for functions, or why not four, two for objects and two for morphisms? One possible answer will be given by Theorem 1.5.16. Another one is the following. A set can be seen as a category in which the only morphisms are the identities. In particular, given $x, x' \in X$, there is an arrow $x \to x'$ if and only if $x = x'$. A function $f : X \to Y$ is therefore injective if and only if *it is surjective on arrows* between any given two objects: If there is an arrow $f(x) \to f(x')$ (i.e. $f(x) = f(x')$), then there is an arrow $x \to x'$ (i.e. $x = x'$). In other words, surjectivity and injectivity can be seen as surjectivity on objects and arrows, respectively. Injectivity on arrows is guaranteed by the fact that between x and x', in a set, there cannot be more than one arrow. In a category, however, there can be many arrows, and so injectivity on arrows becomes an issue too. Again, injectivity on arrows can be seen as surjectivity between "arrows between arrows," and so on. This idea becomes precise and more general in the context of higher category theory.

1.5.4 Equivalences of categories

Definition 1.5.13. Let **C** and **D** be categories. An **equivalence of categories** between **C** and **D** consists of a pair of functors $F : \mathbf{C} \to \mathbf{D}$ and $G : \mathbf{D} \to \mathbf{C}$ and natural isomorphisms $\eta : G \circ F \Rightarrow \mathrm{id}_{\mathbf{C}}$ and $\varepsilon : F \circ G \Rightarrow \mathrm{id}_{\mathbf{D}}$.

In that case, we call G a **pseudoinverse** of F (and vice versa).

Compare this with the definition of isomorphism in a category (Definition 1.1.24). In some sense, here we require G to be the inverse of F "only up to isomorphism." Readers familiar with topology may find this analogous to a *homotopy equivalence* between two spaces (rather than a homeomorphism).

Similarly to isomorphisms, we sometimes denote the equivalence simply by F (or G). The other functor, if it exists, is unique up to natural isomorphism (why?).

Example 1.5.14 (sets and relations, group theory). The following are examples of equivalences of categories:

- Two posets are equivalent as categories if and only if they are isomorphic as posets.
- Two sets equipped with equivalence relations are equivalent as categories if and only if they have isomorphic quotients.
- Two groups G and H are isomorphic if and only if $\mathbf{B}G$ and $\mathbf{B}H$ are equivalent categories. The same is true for monoids.

Exercise 1.5.15. If any of the statements above do not convince you, try to prove them explicitly by plugging in the definition.

In the category of sets, a function between sets is a bijection (i.e. it is invertible) if and only if it is both injective and surjective. We have a similar statement for functors, where instead of injectivity and surjectivity, we have the properties of Section 1.5.3.

Theorem 1.5.16. *A functor $F : \mathbf{C} \to \mathbf{D}$ defines an equivalence of categories if and only if it is fully faithful and essentially surjective.*

This theorem is quite convenient since it is often much easier to check the specified properties for F than to find a specific pseudoinverse. Following the approach of [Rie16], we prove the theorem using the following lemma, which is itself interesting.

Lemma 1.5.17 ([Rie16, Lemma 1.5.10]). *Let $f : X \to Y$ be a morphism in a category* **C**. *Consider isomorphisms $\phi : X \to X'$ and $\psi : Y \to Y'$. Then, there is a unique morphism $X' \to Y'$ such that the following diagram commutes:*

$$\begin{array}{ccc} X & \xrightarrow{f} & Y \\ {\scriptstyle \phi}\downarrow{\scriptstyle \cong} & & {\scriptstyle \psi}\downarrow{\scriptstyle \cong} \\ X' & \longrightarrow & Y', \end{array}$$

and this morphism is given by $\psi \circ f \circ \phi^{-1}$.

Proof of Lemma 1.5.17. First of all, the morphism $\psi \circ f \circ \phi^{-1}$ makes the diagram commute since $\psi \circ f \circ \phi^{-1} \circ \phi = \psi \circ f$. Now, suppose that there is any other morphism $g : X' \to Y'$ that makes the diagram commute. That is, $g \circ \phi = \psi \circ f$. By applying ϕ^{-1} on the right on both sides, we get $g = \psi \circ f \circ \phi^{-1}$. □

We are now ready to prove the theorem.

Proof of Theorem 1.5.16. Let $F : \mathbf{C} \to \mathbf{D}$ be a functor with pseudoinverse $G : \mathbf{D} \to \mathbf{C}$ and with natural isomorphisms $\eta : G \circ F \Rightarrow \mathrm{id}_{\mathbf{C}}$ and $\varepsilon : F \circ G \Rightarrow \mathrm{id}_{\mathbf{D}}$. Let's prove that F is faithful. Let C and C' be objects of **C**. Suppose that $f, f' : C \to C'$ are morphisms of **C** such that $Ff = Ff'$ in **D**. By the naturality of η, these two diagrams commute:

$$\begin{array}{ccc} GFC & \xrightarrow{GFf} & GFC' \\ {\scriptstyle \eta_C}\downarrow{\scriptstyle \cong} & & {\scriptstyle \eta_{C'}}\downarrow{\scriptstyle \cong} \\ C & \xrightarrow{f} & C' \end{array} \qquad \begin{array}{ccc} GFC & \xrightarrow{GFf'} & GFC' \\ {\scriptstyle \eta_C}\downarrow{\scriptstyle \cong} & & {\scriptstyle \eta_{C'}}\downarrow{\scriptstyle \cong} \\ C & \xrightarrow{f'} & C'. \end{array}$$

Now, if $Ff = Ff'$ then $GFf = GFf'$, which are the arrows at the top of the two diagrams, and by uniqueness (Lemma 1.5.17), it must be that $f = f'$. So, F is faithful (and by the same argument, G is faithful too). Let's now prove that F is full. Let $g : FC \to FC'$ be a morphism of $\mathbf{D}$. We have to prove that there is a morphism $f : C \to C'$ of $\mathbf{C}$ such that $Ff = g$. So, consider the morphism Gg of $\mathbf{C}$, fitting into the following diagram:

$$\begin{array}{ccc} GFC & \xrightarrow{Gg} & GFC' \\ {\scriptstyle \eta_C}\downarrow{\scriptstyle \cong} & & {\scriptstyle \eta_{C'}}\downarrow{\scriptstyle \cong} \\ C & & C'. \end{array}$$

We can construct the morphism $f := \eta_{C'} \circ Gg \circ \eta_C^{-1} : C \to C'$ as the unique arrow that makes the diagram above commute. By the naturality of η, the following diagram commutes:

$$\begin{array}{ccc} GFC & \xrightarrow{GFf} & GFC' \\ {\scriptstyle \eta_C}\downarrow{\scriptstyle \cong} & & {\scriptstyle \eta_{C'}}\downarrow{\scriptstyle \cong} \\ C & \xrightarrow{f} & C'. \end{array}$$

Again by uniqueness (Lemma 1.5.17), we then must conclude that $GFf = Gg$. But since G is faithful, this implies $Ff = g$. Therefore, F is full. Let's now turn to essential surjectivity. Let D be an object of $\mathbf{D}$. We have to find an object C of $\mathbf{C}$ such that $FC \cong D$. So, define $C := GD$. We have $FC = FGD \cong D$ via the map $\varepsilon_D : FGD \to D$. Therefore, F is essentially surjective. We have proven one direction of the theorem.

Now, consider the converse. Let F be fully faithful and essentially surjective. We need to construct $G : \mathbf{D} \to \mathbf{C}$ with $G \circ F \cong \mathrm{id}_{\mathbf{C}}$ and $F \circ G \cong \mathrm{id}_{\mathbf{D}}$. Since F is essentially surjective, for every object D of $\mathbf{D}$, we can choose an object C of $\mathbf{C}$ and an isomorphism $\phi_D : D \to FC$. We define then, on objects, $GD := C$. For the morphisms, we proceed as follows. Given $g : D \to D'$ in $\mathbf{D}$, we can apply Lemma 1.5.17 and obtain $\phi_{D'}^{-1} \circ g \circ \phi_D : FGD \to FGD'$. Since F is full, the latter map can be written as Ff for some morphism $f : C_D \to C_{D'}$ of $\mathbf{C}$. We then define, for each $g : D \to D'$, $Gg := f$. This assignment is functorial:

Let $g : D \to D'$ and $g' : D' \to D''$ in $\mathbf{D}$. Then,

$$\begin{aligned} FG(g' \circ g) \;=\; \phi_{C''}^{-1} \circ g' \circ g \circ \phi_C \;&=\; (\phi_{C''}^{-1} \circ g' \circ \phi_{C'}) \circ (\phi_{C'}^{-1} \circ g \circ \phi_C) \\ &=\; FGg' \circ FGg \;=\; F(Gg' \circ Gg), \end{aligned}$$

and since F is faithful, this implies $G(g' \circ g) = Gg' \circ Gg$. Now, let C be an object of $\mathbf{C}$. By the construction of G, we have that GFC is an object of $\mathbf{C}$ admitting an isomorphism $\phi_{FC} : FGFC \to FC$. By Proposition 1.5.7, since F is fully faithful, we have an isomorphism $\tilde{\phi}_{FC} : GFC \to C$. This isomorphism is natural: Let $f : C \to C'$ be a morphism of $\mathbf{C}$, and consider the following diagrams:

$$\begin{array}{ccc} GFC & \xrightarrow{GFf} & GFC' \\ \big\downarrow{\scriptstyle \tilde{\phi}_{FC}} & & \big\downarrow{\scriptstyle \tilde{\phi}_{FC'}} \\ C & \xrightarrow{f} & C' \end{array} \qquad\qquad \begin{array}{ccc} FGFC & \xrightarrow{FGFf} & FGFC' \\ \big\downarrow{\scriptstyle \phi_{FC}} & & \big\downarrow{\scriptstyle \phi_{FC'}} \\ FC & \xrightarrow{Ff} & FC'. \end{array}$$

The diagram on the left commutes exactly if the isomorphism $\tilde{\phi}_{FC} : GFC \to C$ is natural. Since F is fully faithful, the diagram commutes if and only if it does after applying F, which gives the diagram on the right. But the diagram on the right commutes since by the construction of G, $FGFf = \phi_{FC'}^{-1} \circ Ff \circ \phi_{FC}$. Therefore, we have a natural isomorphism $G \circ F \cong \mathrm{id}_{\mathbf{C}}$. Similarly, let D be an object of $\mathbf{D}$. We have by the construction of G that GD is an object of $\mathbf{C}$ such that FGD is isomorphic to D via the map ϕ_D. Again, this isomorphism is natural: Let $g : D \to D'$ be a morphism. By the construction of G, the map FGg is equal to $\phi_{D'}^{-1} \circ g \circ \phi_D$, so the following diagram commutes:

$$\begin{array}{ccc} FGD & \xrightarrow{FGg} & FGD' \\ \big\downarrow{\scriptstyle \phi_D} & & \big\downarrow{\scriptstyle \phi_{D'}} \\ D & \xrightarrow{g} & D'. \end{array}$$

Therefore, we have a natural isomorphism $\Gamma \circ G \cong \mathrm{id}_{\mathbf{D}}$ and hence an equivalence of categories between $\mathbf{C}$ and $\mathbf{D}$. □

Here is a first example, written as a corollary.

Corollary 1.5.18. *A monotone function $f : (X, \leq) \to (Y, \leq)$ is an isomorphism of partial orders if and only if it is order-reflecting and surjective.*

A more suggestive example comes from linear algebra. Some people consider a vector in finite dimensions "just an array of numbers" and a linear map "just a matrix." Of course, they are not *exactly* the same, but one thing is true: Everything that can be done with abstract finite-dimensional vector spaces can also be done with their representation as arrays of numbers, and vice versa, if one is careful. This can be made precise in the following way: *There is an equivalence of categories between* **FVect** *and the category of matrices.* In the following example, we show it for vector spaces over the reals, but the same can be done for complex numbers or any other field.

Example 1.5.19 (linear algebra). First of all, we want to define a category **Mat** whose morphisms are matrices with their usual matrix multiplication. As you know from linear algebra, an $m \times n$-matrix A (i.e. a matrix with m rows and n columns) can only be multiplied on the left with a matrix B that has m columns (the same m as the rows of A) and an arbitrary number of rows. Therefore, as objects, we take natural numbers, which keep track of the numbers of rows and columns in such a way that composable pairs of arrows correspond to multipliable pairs of matrices. That is:

- The objects of **Mat** are natural numbers $n \in \mathbb{N}$, including zero.
- For objects (natural numbers) $m, n \neq 0$, the morphisms $M : n \to m$ are the $m \times n$ matrices M with real entries.
- If m or n is zero, we just assign a unique morphism $m \to n$, which we can see as a "zero-dimensional matrix."
- The identity of n is just the $n \times n$ identity matrix, and the composition is given by the usual multiplication of matrices.

Let's now construct a functor $F : \mathbf{Mat} \to \mathbf{FVect}$ as follows:

- F maps each object (natural number) n of **Mat** to the vector space $\mathbb{R}^n$ and 0 to the zero vector space;
- F maps each $m \times n$ matrix $M : n \to m$ to the linear map $\mathbb{R}^n \to \mathbb{R}^m$ represented by the matrix M.

This preserves identities and composition, and so it is a functor. Now:

- F is faithful: different $m \times n$ matrices give different linear maps $\mathbb{R}^n \to \mathbb{R}^m$;
- F is full: all linear maps $\mathbb{R}^n \to \mathbb{R}^m$ arise in this way, i.e. can be represented by a matrix;
- F is essentially surjective: every finite-dimensional vector space is isomorphic to $\mathbb{R}^n$, for some n.

Therefore, by Theorem 1.5.16, F defines an equivalence of categories.

In particular, as in the proof of Theorem 1.5.16, Lemma 1.5.17 implies the following: Let V and W be vector spaces of dimension n and m, respectively. Then, there are isomorphisms $\phi_V : \mathbb{R}^n \to V$ and $\phi_W : \mathbb{R}^m \to V$. These can be seen as *choices* of bases for V and W, respectively. There are actually many more possible choices of isomorphism than these two, which correspond to different choices of bases, but let's keep these fixed for the moment. Now, given a linear map $f : V \to W$, by Lemma 1.5.17, there exists a *unique* linear map $\mathbb{R}^n \to \mathbb{R}^m$ making this diagram commute:

$$\begin{array}{ccc} \mathbb{R}^n & \longrightarrow & \mathbb{R}^m \\ {\scriptstyle \phi_V}\downarrow{\scriptstyle \cong} & & {\scriptstyle \phi_W}\downarrow{\scriptstyle \cong} \\ V & \xrightarrow{f} & W, \end{array}$$

namely, the map $\phi_W^{-1} \circ f \circ \phi_V$. This map can be seen as the *representation of f as a matrix given the chosen bases of V and W*. If we keep ϕ_V and ϕ_W fixed, this representation is unique, and it uniquely specifies f. But of course, if we let ϕ_V and ϕ_W vary, then the representation of f will be different.

As the example above shows, an equivalence of categories between **C** and **D** can be interpreted, intuitively, as the fact that *whatever can be done in* **C** *can be equivalently done in* **D** *too, and vice versa*. At first glance, the two categories may appear different, but they essentially encode the same mathematical methods and possess the same expressive power.

Exercise 1.5.20. A category is called **connected** if between any two objects X and Y, there is a **zig-zag**, i.e. an *undirected* path of arrows between them (e.g. $X \to A \leftarrow Y$ and $A \leftarrow B \to Y$ are both zig-zags of length 2). Prove that a connected groupoid is equivalent to a category in the form $\mathbf{B}G$ for some group G.

Is an analogous statements true for connected categories and monoids? (*Hint*: No. Can you give a counterexample?)

2

The Yoneda Lemma

In this chapter, we study the notions of representable functors and of universal property, and the Yoneda lemma. These ideas turn category theory from a mere language into a mathematical theory of its own.

Some of these concepts, such as universal properties, will be very abstract — if you are a beginner in category theory, the content of this chapter is likely to be *the most abstract mathematics you have ever seen*. It may be that initially, one has no clear mental picture for this abstract ideas. If this is the case, start with the more concrete examples later on in the chapter: they usually help a lot. You can also first check out the more concrete Sections 3.1 and 3.2 in the following chapter (skipping the references to Yoneda) and then come back to this material for the general theory.

2.1 Representable Functors and the Yoneda Embedding Theorem

2.1.1 Extracting sets from objects

We have said in Section 1.3.7 that the category **Set** plays a special role in (ordinary) category theory because the arrows between two given objects (in a locally small category) form a set. Given an object X in a category **C**, we can usually extract many sets out of X, which can be considered as "capturing part of the structure" whenever they are functorial. Let's see some examples.

Example 2.1.1 (topology). Let $\mathbf{C} = \mathbf{Top}$. Given a topological space X, we can consider, for example, the following sets:

(a) the underlying set $U(X)$, which is the set of *points* of X;

(b) the set $Curve(X)$ of continuous parameterized curves in X, with or without endpoints;

(c) the set $Loop(X)$ of continuous parameterized loops, or closed curves, in X;

(d) the set $\pi_0(X)$ of path-connected components of X;

(e) the set $O(X)$ of open sets of X, i.e. the topology of X.

All these constructions are functorial in X in the following ways. Let $f : X \to Y$ be a continuous function:

(a) There is an "underlying function" Uf between the underlying sets of points $U(X)$ and $U(Y)$. So, we have a functor $U : \mathbf{Top} \to \mathbf{Set}$ (see also Example 1.3.20).

(b) If c is a continuous curve in X, then composing with a continuous function f, we get a continuous curve $f_*c = f \circ c$ in Y. Therefore, we get an induced map $f_* : Curve(X) \to Curve(Y)$ (we can set $Curve(f) := f_*$). So, we have a functor $Curve : \mathbf{Top} \to \mathbf{Set}$.

(c) Similarly, we get an induced map $f_* : Loop(X) \to Loop(Y)$. Again, we have a functor $Loop : \mathbf{Top} \to \mathbf{Set}$.

(d) If $x, y \in X$ are in the same path component of X, then f, by continuity, has to map them to the same path component of Y. This induces a well-defined map $\pi_0(X) \to \pi_0(Y)$ (compare with the graph case in Example 1.3.36). Therefore, we have a functor $\pi_0 : \mathbf{Top} \to \mathbf{Set}$.

(e) Let U be an open set of Y. Then, its preimage $f^{-1}(U)$ is an open set of X. Therefore, there is a well-defined function $f^{-1} : O(Y) \to O(X)$ (note the direction of the arrow!). So, we have a *presheaf* $O : \mathbf{Top}^{\mathrm{op}} \to \mathbf{Set}$. (In general, the *image* of an open set under a continuous map is not open — we really need to take preimages here.)

The cases of graphs and multigraphs are analogous.

Example 2.1.2 (graph theory). Let **C** be the category **MGraph** of directed multigraphs and their morphisms (Example 1.4.8). Given a multigraph G, the following are functorial assignments into **Set**: (How are they functorial?)

(a) the set $Vert(G)$ of its vertices;

(b) the set $Edge(G)$ of its edges;

(c) the set $Chain_2(G)$ of 2-chains in G, i.e. pairs of edges which are head-to-tail (like composable morphisms in a category), or, more generally, the set $Chain_n(G)$ of n-chains, or n-walks, in G, i.e. a series of n consecutive edges.

(d) The set $\pi_0(G)$ of connected components (see also Example 1.3.36).

(Note that the word "path" in graph theory is reserved for those chains which do not cross the same vertices more than once.)

Here are some examples of the category of groups.

Example 2.1.3 (group theory). The following are functorial assignments **Grp** $\to$ **Set**: Given a group G,

(a) the underlying set $U(G)$;

(b) the set of elements of G of order 2, or, more generally, of elements whose order divides n, for fixed n. This means the elements $g \in G$ such that $g^n = e$. For example, -1 has order 2 in $(\mathbb{R}_{\neq 0}, \cdot)$.

Exercise 2.1.4 (group theory). Can you give more examples of functors **Grp** $\to$ **Set**? What about presheaves $\mathbf{Grp}^{\mathrm{op}} \to$ **Set**?

Exercise 2.1.5. Take a category **C** of your choice, useful to your field. Can you give examples of functors **C** $\to$ **Set** or presheaves $\mathbf{C}^{\mathrm{op}} \to$ **Set** which extract useful information from the objects?

2.1.2 Representable functors

In many of the examples in the section above, the given functors are actually of a very simple form: They are in the form $\mathrm{Hom}_{\mathbf{C}}(S,-) : \mathbf{C} \to \mathbf{Set}$, for some object S.

Definition 2.1.6. Let $\mathbf{C}$ be a category. A functor $F : \mathbf{C} \to \mathbf{Set}$ is called **representable** if it is naturally isomorphic to the functor $\mathrm{Hom}_{\mathbf{C}}(S,-) : \mathbf{C} \to \mathbf{Set}$, for some object S of $\mathbf{C}$. In that case, we call S a **representing object.**

A presheaf $F : \mathbf{C}^{\mathrm{op}} \to \mathbf{Set}$ is called **representable** if it is naturally isomorphic to the functor $\mathrm{Hom}_{\mathbf{C}}(-,S) : \mathbf{C}^{\mathrm{op}} \to \mathbf{Set}$, for some object S of $\mathbf{C}$. Again, in that case, we call S a **representing object.**

Here is a way to interpret representable functors and representable presheaves. A representable functor $\mathrm{Hom}_{\mathbf{C}}(S,-) : \mathbf{C} \to \mathbf{Set}$ takes an object X of $\mathbf{C}$ and gives the set $\mathrm{Hom}_{\mathbf{C}}(S,X)$ of arrows from S to X. Intuitively, the features that it extracts from X are exactly all the possible ways of mapping S into X, as if S were a sort of *probe* that we use to explore the structure of X.

Dually, a representable presheaf $\mathrm{Hom}_{\mathbf{C}}(-,S) : \mathbf{C}^{\mathrm{op}} \to \mathbf{Set}$ takes an object X of $\mathbf{C}$ and gives the set $\mathrm{Hom}_{\mathbf{C}}(X,S)$ of arrows from X to S. Intuitively, the features that it extracts from X are all the possible ways of mapping X into S, or all the possible *observations* of X with values in S (think of S as a "screen" onto which X can be projected in many possible ways).

Example 2.1.7 (topology). Let's see which functors from Example 2.1.1 are representable. Let X be a topological space:

(a) Let x be a point of X. If we denote by 1 the one-point space, then there exists a (continuous) map $1 \to X$ which picks out exactly x, i.e. which maps the unique point of 1 to x. This can be done for every point x of X, and different maps $1 \to X$ pick out necessarily different points. Therefore, there is a bijection between the points of X and the maps $1 \to X$, i.e. $U(X) \cong \mathrm{Hom}_{\mathbf{Top}}(1,X)$.

This bijection is natural in X (why?), and so the functor U is representable, and the representing object is the one-point space 1.

(b) Analogously, curves in X with endpoints correspond to continuous maps from $[0,1]$ to X, and curves without endpoints correspond to continuous maps from $(0,1)$ (or equivalently, $\mathbb{R}$) to X. Therefore, the functor *Curve* is representable.

(c) In the same way, the functor *Loop* is representable, and the representing object is the circle S^1. That is, loops in X correspond to continuous maps $S^1 \to X$.

(d) The functor π_0 is not representable. (It is, however, a sort of homotopy version of the forgetful functor U.)

(e) Consider now the topology presheaf $O : \mathbf{Top}^{\mathrm{op}} \to \mathbf{Set}$, which assigns to each set X its topology. Denote by S the Sierpinski space, which is a space with two elements 0 and 1, with the topology given by $\{\varnothing, \{1\}, \{0,1\}\}$, but not $\{0\}$. Given a topological space X, consider a function $f : X \to S$. Necessarily, $f^{-1}(\varnothing) = \varnothing$ and $f^{-1}(S) = X$. Therefore, f is continuous if and only if $f^{-1}(1)$ is open in X. Conversely, given any open set $U \subseteq X$, the function $X \to S$ mapping U to 1 and everything else to 0 is continuous. Therefore, there is a bijection between the open sets of X and the continuous maps $X \to S$, i.e. $O(X) \cong \mathrm{Hom}_{\mathbf{Top}}(X, S)$. This is natural in X, and so the presheaf O is representable by the space S.

In the interpretation given above, for example, the functor *Loop* extracts from a space X the features that can be obtained by "probing" X with the space S^1, i.e. by looking at all the possible ways of mapping S^1 into X. Similarly, the topology presheaf O extracts from a space X all the possible ways in which X can be mapped continuously to the Sierpinski space S, i.e. all possible ways to observe X with "resolution" S.

Example 2.1.8 (graph theory). Let's see which functors from Example 2.1.2 are representable. Let G be a directed multigraph:

(a) Denote by G_0 the graph with one single vertex and no edges. For each vertex v of G, there is a unique morphism $G_0 \to G$ picking out exactly v. Therefore, $Vert(G) \cong \mathrm{Hom}_{\mathbf{MGraph}}(G_0, G)$. This is natural in G, so we have that $Vert$ is representable by G_0.

(b) Denote by G_1 the graph with two vertices and a unique edge between them. For each edge e of G, there is a unique morphism $G_1 \to G$ picking out exactly the edge e. Again, this assignment is natural. Therefore, $Edge$ is representable by the graph G_1.

(c) More generally, for each $n \in \mathbb{N}$, denote by G_n the graph of $n+1$ vertices and n edges forming a chain. The functor $Chain_n$ is representable by the graph G_n.

(d) The set π_0 of connected components is not representable, just as for topological spaces.

Example 2.1.9 (group theory). The functors given in Example 2.1.3 are representable. Let G be a group:

(a) The elements of G are in bijection with group homomorphisms $(\mathbb{Z}, +) \to G$. Every element $g \in G$ defines a group homomorphism $f : \mathbb{Z} \to G$ defined by $f(1) = g$ (necessarily, $f(0) = e$, $f(-1) = g^{-1}$, $f(2) = g^2$, and so on). Conversely, every homomorphism $f : \mathbb{Z} \to G$ gives an element of G by looking at where it maps $1 \in \mathbb{Z}$. Therefore, $U : \mathbf{Grp} \to \mathbf{Set}$ is representable by $\mathbb{Z}$. Note that we cannot, as we did for topological spaces, use the one-point group $\{e\}$ as representing an object: Any group homomorphism from $\{e\}$ to G can only have $e \in G$ as an image.

(b) Similarly, the elements of G whose order divides n are in bijection with group homomorphisms $\mathbb{Z}/n \to G$.

Example 2.1.10 (group theory). Let G be a group. We have seen (Example 1.3.10) that a functor $F : \mathbf{B}G \to \mathbf{Set}$ is a permutation representation of G, i.e. a set X (which is the image $F\bullet$ of the unique object $\bullet$ of $\mathbf{B}G$) equipped with an action of G on X. When is this functor representable?

Let's instantiate the definition: F is representable if for some object of $\mathbf{B}G$ (there is only one possible object, namely $\bullet$), F is naturally isomorphic to $\mathrm{Hom}_{\mathbf{B}G}(\bullet, -)$. This in turn means that for every object of $\mathbf{B}G$ (again, there is only one, $\bullet$), we have that $F\bullet \cong \mathrm{Hom}_{\mathbf{B}G}(\bullet, \bullet)$ and that this bijection is natural. Now, $F\bullet$ is exactly the set X on which G is acting. Further, $\mathrm{Hom}_{\mathbf{B}G}(\bullet, \bullet)$ is the set of arrows of $\mathbf{B}G$ from $\bullet$ to $\bullet$, which by definition is the set of elements of G. Therefore, in order for F to be representable, we need X to be isomorphic to the underlying set $U(G)$ of G. That is, the group has to act on itself ($g \in G$ acts on the element $h \in G$ by mapping it to $gh \in G$). Moreover, the naturality condition says that the isomorphism $X \to U(G)$ has to be G-equivariant; that is, X and $U(G)$ must also be isomorphic as G-sets.

Exercise 2.1.11 (graph theory). We have seen in Example 1.3.46 that multigraphs are exactly presheaves on the category **Par** (defined there). Prove that the representable presheaves on **Par** are, up to natural isomorphism, exactly the multigraphs G_0 and G_1, with a single vertex and a single edge, respectively, appearing in Example 2.1.8. Which object of **Par** represents which presheaf?

Exercise 2.1.12 (sets and relations). Show that the identity $\mathbf{Set} \to \mathbf{Set}$ is representable. What is the representing object?

Exercise 2.1.13 (measure theory). Let **Meas** be the category of measurable spaces and measurable maps. To each measurable space X assign the set $\Sigma(X)$ of its measurable subsets. Show that this is part of a presheaf $\mathbf{Meas}^{\mathrm{op}} \to \mathbf{Set}$. Is this representable?

Exercise 2.1.14 (topology). Let X be a topological space. According to Example 1.3.47, presheaves on X are presheaves on the poset $O(X)$ seen as a category. What do representable presheaves look like?

Exercise 2.1.15. If you have done Exercise 2.1.5, are the functors or presheaves (that you defined) representable?

2.1.3 The Yoneda embedding theorem

The following two natural questions may arise from the examples in the previous section:

- Suppose that a functor (or presheaf) is representable. Then, is the representing object necessarily unique (up to isomorphism)?
- Suppose that given two objects X and Y of $\mathbf{C}$, then, for each object S, we have that $\mathrm{Hom}_{\mathbf{C}}(S, X)$ is naturally isomorphic to $\mathrm{Hom}_{\mathbf{C}}(S, Y)$. That is, suppose that for each S, "what S sees in X and Y is the same." Can we conclude that X and Y are isomorphic? (The same question can be asked with morphisms into S instead of out of S.)

The answer to both questions, which are really one and the same, is *yes*. This follows from one of the most important results in category theory.

Theorem 2.1.16 (Yoneda embedding). *Let* $\mathbf{C}$ *be a category, and let* X *and* Y *be objects of* $\mathbf{C}$. *There is a natural bijection of sets*

$$\mathrm{Hom}_{\mathbf{C}}(X, Y) \;\cong\; \mathrm{Hom}_{[\mathbf{C}^{\mathrm{op}}, \mathbf{Set}]}\big(\mathrm{Hom}_{\mathbf{C}}(-, X), \mathrm{Hom}_{\mathbf{C}}(-, Y)\big)$$

between the morphisms of $\mathbf{C}$ *from* X *to* Y *and the natural transformations from the presheaf* $\mathrm{Hom}_{\mathbf{C}}(-, X) : \mathbf{C}^{\mathrm{op}} \to \mathbf{Set}$ *represented by* X *to the presheaf* $\mathrm{Hom}_{\mathbf{C}}(-, Y) : \mathbf{C}^{\mathrm{op}} \to \mathbf{Set}$ *represented by* Y.

Naturality is meant in both variables X *and* Y.

The dual statement replacing $\mathbf{C}$ by $\mathbf{C}^{\mathrm{op}}$ reads as follows.

Corollary 2.1.17. *Let* $\mathbf{C}$ *be a category, and let* X *and* Y *be objects of* $\mathbf{C}$. *There is a natural bijection of sets*

$$\mathrm{Hom}_{\mathbf{C}}(X, Y) \;\cong\; \mathrm{Hom}_{[\mathbf{C}, \mathbf{Set}]}\big(\mathrm{Hom}_{\mathbf{C}}(Y, -), \mathrm{Hom}_{\mathbf{C}}(X, -)\big)$$

between the morphisms of **C** *from* X *to* Y *and the natural transformations from the functor* $\mathrm{Hom}_{\mathbf{C}}(Y,-) : \mathbf{C} \to \mathbf{Set}$ *represented by* Y *to the functor* $\mathrm{Hom}_{\mathbf{C}}(X,-) : \mathbf{C} \to \mathbf{Set}$ *represented by* X.

Here are some consequences of this theorem, which give its intuitive interpretation.

Corollary 2.1.18. *Let* **C** *be a category, and let* X *and* Y *be objects of* **C**.

- X *and* Y *are isomorphic if and only if the functors (resp., presheaves) that they represent are naturally isomorphic. In particular, if* X *and* Y *represent the same functor (resp., presheaf), then they must be isomorphic.*
- *A more concrete view of the point above is that* X *and* Y *are isomorphic if and only if for every object* S *of* **C**, *the sets* $\mathrm{Hom}_{\mathbf{C}}(S,X)$ *and* $\mathrm{Hom}_{\mathbf{C}}(S,Y)$ *(resp.,* $\mathrm{Hom}_{\mathbf{C}}(X,S)$ *and* $\mathrm{Hom}_{\mathbf{C}}(Y,S)$*) are isomorphic, naturally in* S. *In particular, if* X *and* Y *are indistinguishable by* S *for every* S *in* **C**, *then* X *and* Y *are necessarily isomorphic.*

In other words, *each object of* **C** *is uniquely specified by the arrows into it (resp., out of it)*, up to isomorphism. The objects of a category can be uniquely defined in terms of the *role they play in the category* in terms of their "interaction with the whole."

Warning. This statement can be considered rather "philosophical," and it is similar to axioms in philosophy that one can assume true or not (such as the *identity of indiscernibles*[1]). In category theory, however, this is a *theorem* with a proof. It is true in every category.

> **Exercise 2.1.19 (relations, topology; [difficult!]).** Prove Theorem 2.1.16 for the easier case of **C** being a partial order (for example, the topology of some space). What does the statement of the theorem look like? (*Hint*: Exercise 2.1.14 can help you.)

The proof of the theorem is given in the following section.

[1]https://ncatlab.org/nlab/show/identity+of+indiscernibles.

2.2 Statement and Proof of the Yoneda Lemma

In order to prove Theorem 2.1.16, we make use of the following statement, the **Yoneda lemma**, which is at least as important as the theorem itself.

Lemma 2.2.1 (Yoneda). *Let* $\mathbf{C}$ *be a category,* X *be an object of* $\mathbf{C}$*, and* $F : \mathbf{C}^{\mathrm{op}} \to \mathbf{Set}$ *be a presheaf on* $\mathbf{C}$*. Consider the map*

$$\mathrm{Hom}_{[\mathbf{C}^{\mathrm{op}},\mathbf{Set}]}\big(\mathrm{Hom}_{\mathbf{C}}(-,X),F\big) \longrightarrow FX$$

assigning to a natural transformation $\alpha : \mathrm{Hom}_{\mathbf{C}}(-,X) \Rightarrow F$ *the element* $\alpha_X(\mathrm{id}_X) \in FX$*, which is the value of the component* α_X *of* α *on the identity at* X*.*

This assignment is a bijection, and it is natural both in X *and* F*.*

This is enough to prove the Yoneda embedding theorem already.

Proof of Theorem 2.1.16. In the hypotheses of the Yoneda lemma, set F to be the representable presheaf $\mathrm{Hom}_{\mathbf{C}}(-,Y) : \mathbf{C}^{\mathrm{op}} \to \mathbf{Set}$. □

A possible interpretation of the Yoneda lemma is the following, which is rather "trivial." We have seen that a consequence of the Yoneda embedding theorem is that if all the ways of observing two objects X and Y coincide, then X and Y must be isomorphic. The proof of the Yoneda embedding theorem through the Yoneda lemma explains *why*: We can observe X in such a way that no information is lost, namely, by mapping it to itself via the identity $\mathrm{id}_X : X \to X$. This is an observation that trivially sees the whole of X faithfully. Therefore, this one observation is sufficient to determine X uniquely. Conversely, every other observation, which possibly loses information, is obtainable from this one, i.e. it is a sort of "corruption" of this trivial observation. Now, if the observations of X and Y coincide, we can observe Y in a way that corresponds to this trivial observation of X, and this again will not lose any information. So, it is possible to study Y by means of X without losing anything, and this means that X is isomorphic to Y. Of course, the actual statement is more complicated, and more general, than this interpretation (it is about *all* morphisms, not just isomorphisms). The actual proof is as follows.

2.2.1 Proof of the Yoneda lemma

First, let's see what the statement of Lemma 2.2.1 really says. We are given an object X of $\mathbf{C}$ and a presheaf $F : \mathbf{C}^{\mathrm{op}} \to \mathbf{Set}$. From the object X, we can form the representable presheaf $\mathrm{Hom}_{\mathbf{C}}(-, X) : \mathbf{C}^{\mathrm{op}} \to \mathbf{Set}$. We can now look at the natural transformations $\mathrm{Hom}_{\mathbf{C}}(-, X) \Rightarrow F$. Let α be one such natural transformation. For each object Y of $\mathbf{C}$, the component of α at Y is a map between the sets $\alpha_Y : \mathrm{Hom}_{\mathbf{C}}(Y, X) \to FY$. Moreover, this is natural in Y, meaning that, for every $f : Y \to Z$ in $\mathbf{C}$, the following diagram has to commute (note that both functors reverse the arrows):

$$\begin{array}{ccc} \mathrm{Hom}_{\mathbf{C}}(Z, X) & \xrightarrow{-\circ f} & \mathrm{Hom}_{\mathbf{C}}(Y, X) \\ \downarrow{\scriptstyle \alpha_Z} & & \downarrow{\scriptstyle \alpha_Y} \\ FZ & \xrightarrow{\quad Ff \quad} & FY. \end{array}$$

Now, the component of α at X is a map $\alpha_X : \mathrm{Hom}_{\mathbf{C}}(X, X) \to FX$. The identity at X is an element of the set $\mathrm{Hom}_{\mathbf{C}}(X, X)$, and its image under α_X is an element $\alpha_X(\mathrm{id}_X)$ of FX. We can assign to α the element $\alpha_X(\mathrm{id}_X)$ of FX, and this gives a mapping from the natural transformations $\alpha : \mathrm{Hom}_{\mathbf{C}}(-, X) \Rightarrow F$ to the elements of FX.

The lemma says that this mapping is a bijection, and it is natural. This is what we have to prove. In category theory, the constructions given in the proof of the Yoneda lemma are almost as important as its statement. In particular, the assignment $p \mapsto \nu^p$ given in (2.2.2) will be central to the idea of universal properties (Section 2.3).

Proof of Bijectivity. First of all, let $\alpha : \mathrm{Hom}_{\mathbf{C}}(-, X) \Rightarrow F$. Let $f : Y \to X$ in $\mathbf{C}$. By the naturality of α, the following diagram commutes:

$$\begin{array}{ccc}\mathrm{Hom}_{\mathbf{C}}(X,X) & \xrightarrow{-\circ f} & \mathrm{Hom}_{\mathbf{C}}(Y,X)\\ \big\downarrow{\scriptstyle \alpha_X} & & \big\downarrow{\scriptstyle \alpha_Y}\\ FX & \xrightarrow{\quad Ff \quad} & FY.\end{array}$$

If we start with the identity $\mathrm{id}_X \in \mathrm{Hom}_{\mathbf{C}}(X,X)$, the commutativity of the diagram above says that

$$Ff(\alpha_X(\mathrm{id}_X)) = \alpha_Y(\mathrm{id}_X \circ f) = \alpha_Y(f). \tag{2.2.1}$$

With this in mind, let's try to construct an inverse assignment to $\alpha \mapsto \alpha_X(\mathrm{id}_X)$, taking an element of FX and mapping it to a natural transformation $\mathrm{Hom}_{\mathbf{C}}(-,X) \Rightarrow F$. Given $p \in PX$, we have to define a natural transformation whose components are going to be functions $\mathrm{Hom}_{\mathbf{C}}(Y,X) \Rightarrow FY$, for each object Y of $\mathbf{C}$. Now, given such p and Y, define the function

$$\begin{array}{rcl}\mathrm{Hom}_{\mathbf{C}}(Y,X) & \xrightarrow{\ \nu^p_Y\ } & FY\\ f & \longmapsto & Ff(p).\end{array} \tag{2.2.2}$$

In order to show that these functions, for varying Y, are the components of a natural transformation $\mathrm{Hom}_{\mathbf{C}}(-,X) \Rightarrow F$, we have to show that for each $g : Y \to Z$ of $\mathbf{C}$, the following diagram commutes:

$$\begin{array}{ccc}\mathrm{Hom}_{\mathbf{C}}(Z,X) & \xrightarrow{-\circ g} & \mathrm{Hom}_{\mathbf{C}}(Y,X)\\ \big\downarrow{\scriptstyle \nu^p_Z} & & \big\downarrow{\scriptstyle \nu^p_Y}\\ FZ & \xrightarrow{\quad Fg \quad} & FY.\end{array}$$

This means that for each $h \in \mathrm{Hom}_{\mathbf{C}}(Z,X)$, we should have $Fg(\nu^p_Z(h)) = \nu^p_Y(h \circ g)$. Now, since F is functorial (reversing the direction of composition),

$$Fg(\nu^p_Z(h)) = Fg(Fh(p)) = F(h \circ g)(p) = \nu^p_Y(h \circ g).$$

Therefore, ν^p is a natural transformation.

Let's now show that the assignment $p \mapsto \nu^p$ is inverse to the assignment $\alpha \mapsto \alpha_X(\mathrm{id}_X)$. First of all, given $p \in FX$, the component of ν^p at X maps the identity id_X to

$$\nu^p_X(\mathrm{id}_X) = F(\mathrm{id}_X)(p) = \mathrm{id}_{FX}(p) = p.$$

In the other direction, given $\alpha : \mathrm{Hom}_{\mathbf{C}}(-, X) \Rightarrow F$, for each object Y of $\mathbf{C}$, let's look at the component of Y in the resulting natural transformation. Given $f : Y \to X$, by Equation (2.2.1),

$$\nu_Y^{\alpha_X(\mathrm{id}_X)}(f) = Ff(\alpha_X(\mathrm{id}_X)) = \alpha_Y(f).$$

Therefore, $\nu^{\alpha_X(\mathrm{id}_X)} = \alpha$, and the assignment $\alpha \mapsto \alpha_X(\mathrm{id}_X)$ is a bijection (and so is its inverse, $p \mapsto \nu^p$).

Proof of Naturality. First, we prove naturality in X. Let $h : Y \to X$ be a morphism of $\mathbf{C}$. This induces a natural transformation $h_* : \mathrm{Hom}_{\mathbf{C}}(-, Y) \Rightarrow \mathrm{Hom}_{\mathbf{C}}(-, X)$ given by postcomposition with h (see also Exercise 1.4.27). This means, for each object A, we get a component $(h_*)_A : \mathrm{Hom}_{\mathbf{C}}(A, Y) \to \mathrm{Hom}_{\mathbf{C}}(A, X)$ mapping $g \in \mathrm{Hom}_{\mathbf{C}}(A, Y)$ to $h \circ g \in \mathrm{Hom}_{\mathbf{C}}(A, X)$. Now, the naturality in X of the bijection of Lemma 2.2.1 means that the following diagram has to commute:

$$\begin{array}{ccc}
\mathrm{Hom}_{[\mathbf{C}^{op},\mathbf{Set}]}\big(\mathrm{Hom}_{\mathbf{C}}(-, X), F\big) & \xrightarrow{-\circ h_*} & \mathrm{Hom}_{[\mathbf{C}^{op},\mathbf{Set}]}\big(\mathrm{Hom}_{\mathbf{C}}(-, Y), F\big) \\
\downarrow{\cong} & & \downarrow{\cong} \\
FX & \xrightarrow{\qquad Fh \qquad} & FY.
\end{array}$$

This means, equivalently, that for every natural transformation $\alpha : \mathrm{Hom}_{\mathbf{C}}(-, X) \Rightarrow F$, we have that $Fh(\alpha_X(\mathrm{id}_X))$ is equal to $\alpha_Y \circ h_*(\mathrm{id}_Y)$. The latter is equal to

$$\alpha_Y(h_*(\mathrm{id}_Y)) = \alpha_Y(\mathrm{id}_Y \circ h) = \alpha_Y(h),$$

and this is equal to $Fh(\alpha_X(\mathrm{id}_X))$ by Equation (2.2.1).

Let's now turn to naturality in F. This means that for every presheaf $G : \mathbf{C}^{\mathrm{op}} \to \mathbf{Set}$ and every natural transformation $\beta : F \to G$, the following diagram has to commute:

$$\begin{array}{ccc} \mathrm{Hom}_{[\mathbf{C}^{\mathrm{op}},\mathbf{Set}]}\big(\mathrm{Hom}_{\mathbf{C}}(-,X),F\big) & \xrightarrow{\beta\circ-} & \mathrm{Hom}_{[\mathbf{C}^{\mathrm{op}},\mathbf{Set}]}\big(\mathrm{Hom}_{\mathbf{C}}(-,X),G\big) \\ \downarrow\cong & & \downarrow\cong \\ FX & \xrightarrow{\beta_X} & GX. \end{array}$$

This means, equivalently, that for every natural transformation $\alpha : \mathrm{Hom}_{\mathbf{C}}(-,X) \Rightarrow F$, we have that $\beta_X(\alpha_X(\mathrm{id}_X))$ has to be equal to $(\beta\circ\alpha)_X(\mathrm{id}_X)$. However, this is guaranteed by the definition of (vertical) composition of natural transformations, as given, for example, in Exercise 1.4.19.

2.2.2 Particular cases

Example 2.2.2 (graph theory). Consider the category **Par** from Example 1.3.46, given by the diagram

$$V \underset{t}{\overset{s}{\rightrightarrows}} E.$$

As we have seen, presheaves on **Par** encode directed multigraphs. The representable presheaves are, up to isomorphism, $\mathrm{Hom}_{\mathbf{Par}}(-,V)$ and $\mathrm{Hom}_{\mathbf{Par}}(-,E) : \mathbf{Par} \to \mathbf{Set}$. Which multigraphs do these encode?

- $\mathrm{Hom}_{\mathbf{Par}}(-,V)$ maps the object V to the set of arrows $V \to V$ of **Par**, which contains just the identity of V, and maps E to the set of arrows $E \to V$, which is empty (there are no such arrows in **Par**). Therefore, $\mathrm{Hom}_{\mathbf{Par}}(V,V)$ is a singleton set, and $\mathrm{Hom}_{\mathbf{Par}}(E,V)$ is an empty set. The graph corresponding to $\mathrm{Hom}_{\mathbf{Par}}(-,V)$ is therefore a graph with a single node and no edges. In Example 2.1.8, we called this graph G_0.
- $\mathrm{Hom}_{\mathbf{Par}}(-,E)$ maps V to the set of arrows $V \to E$ of **Par**, which contains the two arrows s and t, and maps E to the set of arrows $E \to E$, which contains only the identity of E. Therefore, $\mathrm{Hom}_{\mathbf{Par}}(V,E)$ is a

two-element set, $\mathrm{Hom}_{\mathbf{Par}}(E,E)$ is a singleton, and the morphisms s and t are mapped to functions $\mathrm{Hom}_{\mathbf{Par}}(E,E) \to \mathrm{Hom}_{\mathbf{Par}}(V,E)$, picking out the elements s and t, respectively. The graph corresponding to $\mathrm{Hom}_{\mathbf{Par}}(-,E)$ is therefore a graph with two distinct nodes and a single edge connecting them in one direction. In Example 2.1.8, we called this graph G_1.

Therefore, G_0 and G_1 are the graphs corresponding to representable presheaves on **Par**. Mind that these are not the same as representable functors *on* the category of multigraphs, as we did in Example 2.1.8. But these constructions are related, as we show in a moment.

Let's instantiate the Yoneda lemma, setting **C** equal to **Par**, so that the category of presheaves $[\mathbf{C}^{\mathrm{op}}, \mathbf{Set}]$ becomes the category **MGraphs** of multigraphs and their morphisms (Example 1.4.8). The resulting statement is that, for each multigraph G, there are natural bijective correspondences

$$\mathrm{Hom}_{\mathbf{MGraph}}\left(\mathrm{Hom}_{\mathbf{Par}}(-,V), G\right) \longrightarrow GV$$

and

$$\mathrm{Hom}_{\mathbf{MGraph}}\left(\mathrm{Hom}_{\mathbf{Par}}(-,E), G\right) \longrightarrow GE.$$

Since, as we have seen above, the representable presheaves $\mathrm{Hom}_{\mathbf{Par}}(-,V)$ and $\mathrm{Hom}_{\mathbf{Par}}(-,E)$ correspond to the graphs G_0 and G_1, respectively, and since GV and GE are the sets of vertices and edges of G, we can equivalently write the bijections as

$$\mathrm{Hom}_{\mathbf{MGraph}}\left(G_0, G\right) \cong \mathrm{Vert}(G)$$

and

$$\mathrm{Hom}_{\mathbf{MGraph}}\left(G_1, G\right) \cong \mathrm{Edge}(G),$$

which are exactly what we had found in Example 2.1.8.

Moreover, the Yoneda lemma says that this bijection is given by looking at where the identities of V and E are mapped to. In particular:

- A multigraph morphism $f : G_0 \to G$ corresponds to a vertex of G, and the correspondence is obtained by looking at where f maps the unique vertex of G_0 to since this vertex of G_0 is the unique element of $\mathrm{Hom}_{\mathbf{Par}}(V, V)$, i.e. the identity of the object V of **Par**. The map f can map the unique vertex of G_0 to any vertex of G, and different maps will pick out different vertices, so we have a bijection between the vertices of G and the multigraph morphisms $G_0 \to G$.
- A multigraph morphism $f : G_1 \to G$ corresponds to an edge of G, and the correspondence is obtained by looking at where f maps the unique edge of G_1 to, since this vertex of G_1 is the unique element of $\mathrm{Hom}_{\mathbf{Par}}(E, E)$, i.e. the identity of the object E of **Par**. The map f can map the unique edge of G_1 to any edge of G, and different maps will pick out different edges, so we have a bijection between the edges of G and the multigraph morphisms $G_1 \to G$.

This again follows the intuition of Example 2.1.8.

Example 2.2.3 (group theory). We know that for a group G, the functors $\mathbf{B}G \to \mathbf{Set}$ are sets equipped with a G-action, or G-sets, and that (up to isomorphism) the only representable functor $\mathbf{B}G \to \mathbf{Set}$ is given by G acting on its underlying set $U(G)$. If we look at presheaves instead of functors, $\mathbf{B}G^{\mathrm{op}} \to \mathbf{Set}$, the situation is analogous, except that G will act on the right instead of on the left. That is, the presheaves $\mathbf{B}G^{\mathrm{op}} \to \mathbf{Set}$ correspond to *right* G-sets, and the unique (up to isomorphism) representable presheaf is given by the action of G on its underlying set by *right* multiplication, i.e. g acts on G by mapping h to hg instead of gh.

Now, the Yoneda embedding theorem (Theorem 2.1.16) says the following, noting that $\mathbf{B}G$ has a single object $\bullet$:

$$\mathrm{Hom}_{\mathbf{B}G}(\bullet, \bullet) \cong \mathrm{Hom}_{[\mathbf{B}G^{\mathrm{op}}, \mathbf{Set}]}\big(\mathrm{Hom}_{\mathbf{B}G}(-, \bullet), \mathrm{Hom}_{\mathbf{B}G}(-, \bullet)\big).$$

Since the set $\mathrm{Hom}_{\mathbf{B}G}(\bullet, \bullet)$ consists by definition of the elements of G, and since the representable functors are given by the underlying set of G, the statement becomes

$$G \cong \mathrm{Hom}_{\mathbf{GSet}}\big(U(G), U(G)\big).$$

The group on the right is a subgroup of the full permutation group of the set $U(G)$. Therefore, G is isomorphic to a subgroup of a permutation group of some set. In other words, we have proven Cayley's Theorem 1.3.12.

The Yoneda lemma says more: Given any (right) G-set X, which is expressed by a presheaf $F : \mathbf{B}G \to \mathbf{Set}$, there is a natural bijection between the elements of X and the G-equivariant maps $f : U(G) \to X$. This bijection is given by looking at where the identity element e of G will be mapped to by f: Every element of X can be chosen to be the image of e as long as the other elements of G are "rigidly" mapped along the orbit. Moreover, different choices of f will necessarily map the identity to different points. To realize the latter statement, note that, just as in the proof of Lemma 2.2.1, by equivariance (i.e. naturality), for every $g \in G$,

$$f(g) = f(eg) = f(e) \cdot g,$$

so that if $f, f' : U(G) \to X$ agree on e, they must agree on all of $U(G)$.

Example 2.2.4 (linear algebra; [Rie16, Corollary 2.2.9]). Consider the matrix category **Mat** of Example 1.5.19. Consider a **linear row operation** of matrices, such as "multiply the second row by 2 and add it to the first one," i.e. a mapping, for example,

$$\begin{pmatrix} a & b & c \\ d & e & f \end{pmatrix} \longmapsto \begin{pmatrix} a+2d & b+2e & c+2f \\ d & e & f \end{pmatrix}.$$

Given any linear operation on rows, in order to apply it to a given matrix M, one can equivalently first apply the operation to the identity matrix and then multiply M by the resulting matrix. For example, we can first apply our operation above to the 2×2 identity matrix,

$$\begin{pmatrix} 1 & 0 \\ 0 & 1 \end{pmatrix} \longmapsto \begin{pmatrix} 1+2\cdot 0 & 0+2\cdot 1 \\ 0 & 1 \end{pmatrix} = \begin{pmatrix} 1 & 2 \\ 0 & 1 \end{pmatrix},$$

and then multiply our given matrix by the image of the identity:

$$\begin{pmatrix}1 & 2\\ 0 & 1\end{pmatrix}\begin{pmatrix}a & b & c\\ d & e & f\end{pmatrix} = \begin{pmatrix}a+2d & b+2e & c+2f\\ d & e & f\end{pmatrix}.$$

This is an instance of the Yoneda embedding theorem. Linear row operations (for example, of matrices of n rows) correspond to natural transformations between the presheaves $\phi : \mathrm{Hom}_{\mathbf{Mat}}(-,n) \Rightarrow \mathrm{Hom}_{\mathbf{Mat}}(-,n)$, since the naturality condition says precisely that the operation ϕ commutes with multiplication on the right by (other) matrices, i.e. forming linear combinations of the rows (why?). The Yoneda embedding theorem for $X = Y = n \in \mathbb{N}$ now says that there is a natural bijection

$$\mathrm{Hom}_{\mathbf{Mat}}(n,n) \cong \mathrm{Hom}_{[\mathbf{Mat}^{\mathrm{op}},\mathbf{Set}]}\left(\mathrm{Hom}_{\mathbf{Mat}}(-,n), \mathrm{Hom}_{\mathbf{Mat}}(-,n)\right).$$

That is, linear operations on n rows are in bijection with the elements of $\mathrm{Hom}_{\mathbf{Mat}}(n,n)$, which are by definition $n \times n$ matrices. Moreover, this correspondence is given by looking at where the identity of n is mapped to, i.e. by applying the operation to the $n \times n$ identity matrix, and then letting naturality (i.e. linearity) do the rest.

2.3 Universal Properties

Thanks to the Yoneda lemma, we have seen that objects are specified, uniquely up to isomorphism, by the functor or presheaf that they represent. We interpreted this as the fact that objects are uniquely specified by the way they interact with the rest of the category. When an object is specified in this way, it is said to satisfy a *universal property*.

Different authors differ slightly in what they mean precisely when they talk about universal properties. In this book, we follow closely the approach, terminology, and (where possible) notation of [Rie16].

Definition 2.3.1. Let $\mathbf{C}$ be a category and X an object of $\mathbf{C}$. A **universal property** of X consists of either a functor $F : \mathbf{C} \to \mathbf{Set}$ or a presheaf

$P : \mathbf{C}^{op} \to \mathbf{Set}$ together with either a specified natural isomorphism $\mathrm{Hom}_{\mathbf{C}}(X, -) \Rightarrow F$ or $\mathrm{Hom}_{\mathbf{C}}(-, X) \Rightarrow P$.

Remark 2.3.2. By definition, the functor or the presheaf given as above is necessarily representable.

Remark 2.3.3. A natural isomorphism, $\mathrm{Hom}_{\mathbf{C}}(-, X) \Rightarrow P$, is in particular a natural transformation. By the Yoneda lemma, such natural transformations are in bijection with the elements of the set PX. Note that not all natural transformations are natural isomorphisms in general, so only *some* of the elements of PX give us a natural isomorphism.

Similarly, replacing $\mathbf{C}$ by $\mathbf{C}^{op}$, we have that natural isomorphisms $\mathrm{Hom}_{\mathbf{C}}(X, -) \Rightarrow F$ correspond to some of the elements of the set FX (the ones corresponding to isomorphisms as opposed to general morphisms).

A universal property for the object X can be considered as a condition for the *existence and uniqueness* of a specified map into X (for presheaves) or out of X (for functors) in the following way. Let $P : \mathbf{C}^{op} \to \mathbf{Set}$ be a representable presheaf, and let $\alpha : \mathrm{Hom}_{\mathbf{C}}(-, X) \Rightarrow P$ be a chosen natural isomorphism, so that we have a universal property for X. Consider now a map $f : Y \to X$. We can form the usual naturality diagram

$$\begin{array}{ccc} \mathrm{Hom}_{\mathbf{C}}(X, X) & \xrightarrow{-\circ f} & \mathrm{Hom}_{\mathbf{C}}(Y, X) \\ \cong \downarrow \alpha_X & & \cong \downarrow \alpha_Y \\ PX & \xrightarrow{Pf} & PY, \end{array}$$

where now the components of α are bijections. Starting as usual with the identity of X in the top-left corner and denoting $\alpha_X(\mathrm{id}_X)$ by p, we get that $Pf(p) = \alpha_Y(f)$. You might recognize the natural transformation ν^p defined in (2.2.2), $\nu^p_Y(f) = Pf(p)$. Let's see in concrete what it means that α_Y, i.e. ν^p_Y, is a bijection: For each element q of PY, there exists a unique $f : Y \to X$ such that $\alpha_Y(f) = q$, i.e. such that $\alpha_Y(\mathrm{id} \circ f) = q$. Since the diagram above commutes, equivalently, α_Y is a bijection if and only if *for each element q of PY there exists a unique $f : Y \to X$ such that $Pf(p) = q$*. That is, there is *exactly one* arrow f into X satisfying the condition $Pf(p) = q$. Now, the entire transformation $\alpha = \nu^p$ is a natural *isomorphism* rather than just a natural transformation $\mathrm{Hom}_{\mathbf{C}}(-, X) \Rightarrow P$ if and only if for every object Y, α_Y (i.e. ν^p_Y) is a bijection. That is, if for every object Y and every element q of PY, there exists a unique $f : Y \to X$ such that $Pf(p) = q$. The universal property can be seen as an existence and uniqueness condition for the maps f, one for each object Y and element $q \in PY$.

Dually, a universal property given by a functor $\mathbf{C} \to \mathbf{Set}$ instead of a presheaf will give a condition for the existence and uniqueness of arrows out of X.

In order to have an interpretation for why we want f to satisfy $Pf(p) = q$, let's see a couple of examples. First, let's fix some notation. In diagrams, we use a dashed arrow, such as

$$X \overset{f}{\dashrightarrow} Y,$$

whenever f satisfies a condition of existence and uniqueness coming from a universal property.

Let's now see two examples of universal properties: the cartesian product of topological spaces, and the tensor product of vector spaces.

2.3.1 The universal property of the cartesian product

Example 2.3.4 (topology). Let $\mathbf{C}$ be the category $\mathbf{Top}$. Let X and Y be objects of $\mathbf{Top}$, i.e. topological spaces. Consider the presheaf P given by

$$\mathrm{Hom}_{\mathbf{Top}}(-, X) \times \mathrm{Hom}_{\mathbf{Top}}(-, Y) \;:\; \mathbf{Top}^{\mathrm{op}} \to \mathbf{Set}.$$

This presheaf maps a space S to the set $\mathrm{Hom}_{\mathbf{Top}}(S,X)\times\mathrm{Hom}_{\mathbf{Top}}(S,Y)$, whose elements are the pairs of arrows

$$\begin{array}{ccccc} & & S & & \\ & \swarrow & & \searrow & \\ X & & & & Y. \end{array}$$

On morphisms, P maps a continuous map $f : S \to T$ to the function

$$\mathrm{Hom}_{\mathbf{Top}}(T,X)\times\mathrm{Hom}_{\mathbf{Top}}(T,Y) \xrightarrow{(-\circ f)\times(-\circ f)} \mathrm{Hom}_{\mathbf{Top}}(S,X)\times\mathrm{Hom}_{\mathbf{Top}}(S,Y)$$

mapping the pair

$$\begin{array}{ccccc} & & T & & \\ & {\scriptstyle a}\swarrow & & \searrow{\scriptstyle b} & \\ X & & & & Y \end{array}$$

to the pair

$$\begin{array}{ccccc} & & S & & \\ & {\scriptstyle a\circ f}\swarrow & & \searrow{\scriptstyle b\circ f} & \\ X & & & & Y \end{array}$$

by precomposition with f in both components. Intuitively, one may think of this presheaf as being a "combined" observation onto X and Y, using two instruments, or two eyes.

Now, is this presheaf representable? The question can be unpacked as follows: Is there an object Z of **Top**, i.e. a topological space, such that maps into Z have a natural bijection with pairs of maps into X and Y? Or, more intuitively, is there a space Z such that observations with instrument Z are the same as the combined observations with the instruments X and Y?

Following the guidelines above, a natural isomorphism,

$$\mathrm{Hom}_{\mathbf{Top}}(-,Z) \Rightarrow \mathrm{Hom}_{\mathbf{Top}}(-,X)\times\mathrm{Hom}_{\mathbf{Top}}(-,Y),$$

is equivalently given by an element p of PZ such that for each object S and each element $q \in PS$, there exists a unique map,

$$S \overset{f}{\dashrightarrow} Z,$$

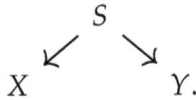

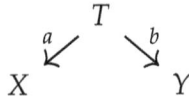

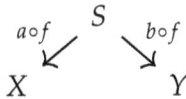

such that $Pf(p) = q$. Let's now unpack this condition. Since $PZ = \mathrm{Hom}_{\mathbf{Top}}(Z, X) \times \mathrm{Hom}_{\mathbf{Top}}(Z, Y)$, an element $p \in PZ$ is given by a pair of maps:

$$\begin{array}{ccc} & Z & \\ {}^{p_1}\swarrow & & \searrow^{p_2} \\ X & & Y. \end{array}$$

Similarly, an element $q \in PS = \mathrm{Hom}_{\mathbf{Top}}(S, X) \times \mathrm{Hom}_{\mathbf{Top}}(S, Y)$ is given by a pair of maps:

$$\begin{array}{ccc} & S & \\ {}^{f_1}\swarrow & & \searrow^{f_2} \\ X & & Y. \end{array}$$

Moreover, the condition $Pf(p) = q$ says that

$$\begin{array}{ccc} & S & \\ {}^{p_1 \circ f}\swarrow & & \searrow^{p_2 \circ f} \\ X & & Y \end{array} \quad = \quad \begin{array}{ccc} & S & \\ {}^{f_1}\swarrow & & \searrow^{f_2} \\ X & & Y, \end{array}$$

which means, equivalently, that $f_1 = p_1 \circ f$ and $f_2 = p_2 \circ f$. In other words, the condition, which is a universal property for Z, reads as follows: Fix the two maps $p_1 : Z \to X$ and $p_2 : Z \to Y$. Then, for every object S and every pair of maps $f_1 : S \to X$ and $f_2 : S \to Y$, there exists a unique map $f : S \to Z$ such that the following diagram commutes:

$$\begin{array}{ccccc} & & S & & \\ & {}^{f_1}\swarrow & \downarrow{\scriptstyle f} & \searrow^{f_2} & \\ X & \xleftarrow{p_1} & Z & \xrightarrow{p_2} & Y. \end{array}$$

In other words, maps $S \to Z$ are "the same" as pairs of maps $S \to X$ and $S \to Y$. Now, can we find such an object Z and maps $p_1 : Z \to X$ and $p_2 : Z \to Y$?

As you probably know, we can. Such an object Z is given by the cartesian product $X \times Y$, equipped with the product topology, and the maps $p_1 : X \times Y \to X$ and $p_2 : X \times Y \to Y$ are the two product projections. The correspondence is given as follows. Given the maps

f_1 and f_2, one can construct the map f as the one mapping $s \in S$ to $(f_1(s), f_2(s)) \in X \times Y$. Different choices of f_1 and f_2 give different maps $f : S \to X \times Y$, and all the maps $S \to X \times Y$ are in this form, for some f_1, f_2, since the elements of $X \times Y$ are uniquely specified by their components. It remains to check that the resulting map f is continuous in order for it to be a morphism of **Top** — for this, see the following exercise.

Exercise 2.3.5 (topology). Show that for every topological space S, the map $f : S \to X \times Y$ is continuous (for the product topology on $X \times Y$) whenever $f_1 : S \to X$ and $f_2 : S \to Y$ are continuous. (*Hint*: Pick convenient generating open sets of the product topology.)

Therefore, the product of two topological spaces is uniquely specified, up to homeomorphism, by its universal property. In some sense, it is the "inevitable" way of combining two spaces X and Y in terms of the maps into them.

Remark 2.3.6. The maps $p_1 : X \times Y \to X$ and $p_2 : X \times Y \to Y$ are part of the universal property of the product. Indeed, in order to have a universal property, it is not enough to say that "there is" a natural isomorphism, one needs to specify which one. In this case, the choice of the natural isomorphism corresponds to the choice of the maps $p_1 : X \times Y \to X$ and $p_2 : X \times Y \to Y$. Intuitively, without a choice of these maps, it may be unclear how to relate $X \times Y$ with X and Y, especially in the presence of symmetries.

We can construct a similar object in categories other than **Top**.

Exercise 2.3.7 (linear algebra). Show that the cartesian product of vector spaces satisfies a similar universal property in **Vect**.

Exercise 2.3.8 (group theory). Show that the direct product of groups satisfies a similar universal property in **Grp**.

Exercise 2.3.9. Show that the product of (small) categories (Definition 1.3.49) satisfies a similar universal property in **Cat**.

This construction, as we will see, is a special case of a very important universal construction called a *limit*. More on this will come in the subsequent chapters.

Exercise 2.3.10 (topology). Given topological spaces X and Y, the space $X \times Y$ is again a topological space. Now, fix X. Show that the assignment $Y \mapsto X \times Y$ is part of a functor $X \times - : \mathbf{Top} \to \mathbf{Top}$. (*Hint*: What could it do on morphisms?)

2.3.2 The universal property of the tensor product

Let V, W, and U be vector spaces. A map $f : V \times W \to U$ is called **bilinear** if it is linear in V and in W separately. That is, whenever we fix $x \in V$, the assignment $y \mapsto f(x, y)$ is linear in y, and whenever we fix $y \in W$, the assignment $x \mapsto f(x, y)$ is linear in x. Note that this is not the same as being linear as a map between the vector spaces $V \times W$ and U. For example, the map $\mathbb{R} \times \mathbb{R} \to \mathbb{R}$ given by $(x, y) \mapsto x + y$ is linear but not bilinear, and the map given by $(x, y) \mapsto xy$ is bilinear but not linear. A geometric example of a bilinear map is the (signed) area enclosed by the parallelogram of two vectors. A scalar product on a vector space is another such example.

Exercise 2.3.11 (linear algebra). Let V, W, U, and U' be vector spaces. Let $f : V \times W \to U$ be a bilinear map and $g : U \to U'$ be a linear map. Show that $g \circ f : V \times W \to U'$ is bilinear.

Let's now look at a new universal property, encoded by bilinearity. Let V, W, and U be vector spaces. Denote by $\mathrm{Bi}(V, W; U)$ the set of bilinear maps $V \times W \to U$. If we keep V and W fixed, this gives a functor $\mathrm{Bi}(V, W; -) : \mathbf{Vect} \to \mathbf{Set}$ in the following way. Given a linear map $g : U \to U'$, we get a function $g \circ - : \mathrm{Bi}(V, W; U) \to \mathrm{Bi}(V, W; U')$

that maps the bilinear function $f : V \times W \to U$ to the bilinear function $g \circ f : V \times W \to U'$, which is again bilinear by Exercise 2.3.11.

Now, is this functor representable? We look for a space Z and a natural isomorphism

$$\mathrm{Hom}_{\mathbf{Vect}}(Z, -) \Rightarrow \mathrm{Bi}(V, W; -).$$

Following the guidelines above, this amounts to an element of the set $p \in \mathrm{Bi}(V, W; Z)$ such that for every vector space S and for each element $q \in \mathrm{Bi}(V, W; S)$, there exists a unique linear map

$$Z \overset{f}{\dashrightarrow} S$$

such that $\mathrm{Bi}(V, W; f)(p) = q$. Let's unpack this definition. The element $p \in \mathrm{Bi}(V, W; Z)$ is a bilinear map $p : V \times W \to Z$, and this map has to satisfy the property that for each vector space S and for each bilinear map $q : V \times W \to S$, there exists a unique linear map f such that $f \circ p = q$, i.e. such that the following diagram commutes[2]:

$$\begin{array}{ccc} V \times W & & \\ {\scriptstyle p}\downarrow & \searrow{\scriptstyle q} & \\ Z & \underset{f}{\dashrightarrow} & S. \end{array}$$

You might know that such an object Z exists: It is the tensor product of vector spaces $V \otimes W$. For readers who are unfamiliar with this product, here is the basic idea for a finite-dimensional case. Take a three-dimensional vector space V, for example $\mathbb{R}^3$, and a two-dimensional vector space W, for example $\mathbb{R}^2$. We would like now to form a six-dimensional vector space $V \otimes W$ out of V and W, with the following intuition: If a vector of V is represented (after picking a basis) by a column of two numbers and a vector of W by a column of

[2]Note that the solid maps are bilinear, not linear, so this is not a diagram in **Vect**, but in **Set**.

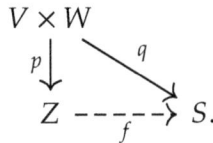

three numbers, a vector of $V \otimes W$ should be a 3×2 matrix.[3] (Note that this is different from the cartesian product, or the direct sum, which would instead have $3 + 2 = 5$ dimensions.) There is, moreover, a map $V \times W \to W \otimes W$, usually denoted again by $\otimes$, which multiplies the entries of the vectors forming all the possibilities; here is an example in coordinates:

$$\begin{array}{ccc} V \times W & \xrightarrow{\quad\otimes\quad} & V \otimes W \\ \left(\begin{pmatrix} a_1 \\ a_2 \\ a_3 \end{pmatrix}, \begin{pmatrix} b_1 \\ b_2 \end{pmatrix}\right) & \longmapsto & \begin{pmatrix} a_1\, b_1 & a_1\, b_2 \\ a_2\, b_1 & a_2\, b_2 \\ a_3\, b_1 & a_3\, b_2 \end{pmatrix}. \end{array}$$

Here is an informal definition for the general case.

Definition 2.3.12 (informal). Let V and W be vector spaces, with bases $E = (e_i)_{i\in I}$ and $E' = (e'_j)_{j\in J}$. Consider the set of symbols of the form $e_i \otimes e'_j$ for $e_i \in E$ and $e'_j \in E'$. (Note that this set is in bijection with the cartesian product of sets $E \times E'$.)

The **tensor product of** V **and** W is the vector space $V \otimes W$ which has $e_i \otimes e'_j$ as a basis. That is, its elements are in the form

$$\sum_{i\in I} \sum_{j\in J} t^{ij}\, e_i \otimes e'_j,$$

where t^{ij} are real numbers which are nonzero only for finitely many i and j.

We also define the following map in coordinates,

$$\begin{array}{ccc} V \times W & \xrightarrow{\quad\otimes\quad} & V \otimes W \\ \left(\sum_i v^i\, e_i, \sum_j w^j\, e'_j\right) & \longmapsto & \sum_{i\in I} \sum_{j\in J} v^i\, w^j\, e_i \otimes e'_j, \end{array}$$

called the **tensor product of vectors.**[4]

[3] In this particular case, we don't consider a matrix as a linear map, or as something to apply or compose, but just as a convenient way to arrange the components of a $3 \times 2 = 6$-dimensional vector.

[4] Alternative name: *Kronecker product*. Note that we are using the same symbol $\otimes$ for two different purposes, as in $V \otimes W$ and $v \otimes w$. Can you tell the difference?

(Why is the finite-dimensional example above an instance of this construction?) This informal definition can be turned into a formal one using the ideas in Chapter 4, see Exercise 4.2.31, but an even more formal definition can be given in terms of the universal property we are about to see. In Section 6.4.4, we will give an even more formal construction.

Let's now show why $V \otimes W$ represents the functor $\mathrm{Bi}(V, W; -)$:

$$
\begin{array}{ccc}
V \times W & & \\
{\scriptstyle p}\downarrow & \searrow^{q} & \\
V \otimes W & \dashrightarrow[f] & S.
\end{array}
$$

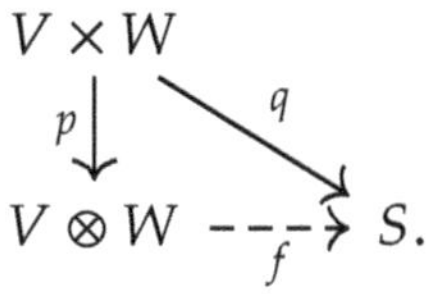

The map $p : V \times W \to V \otimes W$ is the tensor product of vectors $(v, w) \mapsto v \otimes w$, which is bilinear (why?). Here is how the bijection works. The elements of $V \otimes W$ are linear combinations of vectors in the form $v \otimes w$, for some $v \in V$ and $w \in W$. Therefore, a linear map $f : V \otimes W \to S$ is uniquely specified by what it does on vectors in the form $v \otimes w$, i.e. in the image of p. Given a bilinear map $q : V \times W \to S$, define $f : V \otimes W \to S$ to be the linear map specified by $f(v \otimes w) = q(v, w)$ and then extend linearly. This is possible since q is bilinear (for example, $f(2v \otimes w) = 2f(v \otimes w) = 2q(v, w) = q(2v, w)$).

This gives the universal property of the tensor product, and it can be considered an alternative definition, specifying $V \otimes W$ up to a unique isomorphism. Note that, similar to the product projections in the previous section, the map p is part of the universal property, and it is part of the tensor product structure. Intuitively, the tensor product is a "unique way to combine V and W in terms of bilinear maps on them."

Exercise 2.3.13 (linear algebra; [difficult!]). Fix a vector space V. Show that the assignment $W \mapsto V \otimes W$ is part of a functor $V \otimes - :$ **Vect** $\to$ **Vect** in a way that makes the map $p : V \times W \to V \otimes W$ natural in W. (The same thing is true for V.) Conclude that, even if our definition was given in terms of coordinates, the tensor product construction is compatible with changes in the basis.

Exercise 2.3.14 (commutative algebra). If you are familiar with these structures, compare the universal property above with one of the tensor products of:

- vector spaces over a generic (fixed) field;
- modules over a generic (fixed) commutative ring;
- abelian groups.

These tensor products will be studied in more detail in Chapter 6.

3

Limits and Colimits

In Section 1.4, we remarked how natural transformations are a generalization of equivariant maps and how invariant maps are a special case of equivariant maps when one of the actions is trivial. Natural transformations, when one of the functors is trivial, form diagrams that look like "cones," and an "optimal" (universal) choice of cone is called a *limit*. This construction gives one of the most fruitful ideas in category theory, used in almost all areas of mathematics.

3.1 General Definitions

Definition 3.1.1. Let $\mathbf{C}$ be a category. Let $\mathbf{J}$ be a small category and X be an object of $\mathbf{C}$. The **constant diagram** (or **constant functor**) **at** X (indexed by $\mathbf{J}$) is the functor $X : \mathbf{J} \to \mathbf{C}$ assigning:

- to each object of $\mathbf{J}$, always the same object X of $\mathbf{C}$;
- to each morphism of $\mathbf{J}$ the identity morphism id_X.

This is analogous to a constant function.

Definition 3.1.2. Let $\mathbf{J}$ be a small category, $F : \mathbf{J} \to \mathbf{C}$ be a diagram (i.e. a functor), and X be an object of $\mathbf{C}$. A **cone over** F with tip X is a natural transformation from the constant diagram at X to the functor F. A **cone under** F, or **cocone**, with bottom X is a natural transformation from the functor F to the constant diagram at X.

Explicitly, a cone over F is the following assignment: For each object J in $\mathbf{J}$, we have a morphism $\alpha_J : X \to FJ$ of $\mathbf{C}$ in such a way that for every morphism $m : J \to J'$ of $\mathbf{J}$, this triangle commutes:

$$
\begin{array}{ccc}
 & X & \\
\alpha_J \swarrow & & \searrow \alpha_{J'} \\
FJ \xrightarrow{Fm} & & FJ'.
\end{array}
$$

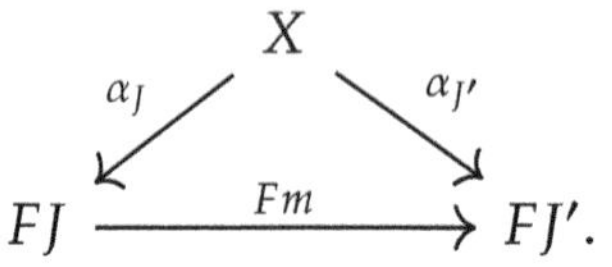

Example 3.1.3. Consider the diagram

$$
\begin{array}{ccc}
 & B & \\
A \nearrow & & \searrow \\
 \searrow & & \nearrow D, \\
 & C &
\end{array}
$$

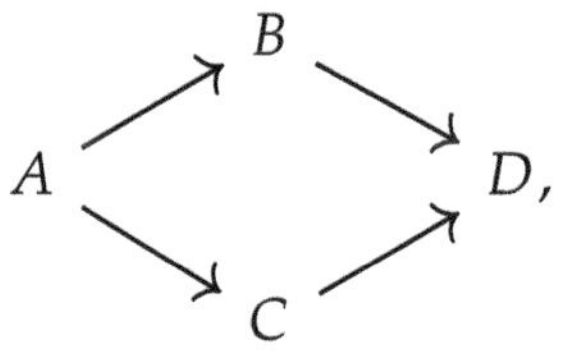

where the objects A, B, C, and D are in the form FJ for some objects J of $\mathbf{J}$, and the arrows in the diagram are in the form $Fm : FJ \to FJ'$, for some arrows $m : J \to J'$ of $\mathbf{J}$. A cone and a cocone over this diagram look as follows, hence their names:

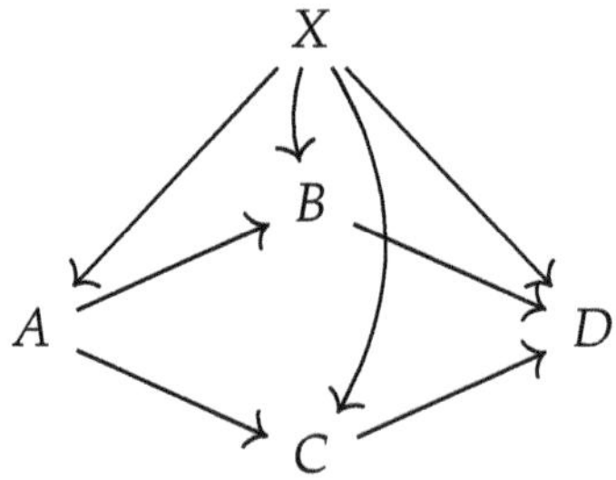

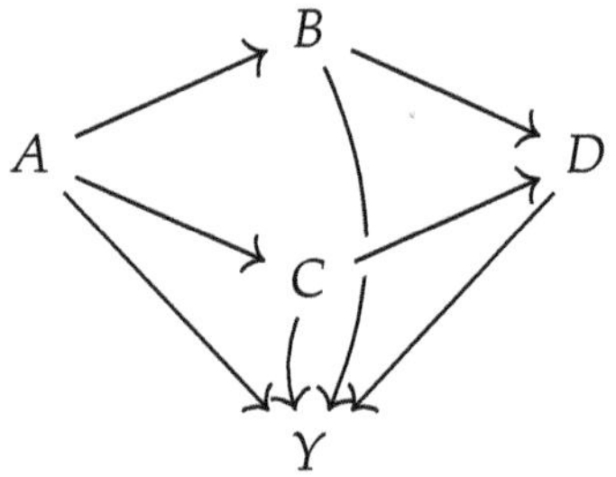

All the triangles involving X and Y are commutative (but the original diagram need not necessarily commute).

The cone construction is functorial, as follows. Given a cone with tip X and a morphism $f : X \to Y$, we get a cone with tip Y via the following composition:

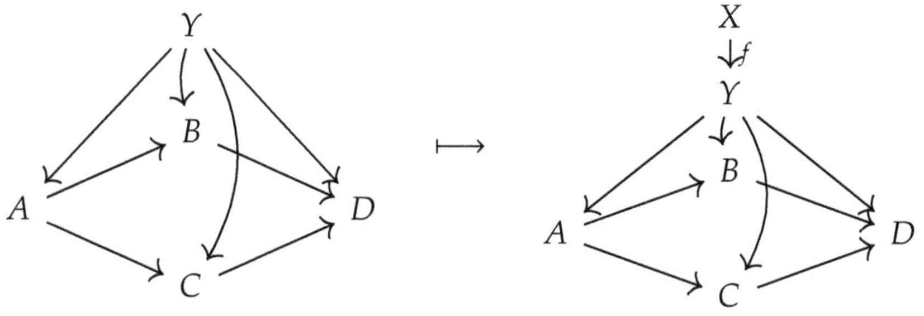

(Note the direction of the arrows.) Here is a more formal definition. You can read it while keeping the picture above in mind.

Definition 3.1.4. Let $F : \mathbf{J} \to \mathbf{C}$ be a diagram. We construct the presheaf $\mathrm{Cone}(-, F) : \mathbf{C}^{\mathrm{op}} \to \mathbf{Set}$ as follows:

- It maps an object X of $\mathbf{C}$ to the set $\mathrm{Cone}(X, F)$ of cones over F with tip X.
- It maps a morphism $f : X \to Y$ to the function $\mathrm{Cone}(Y, F) \to \mathrm{Cone}(X, F)$ given in the following way. It assigns to a cone $\alpha : Y \Rightarrow F$ with components $\alpha_J : Y \to FJ$, for each object J of $\mathbf{J}$, the cone $\alpha \circ f : X \Rightarrow F$ with components $(\alpha \circ f)_J := \alpha_J \circ f$.

Similarly, the functor $\mathrm{Cone}(F, -) : \mathbf{C} \to \mathbf{Set}$ is defined as follows:

- It maps an object X of $\mathbf{C}$ to the set $\mathrm{Cone}(F, X)$ of cocones under F with bottom X.
- It maps a morphism $f : X \to Y$ to the function $\mathrm{Cone}(F, X) \to \mathrm{Cone}(F, Y)$ given in the following way. It assigns to a cocone $\alpha : F \Rightarrow X$ with components $\alpha_J : FJ \to X$, for each object J of $\mathbf{J}$, the cocone $f \circ \alpha : F \Rightarrow Y$ with components $(f \circ \alpha)_J := f \circ \alpha_J$.

Limits and colimits are *universal* cones and cocones.

Definition 3.1.5. Let $F : \mathbf{J} \to \mathbf{C}$ be a diagram. A **limit** of F, if it exists, is an object of $\mathbf{C}$, which we denote $\lim F$, representing the presheaf $\mathrm{Cone}(-, F) : \mathbf{C}^{\mathrm{op}} \to \mathbf{Set}$, together with its universal property.

A **colimit** of F, if it exists, is an object of $\mathbf{C}$, which we denote $\operatorname{colim} F$, representing the functor $\operatorname{Cone}(F,-) : \mathbf{C} \to \mathbf{Set}$, together with its universal property.

Let's see what this means concretely. The objects $\lim F$ and $\operatorname{colim} F$, whenever they exist, are equipped by definition with the natural isomorphisms

$$\operatorname{Hom}_{\mathbf{C}}(-, \lim F) \Rightarrow \operatorname{Cone}(-, F) \quad \text{and}$$
$$\operatorname{Hom}_{\mathbf{C}}(\operatorname{colim} F, -) \Rightarrow \operatorname{Cone}(F, -).$$

As we saw in Section 2.3, by the Yoneda lemma, these natural transformations are uniquely specified by "universal" elements of the sets

$$\operatorname{Cone}(\lim F, F) \quad \text{and} \quad \operatorname{Cone}(F, \operatorname{colim} F),$$

i.e. cones in the following form:

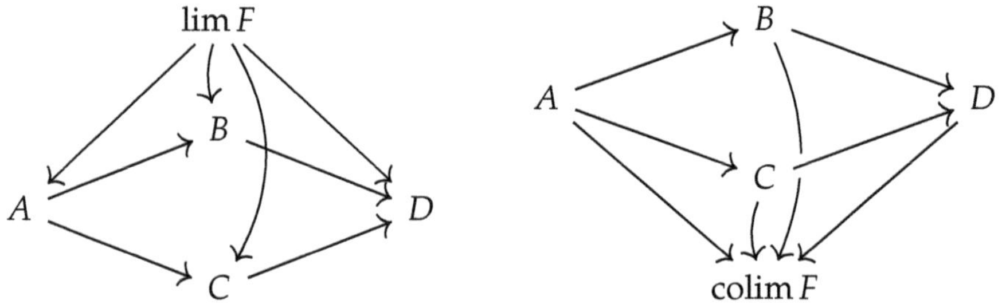

Moreover, these cones are universal in the following sense. For the limiting cone (the diagram on the left), given any (other) cone α with (any) tip X, such as

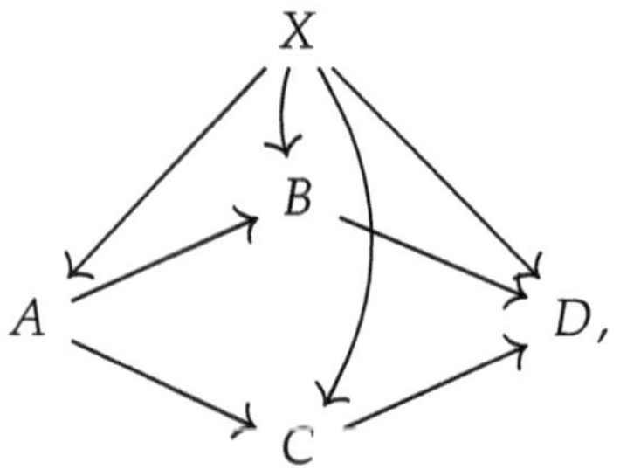

there is a unique map $u : X \to \lim F$ such that for each $J \in \mathbf{J}$, each component $\alpha_J : X \to FJ$ of the cone with tip X factors uniquely through u and through the component $\phi_J : \lim F \to FJ$ of the limiting cone. That is, for each J, this triangle has to commute:

$$
\begin{array}{ccc}
X & & \\
\downarrow u & \searrow^{\alpha_J} & \\
\lim F & \xrightarrow[\phi_J]{} & FJ.
\end{array}
$$

In our example,

$=$

That is,

these triangles commute. ——————

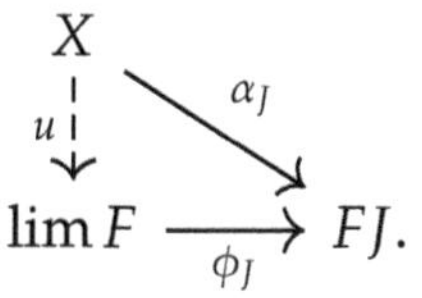

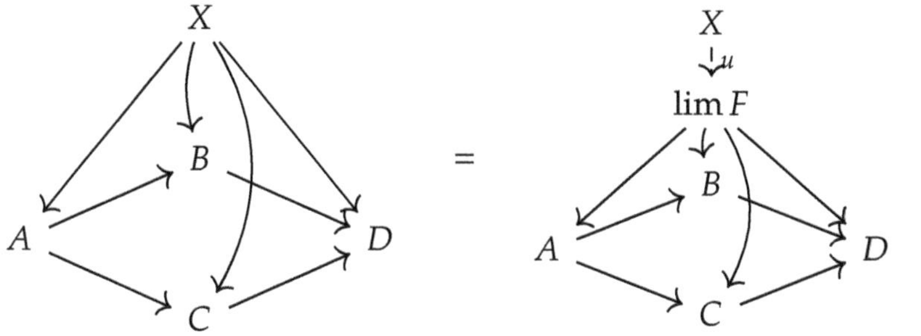

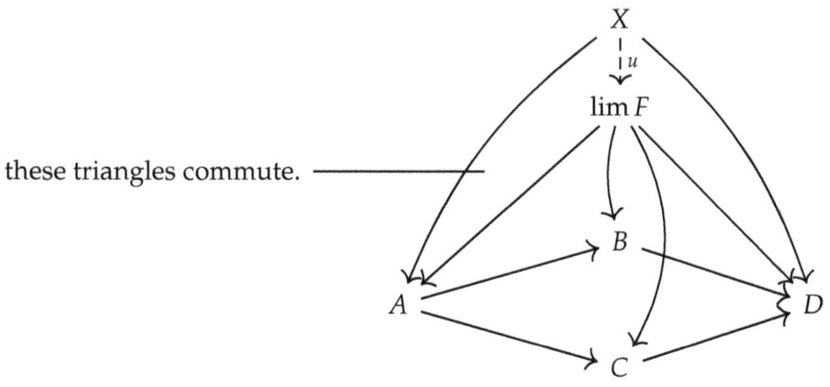

For the colimiting cone, given any cocone β under F with bottom Y, such as

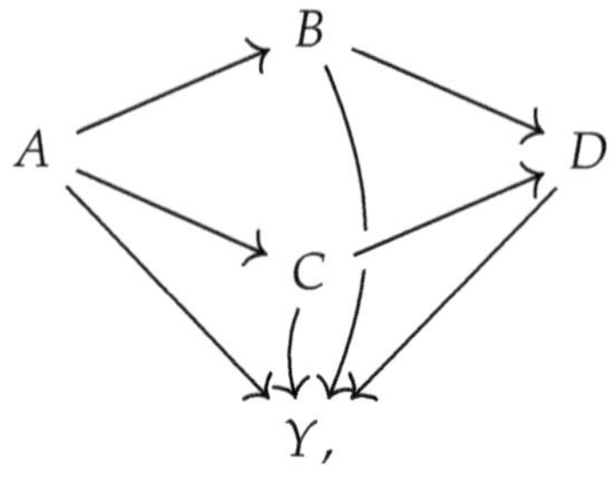

there is a unique map u : colim $F \to Y$ such that for each $J \in \mathbf{J}$, each component $\beta_J : FJ \to Y$ of the cone with bottom Y factors uniquely through u and through the component $\psi_J : FJ \to$ colim F of the colimiting cone. That is, for each J, this triangle has to commute:

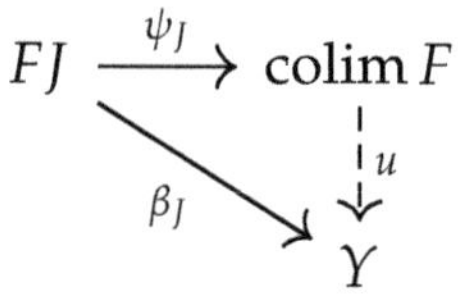

In our example,

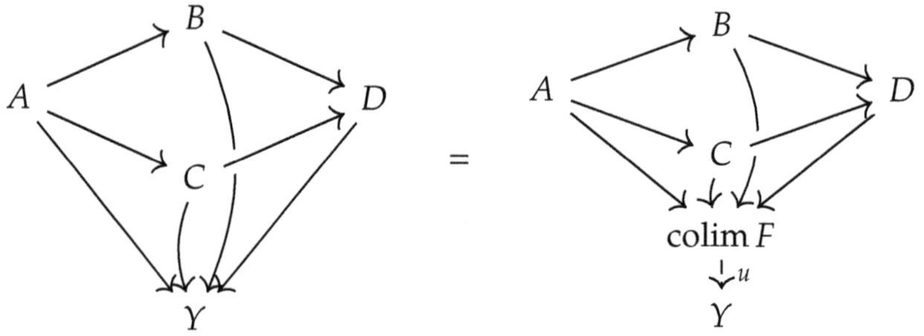

That is,

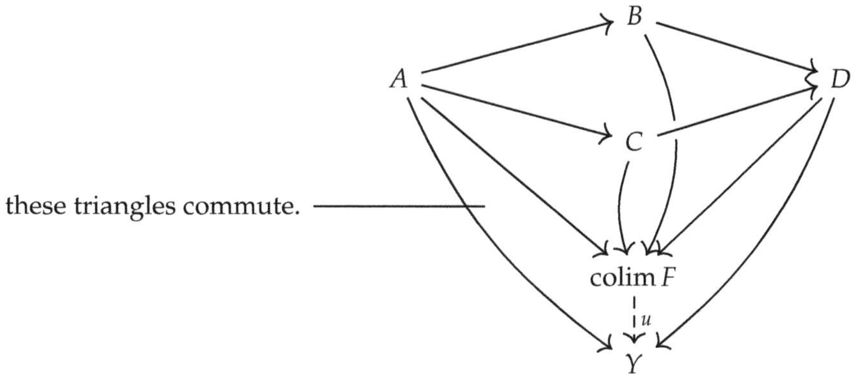

As we know by the Yoneda lemma, moreover, if $\lim F$ and $\operatorname{colim} F$ exist, they are unique up to isomorphism.

Definition 3.1.6. A category **C** is called **complete** if every diagram in **C** has a limit. **C** is called **cocomplete** if every diagram in **C** has a colimit.

Let's now see some practical examples of limits and colimits.

3.2 Particular Limits and Colimits

3.2.1 The poset case: Constrained optimization

Let $(X, \leq)$ be a poset. If we view it as a category, then a diagram in X has the same relevant information as a subset of X because, given any two elements, there is up to one arrow between them. So, let $S \subseteq X$. Plugging in the definitions of cone and cocone, we get the following. A cone over S with tip x is an element x together with an arrow $x \to s$, for each element $x \in S$. That is, it is an element $x \in X$ such that for all $s \in S$, $x \leq s$. This is the same as a **lower bound** for S. A cocone over S with bottom y is an element y together with arrows

$s \longrightarrow y$, i.e. $s \leq y$, for each $s \in S$. Therefore, it is the same as an **upper bound** for S. Mind the possibly confusing direction of the arrow: If x is the tip of a cone *over* S in the sense of category theory, it is actually *below* S in the partial order.

A cone with tip x is a limiting cone if the following universal property holds: Given any (other) lower bound z of S, we have a (necessarily unique) arrow $z \longrightarrow x$, i.e. $z \leq x$. Equivalently, for every $z \in X$, we have that $z \leq s$ for every $s \in S$ if and only if $z \leq x$.

Exercise 3.2.1 (sets and relations). Prove that these two statements are indeed equivalent.

In other words, we are saying that x is the **greatest lower bound**, or **infimum**, or **meet**, of S. This is usually denoted by $\inf S$ or $\wedge S$. Similarly, y defines a colimiting cone if and only if it is the **least upper bound**, or **supremum**, or **join**, of S. This is usually denoted by $\sup S$ or $\vee S$.

As you probably know from analysis, infima and suprema do not always exist, but when they do, they are unique.

In a way, the universal properties of limits and colimits in a poset correspond to solutions of a constrained optimization problem. They are the "optimal" elements satisfying a certain constraint: The limit (meet) of S is the largest element which is still below S, while the colimit (join) is the smallest element which is still above S. Because of this, one may view limits and colimits as a generalization of constrained optimization. This idea that limits and colimits are the largest or smallest objects satisfying a certain property can also help our intuition in most of the following examples.

3.2.2 The group case: Invariants and orbits

Let G be a group. We know that a functor (or a diagram) $F : \mathbf{B}G \longrightarrow \mathbf{Set}$ is the same as a G-set, a set X equipped with an action of G. Let's see what cones, cocones, limits, and colimits are.

Plugging in the definition, a cone over F is a set S together with a map $f : S \to X$ such that for each $g \in G$, the following triangle commutes:

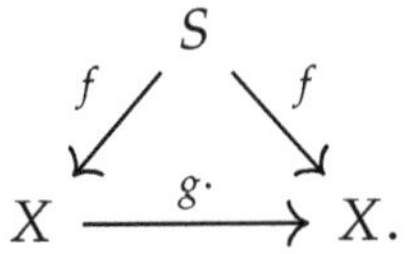

In other words, for every $s \in S$ and $g \in G$, we must have $g \cdot f(s) = f(s)$. This means exactly that every element in the image of f is an invariant element for the action. In particular, if f is injective, S can be seen as a **subset of invariant elements.**

Warning. A *set of invariant elements* is not the same thing as an *invariant set*. If S is a set of invariant elements, then applying g to any $s \in S$ leaves s where it is. Instead, if S is an invariant set, applying g to $s \in S$ may move it, but it will stay inside S. For example, for rotations in a plane, a circle centered at the origin is an invariant set, but not a set of invariant elements (its points move). The origin is an invariant element.

Let us now see what the limit is. Plugging in the definition, $\lim F$ is a cone such that for any (other) cone, say with tip S, there is a unique map $S \to \lim F$ that makes the following diagram commute for each $g \in G$:

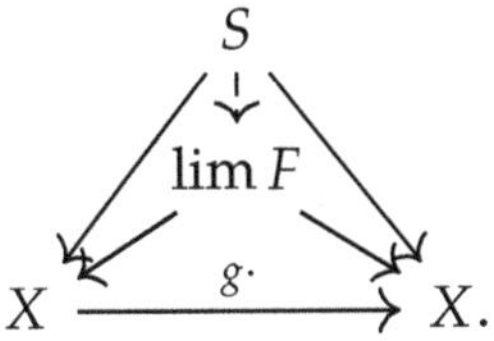

So, first of all, any such map $\lim F \to X$ must be injective. Here is why. Let $I \subseteq X$ be the image of this map. Then, the inclusion $I \to X$ (which is injective) would also give a cone over F (why?), and every map of a cone $S \to X$ would factor uniquely through I, i.e. a composition $S \to I \to X$. But this is precisely the universal property of the limit, and so I would be the limit. Since limits are unique up to isomorphism, we conclude that $\lim F \cong I$, and so the map $\lim F \to X$ must be injective too. Therefore, $\lim F$ is (isomorphic to) a subset of invariant elements. But which subset? Since every subset S of invariant elements gives a cone over F, and by the universal property of the limit, the inclusion $S \to X$ has to factor as $S \to \lim F \to X$, it must necessarily be that $\lim F$ contains all the invariant elements. Therefore, $\lim F \to X$ is, up to isomorphism, exactly the *set of all invariant elements for the action*. In terms of the natural isomorphism, the universal property reads as follows: A function into X picking out only G-invariant elements is the same as a function into $\lim F$. In terms of constrained optimization, it is the largest subset of X whose elements are all invariant for the G-action. Note that this set may be empty, but as a set, it always exists.

Let us now turn to cocones. A cocone under F, plugging in the definition, is a set Y together with a map $p : X \to Y$ such that for each $g \in G$, the following diagram commutes:

$$\begin{array}{ccc} X & \xrightarrow{g\cdot} & X \\ & {\scriptstyle p}\searrow \quad \swarrow{\scriptstyle p} & \\ & Y. & \end{array}$$

In other words, for each $x \in X$ and $g \in G$, we must have $p(g \cdot x) = p(x)$. This is the same thing as an **invariant function**: a function whose value does not change after applying g to its argument.

We can view the function p also in a different way, as follows. Any function $p : X \to Y$ can be seen as a **partition** of X. That is, we can divide X into different regions, according to the values that p assumes. For the value $y \in Y$, we have the region $p^{-1}(y) \subseteq X$ of the elements $x \in X$ such that $p(x) = y$. For a different value $y' \in Y$, we have the region $p^{-1}(y') \subseteq X$ of the elements $x \in X$ such that $p(x) = y'$, and so on. We have that X is the disjoint union

$$X \cong \coprod_{y \in Y} p^{-1}(y).$$

Note that some of the regions may be empty (precisely when p is not surjective). Now, if and only if p is an invariant function, we have that for all $y \in Y$, $p^{-1}(y)$ is an invariant set. If $x \in p^{-1}(y)$, then $p(x) = y$, but then also for each $g \in G$, $p(g \cdot x) = y$, i.e. $g \cdot x \in p^{-1}(y)$. (Note that this is an invariant set, not necessarily a set of invariant elements, see the remark above.) We are saying that p is an invariant function if and only if, if we see it as a partition, all the regions of the partition are invariant sets; that is, G acts on those regions separately, never moving elements from one region to another. Therefore, a cocone under F is the same thing as a **partition of invariant sets,** also called an **invariant partition.**

What about the colimit of F? Plugging in the definition, $\operatorname{colim} F$ is a cocone such that for any (other) cone, say with bottom Y, there is a unique map $\operatorname{colim} F \to Y$ that makes the following diagram commute, for each $g \in G$:

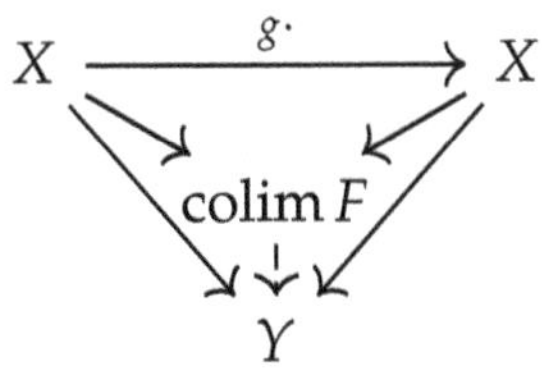

Dually to the case above, the map $X \to \operatorname{colim} F$ must first of all be surjective. Here is why. We can consider its image $I \subseteq \operatorname{colim} F$, and the induced map $X \to I$ is then surjective by construction. Again, any (other) map $X \to Y$ defining a cocone must factor uniquely through $\operatorname{colim} F$ by the universal property of the colimit and, therefore, also through I (why?). We can again conclude that $I \cong \operatorname{colim} F$, and so the map $X \to \operatorname{colim} F$ must be surjective as well. Therefore, in terms of partitions, the map $X \to \operatorname{colim} F$ corresponds to a partition where no region is empty. Now, let O be an orbit of G. Consider the function p from X to the two-element set $\{0, 1\}$ mapping O to 0 and everything else to 1. This function is invariant (why?). If we want this function to factor through $\operatorname{colim} F$, then the partition induced by $\operatorname{colim} F$ must be as fine as the one induced by p; that is, it needs to have O as one of its regions. But this can be done for all orbits O of X; therefore, the partition induced by $\operatorname{colim} F$ must precisely be the partition of X into the orbits of the G-action. In terms of the natural isomorphism, the universal property reads as follows: An invariant function on F is the same thing as a function on $\lim F$, i.e. a function on the set of orbits. In terms of constrained optimization, this is the finest G-invariant partition. Again, note that such a partition always exists.

Exercise 3.2.2 (group theory, linear algebra). Let $F : \mathbf{B}G \to \mathbf{Vect}$ be a linear representation of G. What are its limit and colimit?

Exercise 3.2.3 (group theory, topology). Let $F : \mathbf{B}G \to \mathbf{Top}$ be an action of G on a topological space. What are its limit and colimit?

3.2.3 Products and coproducts

Consider now the category **Set**, and take two sets A and B. They form a discrete diagram F, which simply looks as follows:

$$A \qquad\qquad B.$$

"Discrete" simply means that the only arrows present in the diagram are the identities (not drawn). Now, a cone over F consists of a set X together with maps $f : X \to A$ and $g : X \to B$, i.e.

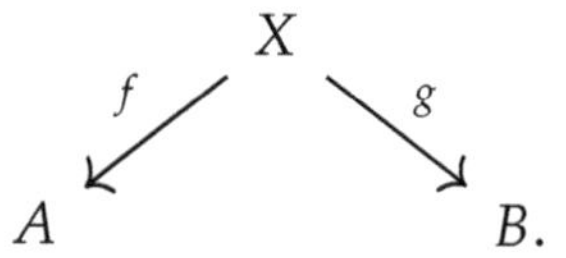

A limiting cone over F consists of a cone with the following universal property. Given any (other) cone over F, say with tip X, there is a unique map $X \to \lim F$ that makes the following diagram commute:

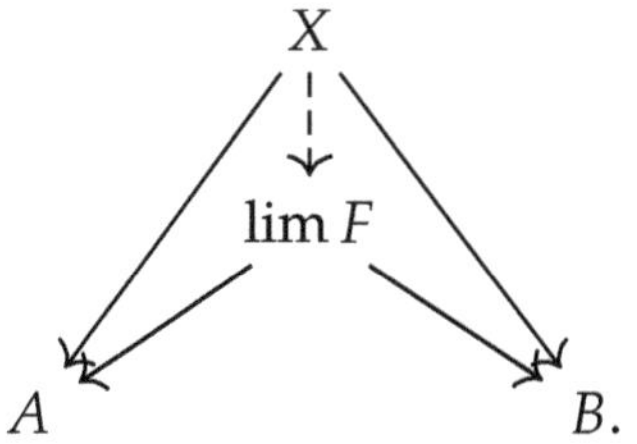

You may recognize that this is exactly the universal property of the cartesian product $A \times B$, as we saw in Section 2.3.1. A way to recognize it is to look at the case of $X = 1$, the one-element set. The maps $1 \to A$ and $1 \to B$ correspond to the elements of A and B, respectively. The universal property says, in this case, that an element of $A \times B$ is equivalently a pair of elements (a, b) of A and B, respectively. More generally, a function into $A \times B$ is the same as a pair of functions into A and B. Therefore, $A \times B$ is (up to isomorphism) the limit of the diagram F, and the maps forming the universal cone are the product projections.

Definition 3.2.4. The limit of a discrete diagram is called the **product**, or **cartesian product**, or **categorical product**. If the diagram contains two objects A, B, their product is usually denoted by $A \times B$, whereas if the diagram contains many objects $\{A_i\}_{i \in I}$, their product is usually denoted by $\prod_{i \in I} A_i$.

Intuitively, products are used to combine objects to form more complex "composite" objects, in such a way that the complex object can always be projected back to its simpler components via the projection maps $A \times B \to A$ and $A \times B \to B$ forming the universal cone. We have seen products of topological spaces already in Section 2.3.1. The exercises in that section involve similar constructions for vector spaces and groups.

Exercise 3.2.5 (graph theory). What is the product of two graphs in the category **MGraph** of multigraphs and their morphisms (Example 1.4.8)?

Exercise 3.2.6. Can we generalize Exercise 2.3.10 to arbitrary categories (assuming that all the necessary products exist)?

What is the colimit of F? Reversing all the arrows, this is a cocone such that for each (other) cocone, say with bottom Y, there exists a unique map $\operatorname{colim} F \longrightarrow Y$ such that the following diagram commutes:

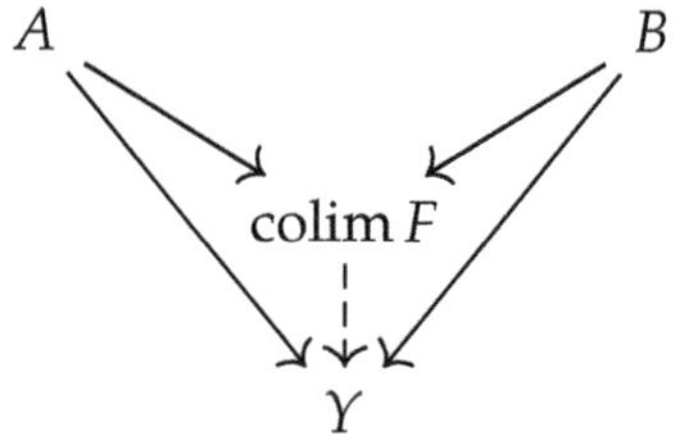

This means that $\operatorname{colim} F$ is a set such that having a function on $\operatorname{colim} F$ is the same as having a function on A and a function on B. In other words, $\operatorname{colim} F$ is the *disjoint union* $A \sqcup B$. A function on the disjoint union of A and B is uniquely specified by its values on A (which give a function on A) and its values on B (which give a function on B), and vice versa. The universal maps forming the colimit cocone are the canonical inclusions $A \longrightarrow A \sqcup B$ and $B \longrightarrow A \sqcup B$.

Definition 3.2.7. The colimit of a discrete diagram is called the **coproduct**, or, sometimes, **sum**. If the diagram contains two objects A, B, their coproduct is usually denoted by either $A + B$ or $A \sqcup B$, whereas if the diagram contains many objects $\{A_i\}_{i \in I}$, their coproduct is usually denoted by $\coprod_{i \in I} A_i$.

Intuitively, coproducts are used to combine objects to form more complex objects by "adding them together" in such a way that the simple objects are contained in the complex object via the inclusion maps $A \longrightarrow A \sqcup B$ and $B \longrightarrow A \sqcup B$ forming the universal cocone.

The symbol $\coprod$ is used because it is the upside-down version of the symbol $\prod$.

Exercise 3.2.8 (linear algebra). Show that in **Vect**, the coproduct of two vector spaces is their direct sum. Conclude that, for two (or finitely many) vector spaces, the product and the coproduct are isomorphic.

Exercise 3.2.9 (topology). Show that the coproduct in **Top** is given by the disjoint union, with the disjoint union topology.

Exercise 3.2.10 (group theory). Show that in **Grp**, the coproduct of two groups is given by their free product.

Exercise 3.2.11 (graph theory). What is the coproduct of two graphs in **MGraph**?

3.2.4 Equalizers and coequalizers

Consider the following diagram in **Set**; let's call it F again, given by a parallel pair of arrows:

$$A \underset{g}{\overset{f}{\rightrightarrows}} B.$$

Note that this is, in general, not a commutative diagram (unless $f = g$).

A cone over F with tip X consists of maps $p : X \to A$ and $q : X \to B$ such that the following triangles commute:

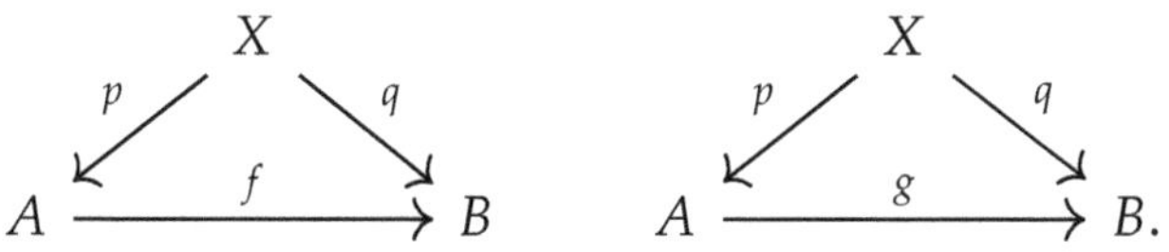

This means, in particular, that $f \circ p = q = g \circ p$. So, since the map q is just a composite, we can omit it from the diagram, and rewrite the condition saying that a cone with tip X over F is a map $p : X \to A$, as in the diagram

$$X \xrightarrow{p} A \underset{g}{\overset{f}{\rightrightarrows}} B,$$

such that $f \circ p = g \circ p$. (Again, note that this does not imply $f = g$ unless p is epi. Therefore, the diagram above is not commutative in general.) This means, in other words, that for every $x \in X$, $f(p(x)) = g(p(x))$, or equivalently, that f and g agree on the image of p. In particular, if p is injective, X can be seen as a subset of A on which f and g agree. Note that this does not imply that f and g necessarily disagree outside of X: X is not necessarily the *largest* subset on which f and g agree.

What is now a limit of the diagram F? First of all, similarly to what we observed in Section 3.2.2, the universal map $\lim F \to A$ must be injective (see Exercise 3.2.16). Therefore, the limit of F must be a subset of A on which f and g agree. Moreover, similarly to what we observed in Section 3.2.2, since any subset of A on which f and g agree gives a cone over F, the limit of F must contain all such subsets. In other words, $\lim F$ is precisely the largest subset of A in which f and g agree, and the universal map $\lim F \to A$ is the inclusion.

Definition 3.2.12. The limit of a pair of parallel arrows f, g is called the **equalizer** of f, g.

Intuitively, equalizers are used to *define subspaces by means of equations*. This is done systematically in algebraic geometry, but it also appears in many other fields of math.

Example 3.2.13 (basic geometry). In the category **Top**, consider the diagram

$$\mathbb{R}^2 \underset{g}{\overset{f}{\rightrightarrows}} \mathbb{R},$$

where f is a function given by $f(x, y) := x^2 + y^2$ and g is a constant function with a value of 1. The equalizer of f and g is the largest subset S of $\mathbb{R}^2$ such that for all of its points $(x, y) \in S$, $x^2 + y^2 = 1$. This is precisely the unit circle S^1.

Example 3.2.14 (topology). Show that, generalizing the example above, the equalizer in **Top** of two maps $f, g : A \to B$ is given by the subset of A on which f and g agree, equipped with the subspace topology inherited by A.

Example 3.2.15 (linear algebra, group theory). Show that in the categories **Vect** and **Grp**, given a map $f : A \to B$ and the constant zero map $0 : A \to B$, the kernel of f is the equalizer of f and 0.

Exercise 3.2.16. Let $m : E \to A$ be the equalizer of $f, g : A \to B$ in a category **C**. Prove that m is necessarily mono. (*Hint*: The idea is similar to what we came across in Section 3.2.2.)

Exercise 3.2.17. In the notation of the exercise above, prove that if and only if m is epi, it is an isomorphism, and $f = g$.

What is a cocone under F? Similarly as above, a cocone under F with bottom Y is given by a map $q : B \to Y$, as in the diagram

$$A \underset{g}{\overset{f}{\rightrightarrows}} B \xrightarrow{q} Y,$$

such that $q \circ f = q \circ g$ (why exactly?). Note that, again, this does not imply that $f = g$ (unless q is mono). In other words, the function q is such that for every $a \in A$, $q(f(a)) = q(g(a))$. Intuitively, q cannot tell whether the points come through f or g — it cannot tell the two maps apart. Let's make this precise. Consider the following relation on B: $b \sim b'$ if and only if there exists $a \in A$ such that $b = f(a)$ and $b' = g(a)$. Then, our condition on q states precisely that q respects the relation, i.e. $b \sim b'$ implies that $q(b) = q(b')$. If we view q as a partition, we are saying that this partition cannot be finer than the relation $\sim$; that is, no two related elements $b \sim b'$ can come from different regions of the partition.

Exercise 3.2.18 (sets and relations). The relation $\sim$ defined above is in general not an equivalence relation. Show that, in any case, q also respects the equivalence relation generated by $\sim$. Such an equivalence relation can be characterized as the smallest (finest) equivalence relation containing $\sim$, or as the one obtained from $\sim$ by enforcing reflexivity, symmetry, and transitivity.

As you might expect, then, a colimiting cone is given by the *finest* function which respects this relation. In other words, analogously to what happens in Section 3.2.2, $\operatorname{colim} F$ is the quotient space of B under the equivalence relation generated by $\sim$, and the universal map $Y \to \operatorname{colim} F$ is the quotient map. In terms of partitions, q corresponds precisely to the partition of B into the equivalence classes generated by $\sim$. This is because, first of all, the universal map is always surjective (Exercise 3.2.21). Moreover, let $C \subseteq B$ be an equivalence class generated by $\sim$, and consider the map $q : B \to \{0, 1\}$ that maps C to 0 and everything else to 1. This map forms a cocone (why?). If we want q to factor through the colimit, as the universal property prescribes, the only possibility is that the universal map $B \to \operatorname{colim} F$, seen as a partition, has C as one of its regions. Again, as in Section 3.2.2, this must be true for all such equivalence classes.

Definition 3.2.19. The colimit of a pair of parallel arrows f, g is called the **coequalizer** of f, g.

Coequalizers are used to create quotient spaces canonically by *identification.*

Exercise 3.2.20 (topology). Consider the following diagram in **Top**:

$$\bullet \overset{1}{\underset{0}{\rightrightarrows}} [0,1],$$

where $\bullet$ is the space of a single point, $[0, 1]$ is the real unit interval, and the maps 0 and 1 map the unique point of $\bullet$, respectively, to 0 and $1 \in [0, 1]$. Show that the coequalizer of these two maps is given by the circle S^1, as the space obtained by identifying 0 and 1 in the unit interval $[0, 1]$.

Exercise 3.2.21. Prove the dual statement to Exercise 3.2.16, that is, the universal map of the coequalizer of f and g is always epi.

Conclude also that if and only if such a map is also mono, then it is an isomorphism, and $f = g$.

3.2.5 Pullbacks and pushouts

Consider the following diagram, called a **cospan**. Let's call the diagram F, as usual:

$$\begin{array}{ccc} & & A \\ & & \downarrow{\scriptstyle f} \\ B & \xrightarrow{g} & C. \end{array}$$

A cone over F with tip X corresponds to an object X together with maps $p : X \to A$ and $q : X \to B$ such that this diagram commutes:

$$\begin{array}{ccc} X & \xrightarrow{p} & A \\ \downarrow{\scriptstyle q} & & \downarrow{\scriptstyle f} \\ B & \xrightarrow{g} & C. \end{array}$$

As before, the map $X \to C$ is implicit since it is just a composition (of which maps?).

A limiting cone is a cone such that for every (other) cone, say with tip X, there is a unique arrow $X \to \lim F$ such that the following diagram commutes:

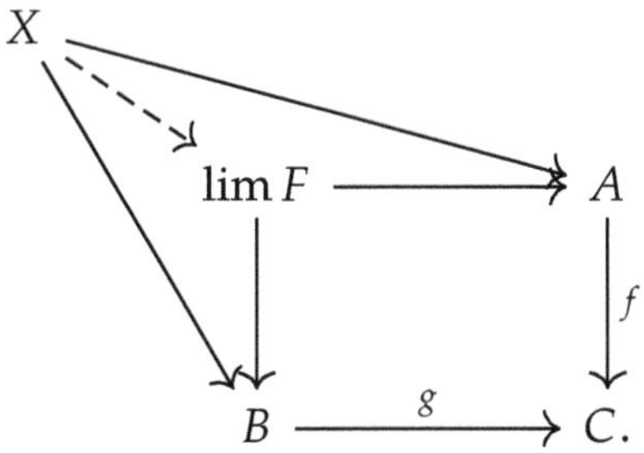

Definition 3.2.22. The limit of a cospan $f : A \to C$, $g : B \to C$ is called **pullback**, or **fibered product**. As an object, it is usually

denoted by $A \times_C B$. The maps giving the universal cone are denoted by $f^*g : A \times_C B \to A$ and $g^*f : A \times_C B \to B$.

The universal cocone is sometimes called a **pullback square**, or **cartesian** square, and it is denoted by putting either the symbol $\ulcorner$ or $\lrcorner$ in the corner of $A \times_C B$, as in one of these diagrams:

$$\begin{array}{ccc} A \times_C B & \xrightarrow{f^*g} & A \\ {\scriptstyle g^*f}\downarrow & \ulcorner & \downarrow{\scriptstyle f} \\ B & \xrightarrow{g} & C \end{array} \qquad \begin{array}{ccc} A \times_C B & \xrightarrow{f^*g} & A \\ {\scriptstyle g^*f}\downarrow & \lrcorner & \downarrow{\scriptstyle f} \\ B & \xrightarrow{g} & C. \end{array}$$

(We use the convention on the left.)

Here are two exercises for people with a background in differential geometry. These also explain the names "pullback" and "fibered product," as well as the notation f^*.

Exercise 3.2.23 (differential geometry, topology). Let $p : E \to X$ be a fiber bundle. Let $f : Y \to X$ be a smooth map. Show that the map $f^*p : E \times_X Y \to Y$ forming the pullback

$$\begin{array}{ccc} E \times_X Y & \xrightarrow{p^*f} & E \\ {\scriptstyle f^*p}\downarrow & \ulcorner & \downarrow{\scriptstyle p} \\ Y & \xrightarrow{f} & X \end{array}$$

is the pullback bundle of p along f in the category **Mfd**.

Exercise 3.2.24 (differential geometry, topology). Let $p : E \to X$ and $q : F \to X$ be fiber bundles on the same base space X. Show that the map $E \times_X F \to X$ obtained from the pullback

$$\begin{array}{ccc} E \times_X F & \xrightarrow{p^*q} & E \\ {\scriptstyle q^*p}\downarrow & \ulcorner & \downarrow{\scriptstyle p} \\ F & \xrightarrow{q} & X \end{array}$$

is a fiber bundle whose fiber is the product of the fibers of p and q in the category **Mfd**.

Warning. The notion of pullback in category theory does not always coincide with the notion of pullback in other fields of mathematics. In the case of bundles, the two notions coincide, but keep in mind that this is not always the case. This is even more evident in the dual case, where in category theory, we have a construction called *pushout* (see the following), while in other fields of mathematics, the dual to pullback is *pushforward*. These are in general very different operations.

To see how the pullback works, consider these two exercises in the category of sets.

Exercise 3.2.25 (sets and relations). In the category **Set**, let S and T be subsets of X together with the inclusion maps

$$\begin{array}{ccc} & & S \\ & & \downarrow \\ T & \hookrightarrow & X. \end{array}$$

Show that the pullback of these diagrams is given by the intersection of S and T. What are the maps of the universal cone?

Exercise 3.2.26 (sets and relations). In the category **Set**, let 1 be a one-element set. Show that $A \times_1 B \cong A \times B$.

Now, we consider more general categories.

Exercise 3.2.27 (topology, graph theory). In the categories **Top** and **Mgraph**, do we have similar results as in the previous two exercises?

Exercise 3.2.28. Consider a morphism $f : X \to Y$ in a category **C**. Let's form the pullback of f with itself (if it exists):

$$\begin{array}{ccc} X \times_Y X & \longrightarrow & X \\ \downarrow & \ulcorner & \downarrow f \\ X & \xrightarrow{f} & Y. \end{array}$$

The two universal maps $X \times_Y X \to X$, which in this case are parallel maps, are called the **kernel pair** of f. Here is now the exercise: prove that f is mono if and only if those maps coincide and are isomorphisms $X \times_Y X \cong X$.

This is where the name "kernel" comes from: The kernel pair of f is trivial if and only if f is mono.

The following exercise can help with the intuition of "what the kernel pair really does."

Exercise 3.2.29 (sets and relations). In the category **Set**, let $m : X \to Y$ be a function, and consider its kernel pair, denoting the resulting two maps $f, g : X \times_Y X \to X$. Consider the relation $\sim$ defined by f and g constructed as in Section 3.2.4; that is, for each $x, x' \in X$, we have $x \sim x'$ if and only if there exists $p \in X \times_Y X$ such that $f(p) = x$ and $g(p) = x'$. Show that:

- in this case, $\sim$ is automatically an equivalence relation;
- $x \sim x'$ if and only if $m(x) = m(y)$, so that as a partition, $\sim$ is exactly the partition induced by m;
- if there exists $p \in X \times_Y X$ such that $f(p) = x$ and $g(p) = x'$, then this p is unique.

Exercise 3.2.30 (difficult!). More generally, suppose that a morphism $f : X \to Y$ in a category **C** has a kernel pair (i.e. it exists), and suppose moreover that f is the coequalizer of *something*. Show that f is then (also) the coequalizer of its kernel pair. Show also that this generalizes the previous exercise for the case where m is surjective.

Let us now consider the dual case. Consider this diagram, called a **span**. Let's call this diagram G:

$$\begin{array}{ccc} A & \xrightarrow{f} & B \\ \big\downarrow{\scriptstyle g} & & \\ C. & & \end{array}$$

A cocone under G of bottom Y consists of an object Y together with maps $B \to Y$ and $C \to Y$ such that the following diagram commutes. Again, there is no need to specify the map $A \to Y$ (why?):

$$\begin{array}{ccc} A & \xrightarrow{f} & B \\ {\scriptstyle g}\downarrow & & \downarrow \\ C & \longrightarrow & Y. \end{array}$$

A colimiting cocone is now a cocone such that for any (other) cocone, say with bottom Y, there is a unique map $\operatorname{colim} G \to Y$ such that the following diagram commutes:

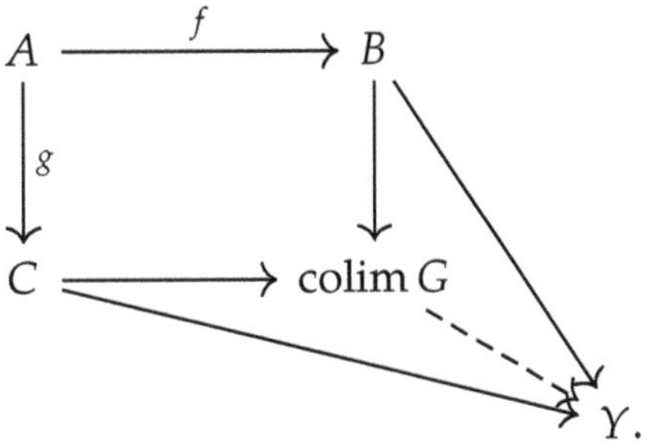

Definition 3.2.31. The colimit of a span is called **pushout**. As an object, it is usually denoted by $B \sqcup_A C$. The maps giving the universal cone are denoted by $f_*g : A \to B \sqcup_A C$ and $g_*f : B \to B \sqcup_A C$.

The universal cocone is sometimes called a **pushout square**, or **cocartesian** square, and it is denoted by putting either the symbol $\ulcorner$ or $\lrcorner$ in the corner of $B \sqcup_A C$, as in one of these diagrams:

$$\begin{array}{ccc} A & \xrightarrow{f} & B \\ {\scriptstyle g}\downarrow & & \downarrow{\scriptstyle f_*g} \\ C & \xrightarrow{g_*f} & \lrcorner\; B \sqcup_A C \end{array} \qquad \begin{array}{ccc} A & \xrightarrow{f} & B \\ {\scriptstyle g}\downarrow & & \downarrow{\scriptstyle f_*g} \\ C & \xrightarrow{g_*f} & \ulcorner\; B \sqcup_A C. \end{array}$$

(We will use the convention on the left.)

Here are some exercises in the category of sets.

Exercise 3.2.32 (sets and relations). In the category **Set**, let $\varnothing$ be the empty set. Show that $A \coprod_{\varnothing} B \cong A \coprod B$. (*Hint*: This is somewhat dual to Exercise 3.2.26.)

Exercise 3.2.33 (sets and relations). More generally, let S be a subset both of A and B, with the inclusion maps forming the following span:

$$\begin{array}{ccc} S & \hookrightarrow & A \\ \big\downarrow & & \\ B. & & \end{array}$$

Show that $A \coprod_S B$ is given by the *non-disjoint* union of A and B, i.e. where the elements of S are counted only once. What are the maps of the universal cocone? Also, why does this generalize the previous exercise? Can you also say why this is somewhat dual to Exercise 3.2.25?

In a generic category, we have the dual statement to Exercise 3.2.28.

Exercise 3.2.34. Consider a morphism $f : X \to Y$ in a category $\mathbf{C}$. Let's form the pushout of f with itself (if it exists). The two universal maps $Y \to Y \coprod_X Y$ are called the **cokernel pair** of f. Prove that f is epi if and only if those maps coincide and are isomorphisms.

Exercise 3.2.35 (difficult!). State and prove (independently) the dual statement to Exercise 3.2.30.

3.2.6 Initial and terminal objects, trivial cases

Let $\mathbf{O}$ be an empty category, i.e. the one with no objects (and no morphisms). This category is small (why?), and there is a unique diagram $E : \mathbf{O} \to \mathbf{C}$ for every category $\mathbf{C}$, which we call the **empty diagram** in $\mathbf{C}$. The empty diagram looks like this:

A cone over the empty diagram E with tip X consists simply of an object X, with no other arrows except the identity of X (why?).

Similarly, a cocone of bottom Y under E is just Y with its identity. So far, everything is trivial; however, the *limit* and *colimit* of E are more interesting.

The *limit* of the empty diagram, following the universal property, is an object $\lim E$ such that for every (other) object X, there is a unique morphism

$$X \dashrightarrow \lim E.$$

Definition 3.2.36. The limit of the empty diagram in **C**, if it exists, is called the **terminal object** of **C**. It is usually denoted by 1.

The terminology comes from the fact that, since every object has a unique morphism to 1, one may view 1 as the "endpoint" of the category. This is particularly visible in the poset case, as follows.

Example 3.2.37 (sets and relations). Let $(X, \leq)$ be a poset. A terminal object of X, if it exists, is an object 1 such that for each $x \in X$, $x \leq 1$. In other words, the terminal object is the top element of X, if it exists. The top element is also denoted by $\top$.

The notation "1" comes instead from the following fact.

Example 3.2.38 (sets and relations). In **Set**, a terminal object is a singleton set 1, i.e. a set with exactly one element. This is because for any other set X, there exists only one function $X \to 1$, namely the one that assigns to each $x \in X$ the unique element of 1.

Exercise 3.2.39 (topology). Show that a terminal object of **Top** is a one-point space.

Exercise 3.2.40 (graph theory). Show that a terminal object of **MGraph** is a graph with a single vertex. Also, how many edges?

Warning. The word "terminal" does *not* mean that, in general, there are no arrows out of 1 except the identity. In a poset, this is true, but not in general, for example, not in **Set**.

The colimit of the empty diagram E is an object $\operatorname{colim} E$ such that for each (other) object Y, there is a unique arrow

$$\operatorname{colim} E \dashrightarrow Y.$$

Definition 3.2.41. The colimit of the empty diagram in **C**, if it exists, is called the **initial** or **coterminal object** of **C**. It is usually denoted by 0.

Example 3.2.42 (sets and relations). Let $(X, \leq)$ be a poset. An initial object of X, if it exists, is an object 0 such that for each $x \in X$, $0 \leq x$. In other words, the terminal object is the bottom element of X, if it exists. The bottom element is also denoted by $\bot$.

The notation "0" comes instead from the following fact.

Example 3.2.43 (sets and relations). In **Set**, the initial object is the empty set $\varnothing$. This is because for any other set X, there exists only one function $\varnothing \to X$, namely the one that doesn't assign anything. Note that this function still exists, even if it's a trivial case (just as zero is a trivial number, but still a number). In particular, there is a unique function $\varnothing \to \varnothing$, the identity function.

Exercise 3.2.44 (topology). Show that the initial object of **Top** is the empty space.

Exercise 3.2.45 (graph theory). What is the initial object of **MGraph**?

Exercise 3.2.46 (linear algebra, group theory). Show that the initial objects of **Vect** and **Grp** coincide (up to isomorphism) with the terminal objects. An object which is both initial and terminal is sometimes called a **zero object**.[1]

[1]This may be confusing since both zero objects and initial objects (which are not necessarily terminal) are usually denoted as 0.

Warning. Just as for terminal objects, the word "initial" does *not* mean that, in general, there are no arrows into 0 except the identity. In a poset, this is true, as well as in **Set** (why?), but not in general, for example, not in **Vect**.

It turns out that *every limit can be seen as a terminal object in some category*. Similarly, every colimit can be seen as an initial object in some category.

Definition 3.2.47. Let $D : \mathbf{J} \to \mathbf{C}$ be a diagram. The **slice category over** D, denoted by $\mathbf{C}/D$, is the category whose:

- objects are pairs (X, α), where X is an object of $\mathbf{C}$ and α is a cone over D with tip X;
- a morphism $(X, \alpha) \to (Y, \beta)$ is given by a morphism $f : X \to Y$ of $\mathbf{C}$ such that for every object $I \in \mathbf{J}$, the following triangle commutes:

$$\begin{array}{ccc} X & \xrightarrow{\quad f \quad} & Y \\ & {\scriptstyle \alpha_I} \searrow \quad \swarrow {\scriptstyle \beta_I} & \\ & DI. & \end{array}$$

One can see the slice category as the category of cones over D with all possible tips, and morphisms between them making the triangles of the cones commute.

Exercise 3.2.48. Show that the limiting cone of D, if it exists, is the terminal object of $\mathbf{C}/D$.

Therefore, complicated limits can be seen as simple limits in a complicated category.

Exercise 3.2.49. Define the **coslice category under** D, denoted by $D/\mathbf{C}$, in such a way that the colimiting cone of D, if it exists, is the initial object of $D/\mathbf{C}$.

Let's conclude this part by giving some trivial cases as exercises.

Exercise 3.2.50. Let X be an object of $\mathbf{C}$, seen as a diagram (with only its identity). Show that its limit and colimit are both (isomorphic to) X.

Exercise 3.2.51. Let now $f : X \to Y$ be a morphism of $\mathbf{C}$, seen as a diagram. Show that its limit is given by X and its colimit is given by Y. Why does this generalize the previous exercise?

Exercise 3.2.52. Consider the following diagrams:

Show that the limit of the diagram on the left is A and the colimit of the diagram on the right is F.

Exercise 3.2.53 (important!). More generally, let $D : \mathbf{J} \to \mathbf{C}$ be a diagram, and suppose that $\mathbf{J}$ has an initial object 0. Prove that $\lim D$ exists and is given by $D(0)$. What is the dual statement? Also, why does this generalize the previous two exercises?

Therefore, limits and colimits can be seen as ways to "complete the diagram" by adding to it an initial or terminal object whenever such an object does not exist already.

3.2.7 Sequential and (co)filtered limits and colimits

In other areas of mathematics, the word "limit" often denotes an approximation of some kind, or encodes an idea of "tending to infinity." There are similar ideas in category theory too — it is probably the reason why categorical limits were named that way.

An **inductive, direct**, or **forward, sequence**, is an infinite diagram in the following form:

$$X_0 \longrightarrow X_1 \longrightarrow X_2 \longrightarrow X_3 \longrightarrow \cdots .$$

The limit of this diagram is not interesting (it is X_0 — why?), but the *colimit* is. Indeed, the colimit of this diagram can be interpreted intuitively as the object to which this sequence "tends." Let's see how. First of all, consider a cocone:

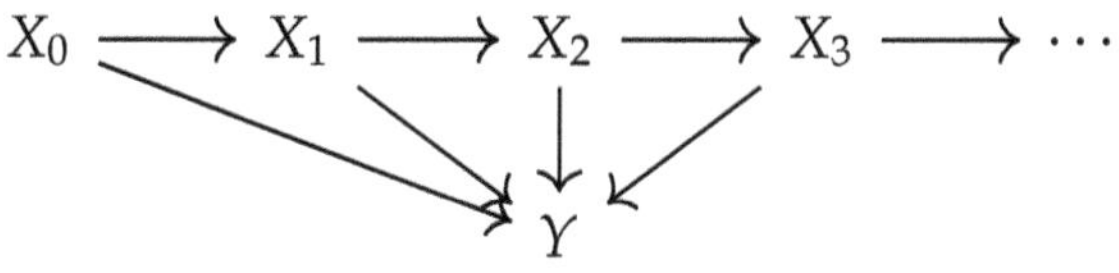

Since this diagram commutes, the arrow $X_0 \to Y$ is equal to the composition $X_0 \to X_1 \to Y$. So, equivalently, that arrow factors through X_1. In turn, the arrow $X_1 \to Y$ decomposes as $X_1 \to X_2 \to Y$, and so on. In other words, each arrow of the cocone equivalently passes through all the objects in the sequence. So, we can equivalently imagine Y at the "far end" of the sequence:

$$X_0 \longrightarrow X_1 \longrightarrow X_2 \longrightarrow X_3 \longrightarrow \cdots \longrightarrow Y.$$

That is, it's as if the arrows $X_i \to Y$ were "infinite compositions" of all the arrows in the sequence, and then some more. (This is just for intuition purposes, it is not a categorical diagram. We have not defined, and we will not define, the composition of an infinite number of arrows.)

Now, a *colimiting* cone is an object that we can interpret as being at the "right end" of the sequence, but not further right than that. That is, it comes "before" all other cocones. Here is the definition.

Definition 3.2.54. A colimit of an inductive sequence is called a **sequential colimit.**

Warning. In other fields of mathematics, a sequential colimit is sometimes called a *direct limit*, but we will not use this terminology here since it conflicts with our usage (this is a *colimit*, not a limit) and since it sounds too similar to *directed colimits*, which are a more general but different notion. (We will not define directed colimits in this text, but we will define the similar notions of *filtered* colimits and *directed sets*, see the following.)

Here is the most prototypical example of a sequential colimit.

Example 3.2.55 (sets and relations). The set $\mathbb{N}$ of natural numbers is the sequential colimit in **Set** of the following sequence of inclusions:

$$\{0\} \hookrightarrow \{0,1\} \hookrightarrow \{0,1,2\} \hookrightarrow \{0,1,2,3\} \hookrightarrow \cdots$$

To prove this, consider a cocone:

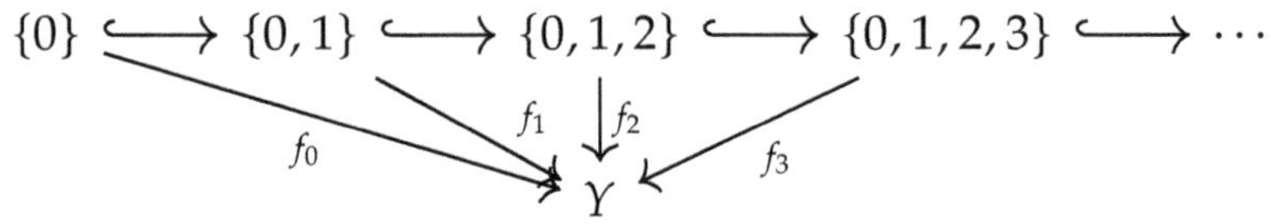

First of all, we see that every $n \in \mathbb{N}$ will be eventually mapped to an element of Y, for example by the map f_n (and all the following ones). Moreover, the maps f_n need to satisfy $f_n(i) = f_{n-1}(i)$, for all $i < n$, for example, $f_2(1) = f_1(1)$, if we want the diagram to commute. That is, each number n will be (eventually) mapped to a *unique* element of Y. Defining now the map $f : \mathbb{N} \to Y$ by $f(n) := f_n(n)$, we see that it is the unique map $\mathbb{N} \to Y$ such that this diagram commutes:

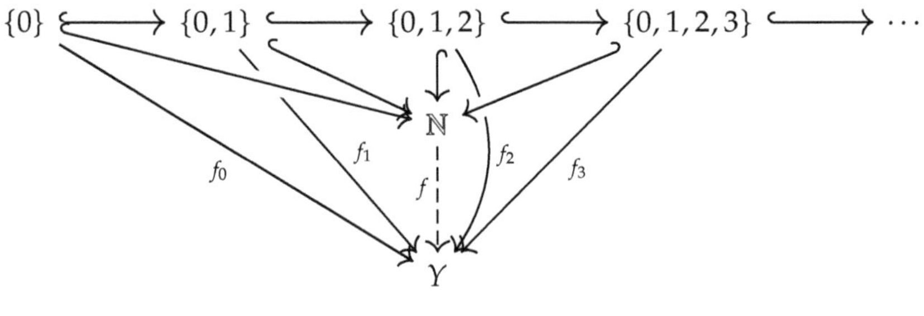

The universal cone is again given by inclusion maps.

We see that, in some sense, the infinite set $\mathbb{N}$ is "approximated" by larger and larger finite subsets. It is a little bit like the limits of analysis and topology. However, one should keep in mind the following important differences:

(a) This notion of "tending" is qualitative and not quantitative: We don't have "epsilons," or "rates of convergence." In a way, this is more analogous to limits in topological spaces than in metric spaces.[2]

(b) This "approximation" is strictly "from below" (or at least *in the direction of the arrows*), not "from above" or "mixed."

(c) We don't just have a sequence of objects; *the arrows matter*.

To illustrate point (b), it is useful to look at the case of posets.

Example 3.2.56 (sets and relations). In a poset, an inductive sequence is an *upward monotone sequence*,[3] which is in the form $x_0 \leq x_1 \leq x_2 \leq \ldots$. Sometimes, it is also called **ω-sequence**. Its limit is exactly its least upper bound. For example, the sequence in $\mathbb{R}$,

$$n \longmapsto -\frac{1}{n} \quad = \quad -1,\ -\frac{1}{2},\ -\frac{1}{3},\ -\frac{1}{4},\ldots,$$

tends to 0, both in the usual sense and as a filtered colimit. Instead, the sequence

$$n \longmapsto \frac{(-1)^n}{n} \quad = \quad -1,\ +\frac{1}{2},\ -\frac{1}{3},\ +\frac{1}{4},\ldots$$

tends to 0 in the usual sense, but *not* as a filtered colimit. (It is not even an inductive sequence.)

As for point (c), the fact that arrows matter is best seen through the following example.

[2]Using *enriched category theory*, one can define notions analogous to "rates of convergence," but we will not do this here. See, for example, https://golem.ph.utexas.edu/category/2017/04/enrichment_and_its_limits.html.

[3]We cannot use the term "nondecreasing" if the order is not total (why?).

Example 3.2.57 (graph theory). Consider the following sequence of simple graphs, which we consider as objects of the category **Graph** of undirected simple graphs and adjacency-preserving functions:

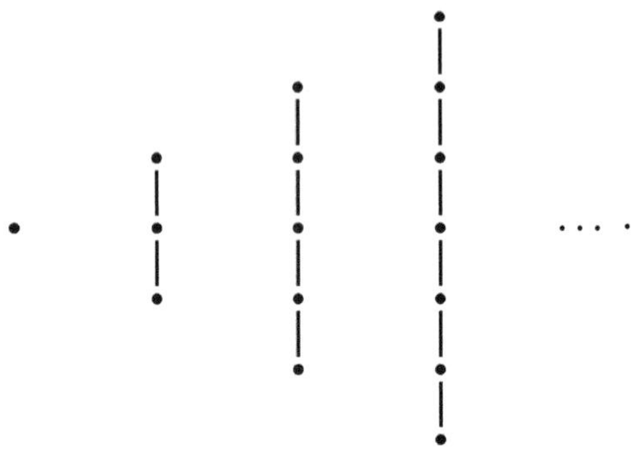

If we take the inductive sequence formed by the inclusion maps

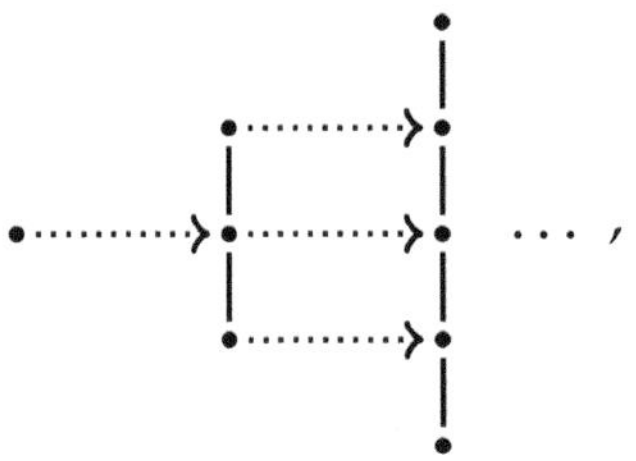

then the colimit is a chain graph which is infinite at both ends:

If instead we take the inclusion maps

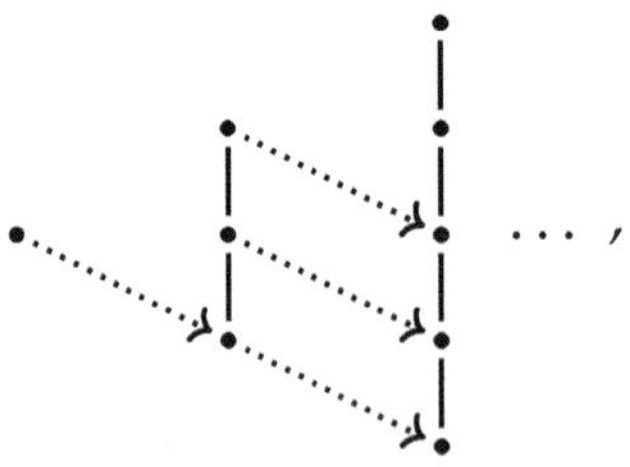

or, equivalently,

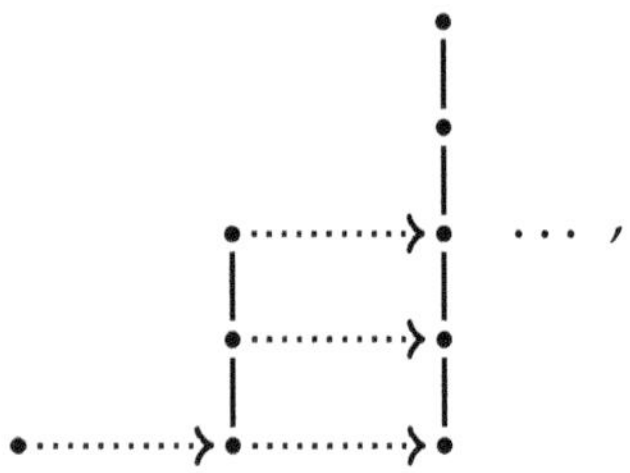

then the colimit is a chain graph which is finite at one end and infinite at the other end:

In other words, the choice of arrows has an effect on "where the sequence is going." (If it is unclear why these graphs are the colimits, try to prove it similarly to how we did in Example 3.2.55.)

Exercise 3.2.58 (graph theory). Construct an example like the one above but where the graphs in the sequence respectively have one vertex, two vertices, three vertices, and so on.

Exercise 3.2.59 (analysis, geometry). In the category of *complete* metric spaces and Lipschitz maps, show that the unit interval $[0,1]$ is the sequential colimit of the following sequence of increasingly fine binary fractions:

$$\{0,1\} \hookrightarrow \{0,\tfrac{1}{2},1\} \hookrightarrow \{0,\tfrac{1}{4},\tfrac{1}{2},\tfrac{3}{4},1\} \hookrightarrow \cdots .$$

Let's now reverse the arrows. A **projective, inverse**, or **backward, sequence** is an infinite diagram as follows:

$$\cdots \longrightarrow X_3 \longrightarrow X_2 \longrightarrow X_1 \longrightarrow X_0 \ .$$

Once again, we can interpret a limit of this diagram as an object at the "far left end" of this sequence.

Definition 3.2.60. A limit of a projective sequence is called a **sequential limit**.

Warning. Just as for direct limits, in other fields of mathematics, sequential limits are often called *inverse limits*; however, once again, we will not follow this convention since it might suggest that they are *co*limits, while they are not.

Projective sequences and their limits are often used in mathematics in two (related) ways:

(a) similarly to sequential colimits, to create larger objects from smaller, increasingly larger ones, but this time in a "product-like way";

(b) to zoom into *finer and finer* structures.

Let's see this through examples.

Example 3.2.61 (sets and relations). Let X be a set. Denote by $X^n = X \times \cdots \times X$ its n-fold cartesian product. Construct now the following sequence:

$$\cdots \longrightarrow X^3 \longrightarrow X^2 \longrightarrow X \longrightarrow 1,$$

where each map $X^n \to X^{n-1}$ is the projection onto the first $n-1$ coordinates (i.e. we discard the last coordinate of the tuple):

$$\begin{array}{ccc} X^n & \longrightarrow & X^{n-1} \\ (x_1, x_2, \ldots, x_{n-1}, x_n) & \longmapsto & (x_1, x_2, \ldots, x_{n-1}). \end{array}$$

The limit of the sequence is given by the countably infinite product $X^{\mathbb{N}}$, i.e. $\prod_{n\in\mathbb{N}} X$.

> **Exercise 3.2.62.** Prove the above statement. (*Hint*: It is somewhat dual to Example 3.2.55.)

In a way, this sequential limit creates a "larger" set, $X^{\mathbb{N}}$, from smaller ones. But we can also see this as going into a limit of *infinitely finer*: Indeed, the elements of $X^{\mathbb{N}}$ are infinite sequences, which have more "details," or "information," than finite ones. (Compare with real numbers, which have infinitely many digits after the decimal point and which can be approximated by decimal numbers of finite, increasing lengths.)

In several areas of mathematics, situations where one has "increasing precision" are often called *filtrations*, or similar words usually containing the word *filter*. Very often, these situations can be modeled categorically by a projective sequence and, if it exists, by its limit.

Here is an example from measure and probability theory.

Example 3.2.63 (measure theory, probability). Let X be a set. A **filtration** on X is a sequence of σ-algebras $\{\mathcal{F}_n\}_{n\in\mathbb{N}}$ on X such that for each n, $\mathcal{F}_n \subseteq \mathcal{F}_{n+1}$, i.e. they get finer and finer. In the poset of

σ-algebras on X and inclusions, this is an upward monotone sequence (i.e. inductive). However, in the category **Meas** of measurable spaces and measurable maps, a filtration can also be written as the following projective sequence:

$$\cdots \longrightarrow (X, \mathcal{F}_3) \longrightarrow (X, \mathcal{F}_2) \longrightarrow (X, \mathcal{F}_1) \longrightarrow (X, \mathcal{F}_0),$$

where the morphisms are the ones induced by the identity map of sets (why are they measurable?). Filtrations are often used in the context of stochastic processes, and the σ-algebras are often the *natural filtrations* of processes; that is, each $\mathcal{F}_n$ is the one generated by the first n steps of a process. The idea is that as the process evolves, by observing what happens, we have more and more information about the state of our system.

Exercise 3.2.64 (measure theory, probability). Prove that the limit in **Meas** of the projective sequence above is given by X together with the join σ-algebra $\vee_{n\in\mathbb{N}}\mathcal{F}_n$, i.e. the σ-algebra generated by the union of all the $\mathcal{F}_n$.

Also for the case of posets, projective sequences (i.e. *downward monotone sequences*) can have an interpretation of "increasing precision."

Exercise 3.2.65 (analysis, topology). Let X be a metric space. Given a point $x \in X$, denote by $B(x, r)$ the open ball of radius r centered at x. The sequence $\{B(x, 1/n)\}_{n\in\mathbb{N}_{>0}}$ is a downward monotone sequence in the power set PX, and it can be interpreted as "zooming in closer and closer to x."

Now, consider a (generic) function $f : X \to Y$ between metric spaces. Given $x \in X$, form the downward sequence $\{B(f(x), 1/n)\}_{n\in\mathbb{N}_{>0}}$ in PY. Its preimages $\{f^{-1}(B(f(x), 1/n))\}_{n\in\mathbb{N}_{>0}}$ give a downward sequence in PX (why?). Show that f is continuous if and only if for each $x \in X$ and for each $n \in \mathbb{N}_{>0}$, we can find an $m \in \mathbb{N}_{>0}$ such that $B(x, 1/m) \subseteq f^{-1}(B(f(x), 1/n))$.

The idea so far has been that sequential limits and colimits allow us to go from a "diagram of finite things" to an "infinite thing" in some way. However, not all "infinite things" are countable or totally ordered, and there are situations where an "infinite thing" does not lend itself well to approximations only by sequences.

A category **J** is called **filtered** if every finite diagram in **J** has a cocone. A **filtered diagram** $D : \mathbf{J} \to \mathbf{C}$ is a diagram where the indexing category **J** is filtered. (Note that the cocones must already exist in **J**, not only in **C**.)

The category indexing a sequential diagram is filtered (why?). Here is another prototypical example of a filtered category. Let X be an infinite set, and let $\mathbf{Fin}(X)$ be the poset of all its *finite* subsets, ordered by inclusion. Given a finite diagram of finite subsets $F_1, \dots, F_n$ and their inclusions, their union is again in $\mathbf{Fin}(X)$ since a union of finitely many finite sets is finite. (There are other cocones over this finite diagram: every finite set which *contains* that union.) We can depict the situation as follows (but of course the real diagram is infinite):

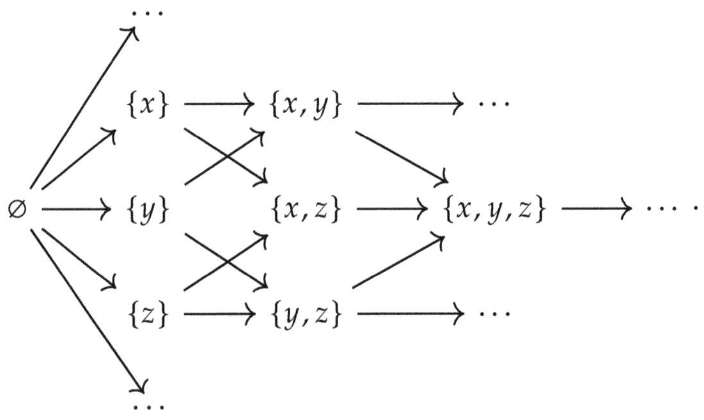

Definition 3.2.66. The colimit of a filtered diagram is called a **filtered colimit.**

Example 3.2.67 (sets and relations). Consider as above the poset $\mathbf{Fin}(X)$ of all finite subsets of an infinite set X. This gives a filtered diagram $\mathbf{Fin}(X) \to \mathbf{Set}$, and X is its colimit, with the universal cone given by the inclusions. To see this, note that if F is a finite subset of X, a cocone consists of maps $f_F : F \to Y$ compatible with the inclusions in the diagram. As in Example 3.2.55, this means that the maps f_F can only map a point of X to a *unique* point of Y. Moreover, every point of X is reached by such a map since $\{x\}$ is a singleton. So, analogously to Example 3.2.55, we can construct a map $f : X \to Y$ by $f(x) := f_{\{x\}}(x)$. (If this is not entirely clear, try to prove this in more detail, as we did with Example 3.2.55.)

Here is the dual notion. A category **J** is called **cofiltered** if every finite diagram has a cone. A **cofiltered diagram** is a diagram indexed by a cofiltered category.

We can generalize Example 3.2.61 to the following case.

Exercise 3.2.68 (sets and relations). Let I be an infinite (small) set, and let $\{X_i\}_{i \in I}$ be a (small) family of objects of a category **C**. Show that the product $\prod_{i \in I} X_i$, if it exists, is the cofiltered limit of a diagram of *finite* products of the X_i. (*Hint*: This is analogous to Example 3.2.61 and somewhat dual to Example 3.2.67.)

The exercise implies, by the way, that if a category has a terminal object, all finite products, and all cofiltered limits, then it has all products. (What's the dual statement?)

A famous theorem of measure theory and probability that can be expressed in terms of cofiltered limits is Kolmogorov's extension theorem.[4] Here is a version.

Theorem 3.2.69 (Kolmogorov extension theorem). *Let X be a standard Borel measurable space. Form the countable cartesian product $X^{\mathbb{N}}$, and consider a probability measure p on it. For each finite subset F of $\mathbb{N}$, denote by $\pi_F : X^{\mathbb{N}} \to X^F$ the projection map onto the components indexed by F,*

[4]See any classical text on probability theory, for example [Kle14, Theorem 14.35].

and denote by p_F the resulting marginal probability measure. Moreover, for each subset inclusion $F \subseteq G$, construct the map $\pi_{F,G} : X^G \to X^F$ which projects to the components indexed by F, discarding the ones in $G \backslash F$. This fits (X^F, p_F) into a cofiltered diagram in the category of probability spaces and measure-preserving maps, and the probability space $(X^{\mathbb{N}}, p)$ is the cofiltered limit of this diagram.

As in the sequential case, we can view this limit as a way either to make things "larger" or to have "finer distinctions." (What is the relation to filtrations, as per Example 3.2.63?)

Let's now turn to the poset case. A cofiltered diagram in a poset X gives (as image) a **downward directed set**, which is a subset $S \subseteq X$ such that if $s, t \in S$, then s and t have a common lower bound in S. A common example of a (downward) directed set is a *filter*, which once again, very often one can interpret as "making finer and finer distinctions."

Definition 3.2.70. Let X be a poset. A **filter** in X is a subset $F \subseteq X$ which is

- **nonempty**: at least one $x \in F$;
- **upward-closed**: if $x \in F$ and $x \leq y$, then $y \in F$;
- **downward-directed**: if $x, y \in F$, then there exists $z \in X$ such that $z \leq x$, $z \leq y$ and $z \in F$.

The dual notion to a filter is called **(order) ideal.**

Using filters, we can in some sense generalize the sequence of open balls of Exercise 3.2.65 to arbitrary topological spaces, as follows.

Example 3.2.71 (topology). Let X be a topological space. The **neighborhood filter** of a point $x \in X$ is the subset $\mathcal{F}_x \subseteq O(X)$ of the topology of X defined by

$$U \in \mathcal{F}_x \iff x \in U.$$

Let's check that it is a filter:

- It is nonempty, since $x \in X$.
- It is upward-closed since if $U \subseteq V$ and $x \in U$, then $x \in V$.
- It is downward-directed since if $x \in U$ and $x \in V$, then $x \in U \cap V$.

As it is downward-directed, the neighborhood filter at X is a cofiltered diagram in the topology $O(X)$, as well as in the power set PX. We can interpret it as a diagram of arrows "tending backward" (or in this case, "downward") to smaller and smaller neighborhoods of x, as if they were tending to the point. This idea is not just for intuition: Filters can be used to give an equivalent definition for continuity of functions.

Exercise 3.2.72 (topology). Generalize Exercise 3.2.65 to give an equivalent definition, in terms of neighborhood filters, of continuous functions between topological spaces.

A very important example of filtered colimit of a neighborhood filter (actually, of its opposite diagram, which is filtered) is the concept of a *stalk*. The following example could be helpful to students with a background (or an interest) in algebraic or differential geometry.

Example 3.2.73 (algebraic geometry, topology). Let X be a topological space, $\Phi : O(X)^{\mathrm{op}} \to \mathbf{Set}$ be a presheaf (recall Example 1.3.47), and $x \in X$. The **stalk of Φ at the point** x is the filtered colimit[5]

$$\operatorname*{colim}_{U \in \mathcal{F}_x} \Phi(U)$$

taken over the neighborhood filter $\mathcal{F}_x$ of x. Note that $\mathcal{F}_x$ is *cofiltered* in $O(X)$; however, since Φ reverses the arrows, the assignment $U \mapsto \Phi(U)$ gives a *filtered* diagram in **Set**.

[5]Note that in algebraic geometry, some authors call this simply a "limit," or even a "contravariant limit."

Intuitively, a stalk is the set corresponding to a "smaller and smaller neighborhood of x." We cannot really talk about the "value of the presheaf Φ at x" because Φ is defined on open sets, and x is just a point (and the singleton $\{x\}$ is usually not open). The idea is then to take increasingly smaller neighborhoods U of x, and see if the resulting diagram of sets $\Phi(U)$ tends to some set (as a colimit). As one can show, this colimit always exists (all colimits of sets exist, see Section 3.4).

For a concrete example of stalk, see the following exercise.

Exercise 3.2.74 (algebraic geometry, topology; difficult!). Let X be a topological space. Consider the presheaf of continuous real-valued functions given in Example 1.3.47, and call it $\Phi : O(X)^{\mathrm{op}} \to \mathbf{Set}$. Given a point $x \in X$, define the set P of pairs (U, f), where U is an open neighborhood of x, and f is a continuous function $f : U \to \mathbb{R}$, i.e. $f \in \Phi(U)$. Define now an equivalence relation on P as follows. Given pairs (f, U) and (g, V) of P, we say that $(f, U) \sim_x (g, V)$ if and only if there exists an open neighborhood W of x, with $W \subseteq U$ and $W \subseteq V$, such that the restrictions $f|_W$ and $g|_W$ are equal. We call the equivalence classes of this relation the (real-valued) **germs** at x, and we denote their set by Φ_x.

Show that the set of germs Φ_x is the stalk at x of the presheaf Φ. (*Hint*: The set P can be seen as the disjoint union $\coprod_{U \in \mathcal{F}_x} \Phi(U)$, so that there are canonical inclusion maps $\Phi(U) \to P$. Also, there is a canonical map from P to the quotient set Φ_x. Composing, we get maps $\Phi(U) \to \Phi_x$. These will form the colimiting cocone.)

A germ is, intuitively, a "minimally small piece of a function at a point." Here is a further exercise that can help with that intuition.

Exercise 3.2.75 (geometry, analysis). Let X be a smooth manifold (or even $\mathbb{R}^n$ if you are not familiar with manifolds). Let $f, g : X \to \mathbb{R}$, and suppose that $f \sim_x g$, i.e. f and g have the same germ at a point $x \in X$. Show that if f is differentiable infinitely many times at x, then so is g, and all partial derivatives of all orders of f are equal to those of g. Can we also say that, conversely, if all partial derivatives at x agree, the germs are equal?

To summarize, then, sequential and filtered colimits are a categorical way of "approximating" things, but unlike the traditional notion of limit, these approximations have a more qualitative flavor (e.g. "finer distinctions" or "larger sets"), are one-sided, and the choice of arrows often matters.

It is now natural to ask: In this light, are functors analogous to *continuous functions*? That is, do functors preserve categorical limits and colimits?

As you might expect, the answer in general is *no*: Already, for the simpler case of posets, consider the following monotone map $f : \mathbb{R} \to \mathbb{R}$ (which we can see as a functor):

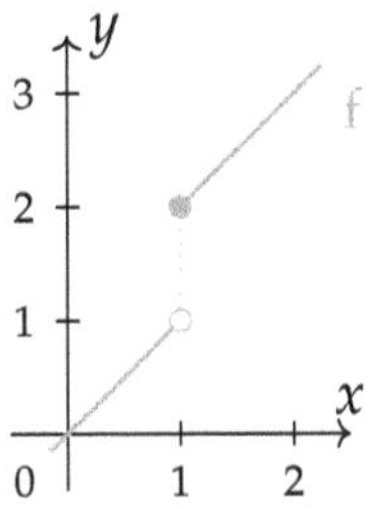

$$f(x) := \begin{cases} x, & x < 1; \\ x+1, & x \geq 1. \end{cases}$$

Given any increasing sequence $\{x_n\}_{n=0}^{\infty}$ tending to 1 from below, we have the following inequality:

$$\lim_{n\to\infty} f(x_n) = 1 \leq 2 = f(1) = f\Big(\lim_{n\to\infty} x_n\Big), \tag{3.2.1}$$

but equality does not hold. Therefore, not all monotone maps preserve sequential suprema.

Exercise 3.2.76 (sets and relations). Show that for every monotone map between posets, whenever the relevant limits exist, an inequality analogous to (3.2.1) holds.

Exercise 3.2.77 (analysis, topology). Show that if $f : \mathbb{R} \to \mathbb{R}$ is monotone and *lower semicontinuous*, it preserves all sequential suprema.

(We saw in Exercise 1.1.18 that lower semicontinuous functions $\mathbb{R} \to \mathbb{R}$ are not closed under composition. What happens if they are monotone?)

An important situation where sequential suprema *are* preserved, and a central result in integration theory, is *Lebesgue's monotone convergence theorem*, as follows.

Theorem 3.2.78 (Lebesgue monotone convergence). *Let (X, μ) be a* ***measure space****, i.e. a measurable space equipped with a measure μ. Let $\{f_n\}_{n=0}^{\infty}$ be a pointwise non-decreasing sequence of non-negative measurable functions $f_n : X \to [0, \infty]$. Then, the function*

$$x \longmapsto \lim_{n\to\infty} f_n(x)$$

is measurable,[6] *and*

$$\int_X \left(\lim_{n\to\infty} f_n \right) d\mu \;=\; \lim_{n\to\infty} \int_X f_n \, d\mu.$$

Indeed, we can see the (Lebesgue) integral as a monotone map between the poset of non-negative measurable functions and $[0, \infty]$, hence, a functor. (Why is the assignment monotone?) The theorem says first of all that sequential suprema (i.e. pointwise monotone limits) exist in that poset and then that integration preserves them.

Exercise 3.2.79. In Theorem 3.2.78, can we replace sequential suprema by cofiltered (i.e. directed) ones? (*Hint*: No. Can you give a counterexample?)

[6]Integrable, if one allows (positive) infinity as a possible result of the integral.

The question of whether a functor preserves a given limit or colimit is interesting also beyond the case of posets, and it has a rich and deep theory, which we introduce in the following section.

3.3 Functors, Limits and Colimits

We have seen that in general functors preserve commutative diagrams, but not the property of being mono or epi. We have also see that monotone maps may fail to preserve suprema. The general situation is that functors do not necessarily preserve limits, colimits, or other universal properties.

Let's see this in detail. Let $D : \mathbf{J} \to \mathbf{C}$ be a diagram with limiting cone $\lambda : \lim D \Rightarrow D$. Given a functor $F : \mathbf{C} \to \mathbf{C}'$, we can form the diagram $F(D) = F \circ D$ in $\mathbf{C}'$, and the image $F\lambda : F(\lim D) \Rightarrow F(D)$ is a commutative cone in $\mathbf{C}'$. Indeed, for each morphism $m : J \to J'$ of $\mathbf{J}$, the commutative triangle on the left gives (by functoriality of F) a commutative triangle on the right:

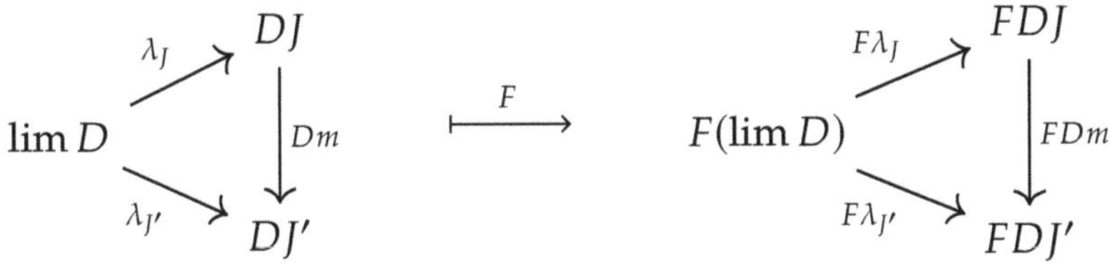

In general, there is no reason why the image cone $F\lambda : F(\lim D) \Rightarrow F(D)$ is a limit in $\mathbf{C}'$. If the image cone *is* limiting, we say that the functor F **preserves the limit of** D.

Note that this, in general, is a stronger requirement than just saying that $F(\lim D)$, as an object, is a limit of $F(D)$: We also need that *the arrows of the limiting cone are the right ones*. It is also stronger than requiring that the limit of $F(D)$ in $\mathbf{C}'$ simply exists. Now, if the limit of the diagram $F(D)$ in $\mathbf{C}'$ *does* exist, then by its universal property and since $F\lambda$ is a cocone, there is a unique morphism $u : F(\lim D) \to \lim F(D)$ that makes the following diagram commute (for all objects J, J' of $\mathbf{J}$), where λ' denotes the limiting cone of $F(D)$ in $\mathbf{C}'$:

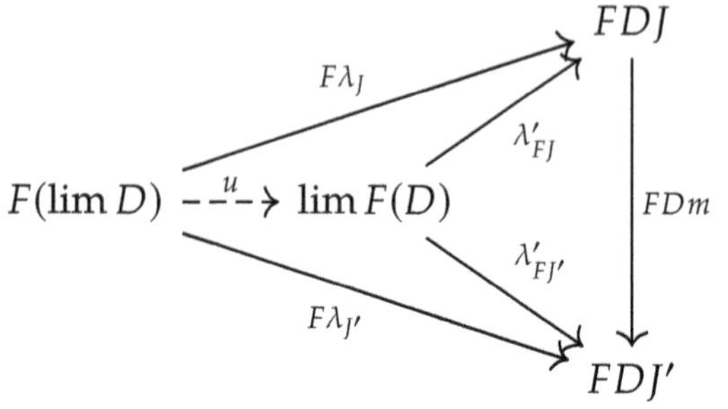

An equivalent way of saying that F preserves the limit of D is to say that $\lim F(D)$ exists and that the map u, as above, is an isomorphism. (Why is this equivalent?) In turn, this is equivalent to requiring that $\lim F(D)$ exists and that there exists *any* isomorphism $F(\lim D) \cong \lim F(D)$ that makes the diagram above commute (for all objects J, J' of $\mathbf{J}$). Indeed, by uniqueness (in the universal property of $\lim F(D)$), this map must be u.

When $\lim F(D)$ exists, the map u is interesting also when it is not an isomorphism. For example, as we will see, the failure of this map from being an isomorphism can be sometimes interpreted in terms of "complexity," see Section 3.3.1. (Note also that the map u is a generalization of the inequality (3.2.1), as well as of Exercise 3.2.76, for general categories instead of posets, and with arrows reversed.)

Dually, suppose that D has a colimiting cone $\kappa : D \Rightarrow \operatorname{colim} D$ in $\mathbf{C}$. The functor $F : \mathbf{C} \to \mathbf{C}'$ maps the colimiting cone of $\mathbf{C}$ to a (not necessarily colimiting) cone of $\mathbf{C}'$:

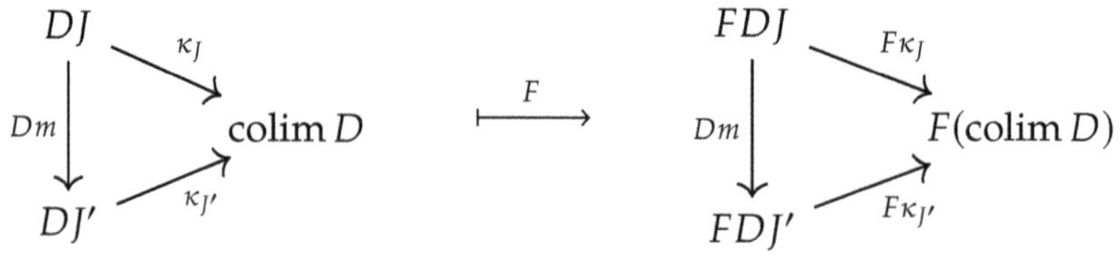

If the image cone $F\kappa$ *is* colimiting, we say that the functor F **preserves the colimit of** D. Analogous remarks as for the case of limits can be made. In particular, this is a stronger condition than just requiring that $F(\mathrm{colim}\, D)$, as an object, is the colimit of $F(D)$.

Definition 3.3.1. A functor $F : \mathbf{C} \to \mathbf{C}'$ is called **continuous** if it preserves all the limits that exist in $\mathbf{C}$.

A functor $F : \mathbf{C} \to \mathbf{C}'$ is called **cocontinuous** if it preserves all the colimits that exist in $\mathbf{C}$.

Here are some examples and nonexamples of functors preserving particular limits.

Example 3.3.2 (topology). Consider the forgetful functor $U : \mathbf{Top} \to \mathbf{Set}$. We have the following (why?):

- The underlying set of the one-point space is the one-point set. Therefore, U preserves terminal objects (which are limits).
- The underlying set of the product of spaces is the cartesian product of sets. Therefore, U preserves products (which are limits).
- The underlying set of the empty space is the empty set. Therefore, U preserves initial objects (which are colimits).
- The underlying set of the disjoint union of spaces is the disjoint union of sets. Therefore, U preserves coproducts (which are colimits).

Exercise 3.3.3 (graph theory). Prove that the same is true for the forgetful functor $\mathit{Vert} : \mathbf{MGraph} \to \mathbf{Set}$.

One may be tempted to ask: Are these functors continuous? What about cocontinuous? A partial answer will be given later on. Not every "forgetful functor" behaves this nicely, as seen in the following.

Example 3.3.4 (linear algebra). Consider the forgetful functor $U : \mathbf{Vect} \to \mathbf{Set}$. We know that binary products and coproducts in $\mathbf{Vect}$ coincide (Exercise 3.2.8), but not in $\mathbf{Set}$. Therefore, U cannot preserve both. Similarly, initial and terminal objects in $\mathbf{Vect}$ coincide (Exercise 3.2.46), but not in $\mathbf{Set}$. Therefore, again, U cannot preserve both. In particular, this forgetful functor preserves products and terminal objects (which are limits), but not coproducts and initial objects (which are colimits).

Exercise 3.3.5 (group theory). Prove that the forgetful functor $U : \mathbf{Grp} \to \mathbf{Set}$ preserves products and terminal objects, but not coproducts and initial objects.

Exercise 3.3.6. Recall that, by Exercise 3.2.28, the property of being mono can be expressed in terms of a universal property: a limit. Give an example of a functor which does not preserve this limit. (*Hint*: In Section 1.3.5, you can find functors which do not preserve monomorphisms.)

Exercise 3.3.7. Can you give an example of the dual case to the previous exercise?

3.3.1 The power set, probability functors, and complexity

Here are two examples of functors which do not preserve products and coproducts for a deep, structural reason.

Example 3.3.8 (sets and relations). Consider the power set functor $P : \mathbf{Set} \to \mathbf{Set}$ (Example 1.3.14). Consider two sets X and Y with two or more elements each.

We can form the coproduct $X \sqcup Y$, i.e. their disjoint union. The set $P(X \sqcup Y)$ has as elements the subsets $S \subseteq X \sqcup Y$. The subsets of $X \sqcup Y$ may be the subsets of X only, of Y only, or mixed, i.e. may contain elements from both. On the other hand, the set $PX \sqcup PY$ is the disjoint union of the set of subsets of X and the set of subsets of Y. The elements of $PX \sqcup PY$, in other words, are either subsets of X or subsets of Y, but not mixed. For example, if $x \in X$ and $y \in Y$, the set $\{x, y\}$ is in $P(X \sqcup Y)$ but not in $PX \sqcup PY$. Therefore, the canonical inclusion map $PX \sqcup PY \to P(X \sqcup Y)$ is not an isomorphism.

Consider now the product $X \times Y$, i.e. their cartesian product. The set $P(X \times Y)$ has as elements the subsets of the cartesian product $X \times Y$.

On the other hand, the set $PX \times PY$ has as elements pairs consisting of a subset of X and a subset of Y. Every subset of $X \times Y$ can be projected onto a subset of X and onto a subset of Y, giving an element of $PX \times PY$. Therefore, we have a canonical map $P(X \times Y) \to PX \times PY$, which is surjective. However, this map is in general far from being injective: Many subsets of $X \times Y$ have the same projections. For example, for $X = Y = \mathbb{R}$, the subsets of $\mathbb{R}^2$ in the following pictures all have the same projections onto the axes:

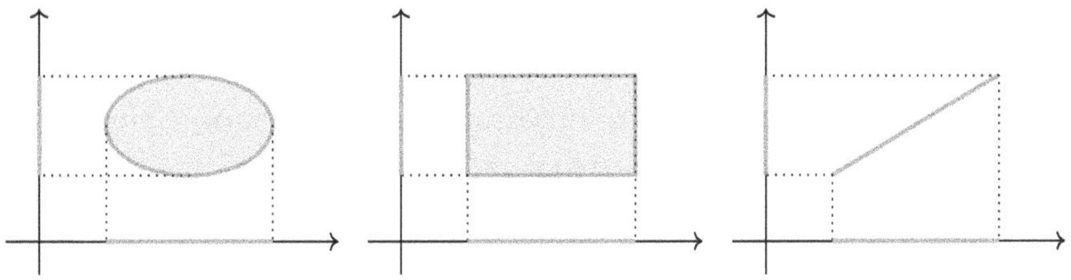

Exercise 3.3.9 (sets and relations). Show that the canonical projection $P(X \times Y) \to PX \times PY$ is an instance of the map u given at the beginning of Section 3.3.

Show that the canonical inclusion $PX \sqcup PY \to P(X \sqcup Y)$ satisfies a dual property to the above.

Example 3.3.10 (probability). Consider the probability functor $\mathcal{P}$: **Set** $\to$ **Set** (Example 1.3.16), or any of its more refined variants in the categories **Meas**, **CHaus**, etc. Consider again two sets X and Y (or spaces) with two or more elements each. The set $\mathcal{P}(X \times Y)$ is the set of *joint* probability distributions on X and Y. On the other hand, the set $\mathcal{P}(X) \times \mathcal{P}(Y)$ is the set of *pairs of marginals*, probability distributions separately on X and on Y. Similarly to the case of the power set, many possible joint distributions correspond to the same marginals, and the different possibilities correspond to the different types of *statistical interactions*, such as *correlation*. For example, for $X = Y$ and given a probability distribution p on X, there are at least two probability distributions on $X \times X$ such that their marginals are both equal to

p: the independent case $p \otimes p$, also called *product probability*, and the perfectly correlated case, supported on the diagonal in $X \times X$.

Just as in the previous exercise, we can construct the map $\mathcal{P}(X \times Y) \to \mathcal{P}(X) \times \mathcal{P}(Y)$ canonically given by the universal property of the product. One can interpret this map as the one "forgetting the statistical interaction": It maps a joint distribution simply to the pair of its marginals.

A possible interpretation for this behavior, for both of these functors, is that *the features extracted by these functors do not respect the composition of objects, either via products or coproducts*. In other words, we are free to compose our objects to obtain more complex ones; however, *observing* the more complex objects is different from observing the parts separately. Observing the parts separately forgets "something," either elements (mixed subsets) or information (statistical interaction). We can think of functors, such as P and $\mathcal{P}$, as functors *detecting* or *exhibiting complexity*.[6] We give a more general version of these ideas in Section 6.3.2.

Exercise 3.3.11 (probability). Does $\mathcal{P} : \mathbf{Set} \to \mathbf{Set}$ preserve coproducts? (*Hint*: No, and this is again because some "mixed" elements are missing, similarly to the power set case.)

Exercise 3.3.12 (sets and relations, probability). Does the power set functor preserve initial and terminal objects? What about the probability functor?

3.3.2 Continuous functors and equivalence

The following exercises show how limits and colimits, and their preservation, behave under equivalences.

[6] The book [FS19] explores this in detail.

Exercise 3.3.13 (important!). Let $F : \mathbf{C} \to \mathbf{D}$ be a functor preserving the limit of a diagram $D : \mathbf{J} \to \mathbf{C}$. Let G be a functor naturally isomorphic to F. Show that G also preserves the limit of D.

Conclude that if F is continuous, then G is continuous too. What is the dual statement?

Exercise 3.3.14. Show that continuous and cocontinuous functors are stable under composition.

Exercise 3.3.15 (important!). Let $F : \mathbf{C} \to \mathbf{D}$ be a functor inducing an equivalence of categories. Show that F is continuous (and hence cocontinuous . . . why?).

Conclude that if a category $\mathbf{C}$ is complete (or cocomplete), then every category equivalent to $\mathbf{C}$ is complete (or cocomplete) too.

3.3.3 The case of representable functors and presheaves

Here is one of the most important properties of representable functors.

Theorem 3.3.16. *Representable functors are continuous.*

Let's see what this theorem means in detail. First of all, every representable functor is by definition naturally isomorphic to a hom-functor, and so, by Exercise 3.3.13, we can equivalently prove the assertion just for hom-functors. Let now $\mathbf{C}$ be a category, J be a small category, and $D : \mathbf{J} \to \mathbf{C}$ be a diagram, and suppose that $\lim D$ exists in $\mathbf{C}$. Let now R be an object of $\mathbf{C}$. Then, we have that the limit in **Set** of the diagram

$$\mathrm{Hom}_{\mathbf{C}}(R, D-) : \mathbf{J} \to \mathbf{Set}$$

exists and is (up to isomorphism) the set $\mathrm{Hom}_{\mathbf{C}}(R, \lim D)$, with limiting cone induced by the one of $\lim D$.

The full proof of this theorem will be given in Section 3.4.2. For now, let's try to understand the result and its significance.

First of all, replacing $\mathbf{C}$ by $\mathbf{C}^{\mathrm{op}}$, we get the following dual statement.

Corollary 3.3.17. *Let $P : \mathbf{C}^{\mathrm{op}} \to \mathbf{Set}$ be a representable presheaf. Then, P turns colimits into limits. That is, if $D : \mathbf{J} \to \mathbf{C}$ is a diagram admitting a colimit, then $P(\operatorname{colim} D)$ is a limit of $P \circ D^{\mathrm{op}}$ existing in* **Set**, *with limiting cone induced by the colimiting cone of* $\operatorname{colim} D$.

Exercise 3.3.18 (important!). Show that this corollary is indeed given by Theorem 3.3.16 if one replaces $\mathbf{C}$ by $\mathbf{C}^{\mathrm{op}}$.

For the case of products, the proof of the theorem is relatively straightforward. Let's consider binary products for simplicity. Take objects X and Y of $\mathbf{C}$, and suppose that their product $X \times Y$ exists. The universal property of the product says that given any (other) object R of $\mathbf{C}$ and maps $R \to X$ and $R \to Y$, there exists a unique map $R \to X \times Y$ that makes the following diagram commute:

$$\begin{array}{ccccc} & \swarrow & R & \searrow & \\ & & \downarrow & & \\ X & \xleftarrow{p_1} & X \times Y & \xrightarrow{p_2} & Y. \end{array} \tag{3.3.1}$$

In particular, there is a bijection between the pairs of maps $R \to X$ and $R \to Y$ and the maps $R \to X \times Y$. Now, the set of pairs of maps $R \to X$ and $R \to Y$ is exactly the cartesian product of the sets $\mathrm{Hom}_{\mathbf{C}}(R, X)$ and $\mathrm{Hom}_{\mathbf{C}}(R, Y)$, so that the map

$$\begin{array}{ccc} \mathrm{Hom}_{\mathbf{C}}(R, X \times Y) & \longrightarrow & \mathrm{Hom}_{\mathbf{C}}(R, X) \times \mathrm{Hom}_{\mathbf{C}}(R, Y) \\ (x \mapsto (f(x), g(x))) & \longmapsto & (x \mapsto f(x), x \mapsto g(x)) \end{array}$$

is a bijection, i.e. an isomorphism in the category **Set**. We now have to check that it preserves the limiting cones, i.e. it commutes with the projection maps,

$$\begin{array}{ccccc} & {\scriptstyle p_1\circ -}\swarrow & \mathrm{Hom}_{\mathbf{C}}(R, X \times Y) & \searrow{\scriptstyle p_2\circ -} & \\ & & \downarrow{\scriptstyle \cong} & & \\ X & \xleftarrow{p_1} & \mathrm{Hom}_{\mathbf{C}}(R, X) \times \mathrm{Hom}_{\mathbf{C}}(R, Y) & \xrightarrow{p_2} & Y, \end{array}$$

where the sides of the triangles are the images under the functor $\mathrm{Hom}_{\mathbf{C}}(R,-)$ of the maps $p_1 : X \times Y \to X$ and $p_2 : X \times Y \to Y$. However, this follows from the way we have constructed the bijection in (3.3.1). Therefore, the representable functor $\mathrm{Hom}_{\mathbf{C}}(R,-) : \mathbf{C} \to \mathbf{Set}$ preserves binary products.

Exercise 3.3.19. Can you work out the case for presheaves and coproducts (independently)?

Here is a possible intuitive interpretation of this statement. We have seen that representable functors can be considered as experiments consisting of probing the objects of $\mathbf{C}$ with a given object R by mapping R into the objects of $\mathbf{C}$ in all possible ways. Now, the statement of Theorem 3.3.16, for products, states that *probing a composite system $X \times Y$ is the same as probing X and Y separately*. Note that this can be seen as a restatement of (or following directly from) the universal property of products.

Dually, the statement with coproducts and presheaves can be interpreted as follows. We can interpret representable presheaves as experiments consisting of observing the objects of $\mathbf{C}$ by mapping them to a given object R. Now, we have that *observing the object $X \sqcup Y$ is the same as observing X and Y separately*. Similarly to above, this can be seen as a restatement of (or following directly from) the universal property of coproducts.

On the other hand, we cannot in general conclude that representable presheaves interact well with products, or that representable functors interact well with coproducts. Probing the coproduct of X and Y via objects R is not the same as probing them separately. For example, a given set R can be embedded into $X \sqcup Y$ in a way which intersects both X and Y partly, and this cannot be obtained by mapping R separately into X and Y. Similarly, observing the product $X \times Y$ is not the same as observing X and Y separately: A function on $X \times Y$ in general does not depend only on the values of X and Y separately. For example, at the end of Example 3.3.8, observing only the projections, we can't distinguish all the possible original subsets of $\mathbb{R}^2$.

Exercise 3.3.20. Can you give other concrete examples of the latter statements?

By the way, by Theorem 3.3.16, the forgetful functors given in Examples 3.3.2 and 3.3.4 are indeed continuous since we have seen that these functors are representable.

3.4 Limits and Colimits of Sets

3.4.1 Completeness of the category of sets

Consider a diagram of sets $D : \mathbf{J} \longrightarrow \mathbf{Set}$. As $\mathbf{J}$ is assumed to be small, the objects of $\mathbf{J}$ form a set (or a small set). Therefore, we can form the cartesian product of all the sets appearing in the diagram D, which is possibly infinite but still well defined:

$$P := \prod_{I \in \mathbf{J}_0} DI.$$

For example, for the diagram

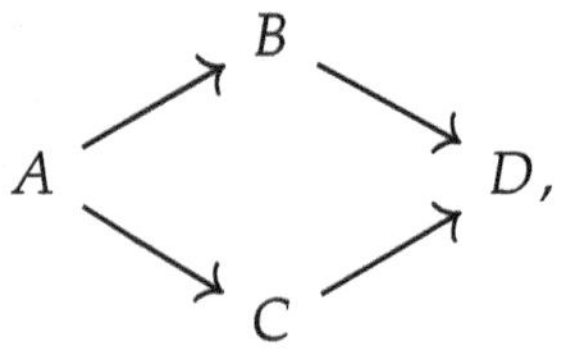

the object P is given by $A \times B \times C \times D$. This cartesian product has projection maps $\pi_I : P \longrightarrow DI$ for each object I of $\mathbf{J}$. For the diagram above, this amounts to the projection maps

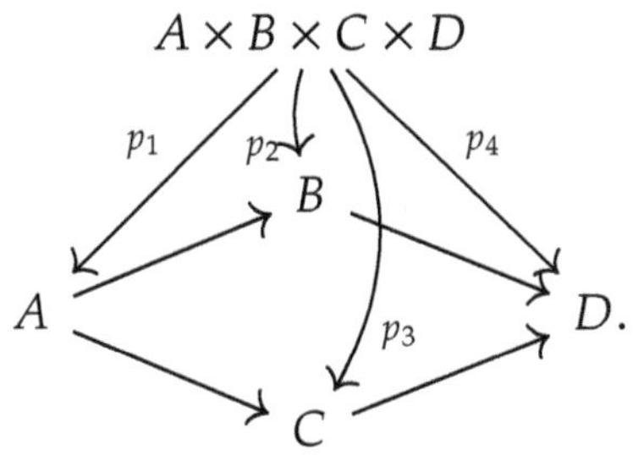

In general, there is no reason to assume that the triangles of the diagram above commute. Similarly, in the general case, it is in general not true that given a morphism $m : I \to I'$ in $\mathbf{J}$, this diagram commutes:

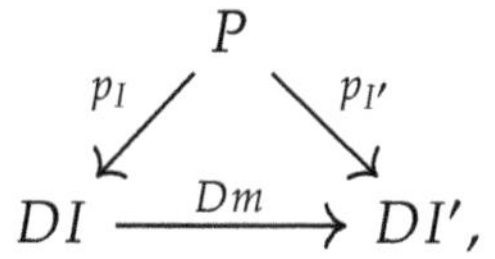

i.e. it may not be true that for all $y \in P$,

$$Dm(p_I(y)) = p_{I'}(y). \tag{3.4.1}$$

However, there is always a *subset* S_m of P whose elements satisfy Equation (3.4.1) — if no element of P satisfies Equation (3.4.1), then S_m is the empty set, which is still a well-defined subset of P. Now, as the category $\mathbf{J}$ is small, also the morphisms of $\mathbf{J}$ form a set (or a small set). Therefore, we can form the following intersection, indexed by all the morphisms appearing in the diagram D, which is possibly empty but is still a well-defined subset of P:

$$S := \bigcap_{m \in \mathbf{J}_1} S_m.$$

More explicitly, S is the subset of P defined by the condition that for each element y of S and each morphism m of $\mathbf{J}$, Equation (3.4.1) has to hold (and it is the largest such subset).

Lemma 3.4.1. *The object S defined above, together with the maps $S \longrightarrow DI$ given by*

$$S \xhookrightarrow{\;i\;} P \xrightarrow{\;p_I\;} DI,$$

where $i : S \to P$ is the inclusion, is a limiting cone over the diagram D.

This gives an explicit way to write the limit of any diagram of sets. In particular, such a limit always exists. Therefore, we have the following.

Theorem 3.4.2. *The category* **Set** *is complete.*

Keep in mind that in the construction of S, we have crucially used the fact that **J** is small.

Proof of Lemma 3.4.1. First of all, by construction, the maps $p_I \circ i : S \to DI$ form a cone over D since by construction, for each morphism $m : I \to I'$ of **J**, the following diagram commutes:

$$\begin{array}{ccc} & S & \\ {\scriptstyle p_I \circ i} \swarrow & & \searrow {\scriptstyle p_{I'} \circ i} \\ DI & \xrightarrow{\;Dm\;} & DI'. \end{array}$$

Take now any (other) cone α over D, say with tip X. So, for every object I of **J**, we have maps $\alpha_I : X \to DI$, and for each morphism $m : I \to I'$ of **J**, this diagram commutes:

$$\begin{array}{ccc} & X & \\ {\scriptstyle \alpha_I} \swarrow & & \searrow {\scriptstyle \alpha_{I'}} \\ DI & \xrightarrow{\;Dm\;} & DI'. \end{array} \qquad (3.4.2)$$

Now, by the universal property of the product, a tuple of maps $\alpha_I : X \to DI$ for each object I of **J** gives a unique map $u : X \to P$ such that for each object I of **J**, this triangle commutes:

$$\begin{array}{ccc} X & & \\ {\scriptstyle u} \downarrow & \searrow {\scriptstyle \alpha_I} & \\ P & \xrightarrow[\;p_I\;]{} & DI. \end{array} \qquad (3.4.3)$$

Moreover, the map u actually factors through S (and uniquely so since $S \subseteq P$). Indeed, for every $x \in X$, we have that $u(x) \in S$ since for all $m : I \to I'$ of $\mathbf{J}$, diagrams (3.4.2) and (3.4.3) state that

$$Dm(p_I(u(x))) = Dm(\alpha_I(x)) = \alpha_{I'}(x) = p_{I'}(u(x)),$$

which means that $u(x)$ satisfies Equation (3.4.1). In conclusion, we have proven that for each limiting cone with tip X over D, there is a unique map $X \to S$ such that this diagram commutes:

$$
\begin{array}{ccc}
X & & \\
\Big\vert & \searrow^{\alpha_I} & \\
\downarrow & & \\
S & \xrightarrow[p_I \circ i]{} & DI,
\end{array}
$$

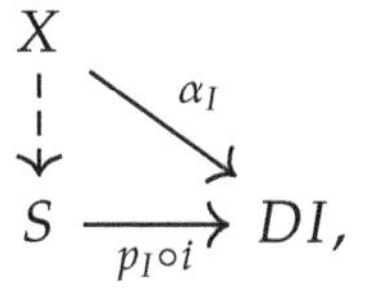

i.e. S is the limit of D. □

Exercise 3.4.3 (sets and relations). Write the subset $S \subseteq P$ as an equalizer of a pair of parallel maps $P \to \prod_{m \in J_1} t(Dm)$, where $t(Dm)$ is the target (codomain) of the morphism Dm.

Exercise 3.4.4 (sets and relations). If you have solved Exercise 3.4.3, rewrite the proof of Lemma 3.4.1 without referring to the elements of the sets involved, using instead the universal properties of the products and equalizers. In other words, you have to show that every limit can be written as an equalizer of products. Conclude that every category where all products and equalizers exist is necessarily complete.

Exercise 3.4.5 (sets and relations; difficult!). In a somewhat dual way to Lemma 3.4.1, give an explicit formula for the colimit of a diagram of sets as a quotient of the disjoint union U of all the sets appearing in the diagram. (*Hint*: This is a generalization of Exercise 3.2.74, if you have solved it.)

Exercise 3.4.6 (sets and relations). If you have solved Exercise 3.4.5, express the quotient of the disjoint union U as the coequalizer of a pair of parallel maps $\coprod_{m\in J_1} s(Dm) \to U$, where $s(Dm)$ is the source (domain) of the morphism Dm.

Exercise 3.4.7. If you have solved the previous two exercises, repeat Exercise 3.4.5 without referring to the elements of the sets involved, using instead the universal properties of the coproducts and coequalizers. In other words, you have to show that every colimit can be written as a coequalizer of coproducts. Conclude that every category where all coproducts and coequalizers exist is necessarily cocomplete.

3.4.2 General proof of Theorem 3.3.16

The statement of Theorem 3.3.16 says the following. Suppose that the diagram $D : \mathbf{J} \to \mathbf{C}$ has a limit. Let R be an object of $\mathbf{C}$. Then, the diagram $\mathrm{Hom}_{\mathbf{C}}(R, D-) : \mathbf{J} \to \mathbf{Set}$ given by

$$\mathbf{J} \xrightarrow{\quad D \quad} \mathbf{C} \xrightarrow{\mathrm{Hom}_{\mathbf{C}}(R,-)} \mathbf{Set}$$

has a limit, and there is an isomorphism

$$\lim_{I\in\mathbf{J}} \mathrm{Hom}_{\mathbf{C}}(R, DI) \;\cong\; \mathrm{Hom}_{\mathbf{C}}\left(R, \lim_{I\in\mathbf{J}} DI\right)$$

commuting with the arrows given by the respective limiting cones.

Proof of Theorem 3.3.16. First of all, note that the action of the functor $\mathrm{Hom}_{\mathbf{C}}(R, D-)$ on the morphisms of $\mathbf{J}$ maps $m : I \to I'$ to the function

$$\begin{aligned} \mathrm{Hom}_{\mathbf{C}}(R, DI) &\xrightarrow{(Dm)\circ -} \mathrm{Hom}_{\mathbf{C}}(R, DI') \\ f &\longmapsto (Dm)\circ f. \end{aligned}$$

Now, let's turn to prove the theorem. By Lemma 3.4.1, we know that the limit of $\mathrm{Hom}_{\mathbf{C}}(R, D-)$ exists since every diagram of sets has a limit. Moreover, Lemma 3.4.1 gives us a way to write down the limit explicitly. Namely, if we form the product

$$P := \prod_{I \in \mathbf{J}_0} \mathrm{Hom}_{\mathbf{C}}(R, DI),$$

then the limit of $\mathrm{Hom}_{\mathbf{C}}(R, D-) : \mathbf{J} \to \mathbf{Set}$ is given by the subset $S \subseteq P$ of all $y \in P$ which satisfy $(Dm) \circ p_I(y) = p_{I'}(y)$, for all $m : I \to I'$ of $\mathbf{J}$, so that the following diagram commutes:

$$\begin{array}{ccc} & S & \\ {\scriptstyle p_I \circ i} \swarrow & & \searrow {\scriptstyle p_{I'} \circ i} \\ \mathrm{Hom}_{\mathbf{C}}(R, DI) & \xrightarrow{(Dm)\circ -} & \mathrm{Hom}_{\mathbf{C}}(R, DI'), \end{array} \tag{3.4.4}$$

where again $i : S \to P$ is the inclusion. But now, P is a product of sets in the form $\mathrm{Hom}_{\mathbf{C}}(R, DI)$. The elements of $\mathrm{Hom}_{\mathbf{C}}(R, DI)$ are exactly the morphisms $R \to DI$, so that the elements of P are exactly the tuples of arrows $R \to DI$ indexed by I (varying over the objects of $\mathbf{J}$). In symbols, a generic element of P is in the form

$$(f_I : R \to DI)_{I \in \mathbf{J}_0},$$

which is a tuple of arrows in $\mathbf{C}$ from R to all the objects in the diagram D:

$$\begin{array}{ccccc} & & R & & \\ & {\scriptstyle f_I} \swarrow & \downarrow {\scriptstyle f_{I'}} & \searrow & \\ DI & & DI' & & \dots \end{array}$$

Does this give a cone over D? Let's see. Condition (3.4.4) says that a tuple $(f_I : R \to DI)_{I \in \mathbf{J}_0}$ is in S if and only if for every $m : I \to I'$ of $\mathbf{J}$,

$$(Dm) \circ f_I = f_{I'};$$

that is, if for every $m : I \to I'$ of $\mathbf{J}$, this diagram commutes:

$$\begin{array}{ccc} & R & \\ {\scriptstyle f_I} \swarrow & & \searrow {\scriptstyle f_{I'}} \\ DI & \xrightarrow{Dm} & DI', \end{array}$$

which means indeed that $(f_I : R \to DI)_{I \in \mathsf{J}_0}$ is a cone over D (and all cones over D are in this form). So, the elements of S are precisely the cones over D with tip R; that is, we have a canonical bijection

$$\begin{aligned} \mathrm{Cone}(R, D) &\xrightarrow{\cong} S = \lim_{I \in \mathsf{J}} \mathrm{Hom}_{\mathsf{C}}(R, DI) \\ (f_I : R \to DI)_{I \in \mathsf{J}_0} &\longmapsto (f_I : R \to DI)_{I \in \mathsf{J}_0}, \end{aligned} \tag{3.4.5}$$

which, in this notation, even looks like the identity since the elements of both S and $\mathrm{Cone}(R, D)$ are tuples of arrows from R (making the relevant triangles commute). But now, by definition , the limit of D is the object representing the functor $\mathrm{Cone}(-, D) : \mathbf{C} \to \mathbf{Set}$, so that we have an isomorphism

$$\mathrm{Hom}_{\mathsf{C}}\left(R, \lim D\right) \xrightarrow{\cong} \mathrm{Cone}(R, D). \tag{3.4.6}$$

By the Yoneda lemma, this isomorphism is determined by where its component at $R = \lim D$ maps the identity to, and the resulting cone is exactly the limiting cone for the limit $\lim D$. That is, if we denote the arrows of the limit cone by λ_I, we have

$$\begin{aligned} \mathrm{Hom}_{\mathsf{C}}\left(\lim D, \lim D\right) &\xrightarrow{\cong} \mathrm{Cone}(\lim D, D) \\ \mathrm{id} &\longmapsto (\lambda_I : \lim D \to DI)_{I \in \mathsf{J}_0}, \end{aligned}$$

so that the bijection (3.4.6) is given by composing with the limiting cone:

$$\begin{aligned} \mathrm{Hom}_{\mathsf{C}}\left(R, \lim D\right) &\xrightarrow{\cong} \mathrm{Cone}(R, D) \\ f &\longmapsto (\lambda_I \circ f : R \to DI)_{I \in \mathsf{J}_0}. \end{aligned}$$

Composing with the bijection (3.4.5), we get the desired bijection:

$$\begin{aligned} \mathrm{Hom}_{\mathsf{C}}\left(R, \lim D\right) &\xrightarrow{\cong} \mathrm{Cone}(R, D) \xrightarrow{\cong} \lim_{I \in \mathsf{J}} \mathrm{Hom}_{\mathsf{C}}(R, DI) \\ f &\longmapsto (\lambda_I \circ f : R \to DI)_{I \in \mathsf{J}_0} \longmapsto (f_I : R \to DI)_{I \in \mathsf{J}_0}. \end{aligned}$$

To see that it commutes with the arrows of the respective limiting cones, by setting again $R = \lim D$, we get that

$$\begin{array}{ccccc} \mathrm{Hom}_{\mathbf{C}}\left(\lim D, \lim D\right) & \xrightarrow{\cong} & \mathrm{Cone}(\lim D, D) & \xrightarrow{\cong} & \lim_{I \in \mathbf{J}} \mathrm{Hom}_{\mathbf{C}}(\lim D, DI) \\ \mathrm{id} & \longmapsto & (\lambda_I : \lim D \to DI)_{I \in \mathbf{J}_0} & \longmapsto & (\lambda_I : \lim D \to DI)_{I \in \mathbf{J}_0}. \end{array}$$

Therefore, the arrow at I of the limit cone in **Set**, given by projecting onto the Ith component of the tuple $(\lambda_I : \lim D \to DI)_{I \in \mathbf{J}_0}$, corresponds exactly to λ_I, which is the arrow at I of the limit cone in **C**. □

Exercise 3.4.8. Is the isomorphism $\lim_{I \in \mathbf{J}} \mathrm{Hom}_{\mathbf{C}}(R, DI) \to \mathrm{Hom}_{\mathbf{C}}\left(R, \lim_{I \in \mathbf{J}} DI\right)$ natural? (By the way, in which argument?)

4

Adjunctions

In this chapter, we look at another cornerstone concept in category theory: adjoint functors. Whenever a functor is not invertible, very often we have a "canonical choice of coming back," but different from a section or a retraction. This notion is central to the theory of monads (Chapter 5) and to the theory of closed monoidal categories (Section 6.5).

An alternative way of reading this book, which could be helpful to some readers, is to read the following chapter on monads before this one and to come back here having in mind that monads give rise to adjunctions.

4.1 General Definitions

Definition 4.1.1. Let $\mathbf{C}$ and $\mathbf{D}$ be categories, and consider functors $F : \mathbf{C} \to \mathbf{D}$ and $G : \mathbf{D} \to \mathbf{C}$. An **adjunction** between F and G is a bijection

$$\mathrm{Hom}_{\mathbf{D}}(FC, D) \xrightarrow{\cong} \mathrm{Hom}_{\mathbf{C}}(C, GD),$$

for each object C of $\mathbf{C}$ and D of $\mathbf{D}$, natural both in C and in D.

Note that the notion is not symmetric in F and G, or in $\mathbf{C}$ and $\mathbf{D}$. We call F the **left-adjoint** and G the **right-adjoint**, and denote the adjunction by $F \dashv G$.

Two maps $FC \to D$ and $C \to GD$ related by the bijection above are called **transpose** or **adjunct** to each other.

Notation 4.1.2. It is useful to use the symbols $\flat$ (flat) and $\sharp$ (sharp) to denote the transposes.[1] We denote by $f^\flat$ a map $FC \to D$ and by $f^\sharp$ its transpose $C \to GD$. Slightly overloading the notation, we can also write the specific bijections as

$$\mathrm{Hom}_{\mathsf{D}}(FC, D) \underset{\flat}{\overset{\sharp}{\rightleftarrows}} \mathrm{Hom}_{\mathsf{C}}(C, GD).$$

Let's now see some more concrete examples of adjunctions.

4.1.1 Free-forgetful adjunctions

Many adjunctions appearing in mathematics are the so-called *free-forgetful adjunctions*, where the right-adjoint is a forgetful functor and the left-adjoint is a functor giving a "free construction." While the notion of "forgetful functor" does not have a formal meaning, the idea of "free," as we will show, can be made precise in terms of left-adjoints.

Example 4.1.3 (topology). Consider the forgetful functor $U : \mathbf{Top} \to \mathbf{Set}$. This takes a topological space and returns the underlying set "forgetting some structure." In general, we cannot go back: Given a set, we cannot assign a topological space uniquely, and many topological spaces have the same underlying set up to isomorphism (for example, $\mathbb{R}$ and $\mathbb{R}^2$). However, given a set X, there is always a "trivial" topology that we can put on X, the discrete topology, for which each subset of X is declared open. Let's denote this space by FX. The discrete topology has the following property: Given any topological space S, every function $f : FX \to S$ is continuous. Indeed, let $O \subseteq S$ be open. Then, $f^{-1}(O) \subseteq FX$ is open because every subset of FX is. Now, given a function between sets $g : X \to Y$, the induced function $FX \to FY$ is continuous (because every function on FX is). Let's denote this continuous map by $Ff : FX \to FY$. Thus, F is a functor $\mathbf{Set} \to \mathbf{Top}$.

---

[1] Note that different authors use different conventions. In particular in [Rie16], the two symbols are switched compared to here.

Let us now prove that there is an adjunction $F \dashv U$. This amounts to a natural bijection,

$$\mathrm{Hom}_{\mathbf{Top}}(FX, S) \xrightarrow{\cong} \mathrm{Hom}_{\mathbf{Set}}(X, US),$$

for each set X and each topological space S. Here, we construct the bijection; the naturality of this construction is considered in the following exercise. Now, consider a continuous function $FX \to S$. As simply a function, this is a map from the set X to the underlying set of S; in other words, it is a function $X \to US$. Conversely, given a function $X \to US$, this gives a function $FX \to S$, which is necessarily continuous since FX is equipped with the discrete topology. Therefore, *continuous functions $FX \to S$ correspond bijectively to functions between the sets $X \to US$.*

Exercise 4.1.4 (topology). Show that the bijection above is natural in both arguments.

A possible interpretation of this adjunction is the following: The functor U "forgets" the structure of topological spaces, and the functor F "tries to recover it" in some default way — in this case, by equipping our sets with the discrete topology. The functors F and U are not inverse to each other, not even pseudoinverse — an adjunction is a more nuanced relation between functors.

Exercise 4.1.5 (topology). Show that the forgetful functor U : **Top** $\to$ **Set** also has a *right*-adjoint. (*Hint*: Which topology is somewhat dual to the discrete topology?)

Example 4.1.6 (linear algebra). Consider now the forgetful functor U : **Vect** $\to$ **Set**. This takes a vector space V and forgets its vector space structure, returning its underlying set. In general, there is no way to go back, and not every set even admits a vector space structure. However, given a set X, we can always canonically form a vector space from X: the set of **formal linear combinations** of the elements of X. Let's denote this latter set by FX. Its elements are expressions in the form

$$a_1x_1 + \cdots + a_nx_n,$$

for all $a_i \in \mathbb{R}$ and $x_i \in X$, and n is finite (but arbitrarily large). Note that we are not *actually computing* the expression above (X is not a vector space); we are simply *writing it* formally.[2] The vector space structure on FX is given by the usual rules of addition and scalar multiplication of expressions by *expanding*, such as

$$\begin{aligned} a_1(b_1x_1 + b_2x_2) + a_2(c_1y_1 + c_2y_2) \\ = (a_1b_1)x_1 + (a_1b_2)x_2 + (a_2c_1)y_1 + (a_2c_2)y_2, \end{aligned}$$

and the set X is a basis of FX (why?). Now, let's make this construction functorial. Given a function between sets $f : X \to Y$, we can obtain the linear function $Ff : FX \to FY$ as

$$Ff(a_1x_1 + \cdots + a_nx_n) \;:=\; a_1f(x_1) + \cdots + a_nf(x_n).$$

(Why is this function linear?) Therefore, we have the functor F : **Set** $\to$ **Vect**.

Let's now show that $F \dashv U$. We have to prove that for all sets X and vector spaces V, there is a natural bijection

$$\mathrm{Hom}_{\mathbf{Vect}}(FX, V) \xrightarrow{\cong} \mathrm{Hom}_{\mathbf{Set}}(X, UV)$$

between linear maps $FX \to V$ and functions from X to the set underlying the vector space V. Now, if we have a linear map l : $FX \to V$, we can restrict it to the basis X and obtain a function from X to UV (i.e. V, but considered as a set). Conversely, given a map $f : X \to UV$, there is a **linear extension** $FX \to V$ that extends f linearly on all linear combinations of elements of X, and this

---

[2] A rigorous definition of *formal linear combination* can be given by a finitely supported function $X \to \mathbb{R}$. For example, the formal sum $2x + 3y$ can be interpreted as the function sending x to 2, y to 3, and everything else to zero.

linear extension is unique because X is a basis of FX (why exactly?). Therefore, we have a bijection. Again, naturality is left as exercise as follows.

Exercise 4.1.7 (linear algebra). Show, again, that the bijection above is natural in both arguments.

Exercise 4.1.8 (linear algebra). This idea of *linear extension* is akin to that of a universal property. Can you make this precise? If not, just wait until the following section.

Exercise 4.1.9 (group theory). What is the left-adjoint to the forgetful functor **Grp** $\to$ **Set**?

4.1.2 Galois connections

If the categories in question are posets, analogously to most category-theoretical concepts, adjunctions looks particularly simple, and they were introduced much earlier than adjunctions (the first example was given by Évariste Galois).

Definition 4.1.10. Let $(X, \leq)$ and $(Y, \leq)$ be posets, and let $f : X \to Y$ and $g : Y \to X$ be monotone maps. An adjunction $f \dashv g$ is called a **Galois connection** or **Galois correspondence**. The left-adjoint f is also called the **lower adjoint**, and the right-adjoint g is also called the **upper adjoint**.

Concretely, an adjunction $f \dashv g$ amounts to the following condition (can you see why?): For every $x \in X$ and $y \in Y$,

$$f(x) \leq y \quad \text{if and only if } x \leq g(y),$$

and the naturality condition is trivial.

We have seen that in many cases, adjunctions between categories can be interpreted in terms of "forgetting structure" and "trying to recover the structure in a default way." Analogously, for posets, Galois connections can be often interpreted in terms of "forgetting properties" and "trying to recover them in a default way." (In category theory, often, structures and properties behave in similar way. If you are interested in this, see for example the nLab page on "stuff, structure, property."[3])

Example 4.1.11 (sets and relations, analysis). Let X be the poset of subsets of $\mathbb{R}^2$, ordered by inclusion. Let Y be the poset of *convex* subsets of $\mathbb{R}^2$ (disks, squares, triangles,. . .). We have a canonical inclusion map $i : Y \to X$, which we can think of as "forgetful": It forgets the convexity. Let's now show that i has a lower adjoint. The **convex hull** of a subset $S \subseteq \mathbb{R}^2$ is defined as

- the smallest convex subset of $\mathbb{R}^2$ containing S,
- the intersection of all convex subsets of $\mathbb{R}^2$ containing S, or
- the set obtained by closing S under all possible convex combinations.

As for why these construction are equivalent, see the following exercise. Define now $c : X \to Y$ to be the map assigning to each $S \in X$, i.e. to each subset of $\mathbb{R}^2$, its convex hull. This map is monotone (why?).

Let us now show that c is left-adjoint (or lower adjoint) to i. We have to show that for each subset $S \subseteq \mathbb{R}^2$ and each *convex* subset $C \subseteq \mathbb{R}^2$,

$$c(S) \subseteq C \quad \text{if and only if } S \subseteq C$$

(note that we have omitted the inclusion map i, although technically it is still there). Now, suppose that the convex hull $c(S)$ is contained in the convex set C. Then, since $c(S)$ contains S, clearly S is contained in C as well. Conversely, suppose that S is contained in C. Then, since C is convex and $c(S)$ is the *smallest* convex set containing S, necessarily, $c(S)$ is contained in C. Therefore, c is left adjoint to i.

[3]https://ncatlab.org/nlab/show/stuff%2C+structure%2C+property.

Exercise 4.1.12 (sets and relations, analysis). Show that the constructions of the convex hull given above are all equivalent. (*Hint*: the first and the second are basically the same.)

Here are other examples of Galois connections with the same interpretation of "forgetting properties."

Exercise 4.1.13 (topology). Show that there is a Galois connection between the subsets of $\mathbb{R}^2$ and the *closed* subsets of $\mathbb{R}^2$.

Exercise 4.1.14 (group theory). Show that there is a Galois connection between the *subsets* of a given group G and the *subgroups* of G.

Here are more general exercises.

Exercise 4.1.15 (sets and relations). Show that the inclusion $i : \mathbb{Z} \to \mathbb{R}$ (with the usual order of numbers) has both an upper and a lower adjoint.

Exercise 4.1.16 (sets and relations, analysis). Show that, for all the examples above, the result of applying the lower adjoint is always the solution to some constrained optimization problem, namely, the *smallest element satisfying a particular constraint*. Express this in terms of a universal property, and show that this universal property holds in general for any lower adjoint of a Galois connection.

(Of course, the upper adjoint satisfies a dual property.)

Exercise 4.1.17 (difficult!). If you have solved the exercise above, show that a similar universal property holds for all left-adjoint functors, not necessarily between posets. (*Hint*: The "smallest element" in a poset corresponds to the "initial object" of some category … *which category*?)

4.2 Unit and Counit

Let's now see in detail the universal arrows defined by an adjunction. Let $\mathbf{C}$ and $\mathbf{D}$ be categories, let $F : \mathbf{C} \to \mathbf{D}$ and $G : \mathbf{D} \to \mathbf{C}$ be functors, and consider an adjunction $F \dashv G$, i.e. a natural bijection,

$$\mathrm{Hom}_{\mathbf{D}}(FC, D) \xrightarrow{\cong} \mathrm{Hom}_{\mathbf{C}}(C, GD),$$

for all objects C of $\mathbf{C}$ and D of $\mathbf{D}$. Now, fix the object C. The bijection above, which is natural in D, can be seen as a natural isomorphism between the functors

$$\mathrm{Hom}_{\mathbf{D}}(FC, -) \quad \text{and} \quad \mathrm{Hom}_{\mathbf{C}}(C, G-) : \mathbf{D} \to \mathbf{Set}.$$

In other words, for all objects C of $\mathbf{C}$, the functor $\mathrm{Hom}_{\mathbf{C}}(C, G-) : \mathbf{D} \to \mathbf{Set}$ is representable, and it is represented by the object FC.

How can we understand this intuitively? In terms of representable functors as "probes," we can interpret the situation as follows. The functor $\mathrm{Hom}_{\mathbf{C}}(C, G-)$ takes an object D of $\mathbf{D}$ and extracts some of its features by first mapping it to $\mathbf{C}$ via the functor G and then probes the resulting object GD by looking at all possible ways of mapping C into it. The natural isomorphism above says that the features extracted this way from D are the same as those that can be extracted by directly probing D with FC. So, for example, in the adjunction between sets and vector spaces of Example 4.1.6, given a set X and a vector space V, all the possible ways of mapping X to V (taken as a set) correspond exactly to all the possible ways of mapping FX linearly to V (taken as a vector space).

Let's now see what is happening in terms of universal properties. By the Yoneda lemma, we know that the natural isomorphism above, for fixed C, is specified uniquely by an element of the set

$$\mathrm{Hom}_{\mathbf{C}}(C, GFC),$$

i.e. a particular ("universal") morphism $C \to GFC$ of $\mathbf{C}$. As usual (Section 2.3), this morphism has to be the image of the identity of FC

under the bijection. That is, set D to be exactly FC, so that we have a bijection:

$$\mathrm{Hom}_{\mathbf{D}}(FC, FC) \xrightarrow{\cong} \mathrm{Hom}_{\mathbf{C}}(C, GFC).$$

In the set on the left, there is a distinguished element, namely id_{FC}. The universal morphism $C \to GFC$ is the image of id_{FC} under the isomorphism above; that is, it is exactly $(\mathrm{id}_{FC})^\sharp : C \to GFC$.

Definition 4.2.1. The universal map $C \to GFC$ induced by the adjunction at the object C is called the **unit** of the adjunction (at C) and denoted by $\eta_C : C \to GFC$.

So, $\eta_C = (\mathrm{id}_{FC})^\sharp$ and $(\eta_C)^\flat = \mathrm{id}_{FC}$. Here is how the universal property of η_C works in practice, following the ideas in Section 2.3. Let D be an object of $\mathbf{D}$, and apply the functor G to it to get the object GD of $\mathbf{C}$. Then, the bijection $\mathrm{Hom}_{\mathbf{D}}(FC, D) \to \mathrm{Hom}_{\mathbf{C}}(C, GD)$ says that, given any morphism $f^\sharp : C \to GD$ of $\mathbf{C}$, there is a unique morphism $f^\flat : FC \to D$ of $\mathbf{D}$ such that the following triangle in $\mathbf{C}$ commutes:

$$\begin{array}{ccccc} C & \xrightarrow{\eta_C} & GFC & & FC \\ & \searrow^{f^\sharp} & \downarrow{\scriptstyle Gf^\flat} & \xleftarrow[G]{} & \downarrow{\scriptstyle f^\flat} \\ & & GD & & D. \end{array} \tag{4.2.1}$$

Example 4.2.2 (linear algebra). Consider again the adjunction between sets and vector spaces of Example 4.1.6. Given a set X, the unit $\eta_X : X \to UFX$ is a function between the set X and the underlying set of FX, i.e. the *set* of formal linear combinations of the elements of X (seen as a set, not as a vector space). Which map is now η_X? We know that $(\eta_X)^\flat = \mathrm{id}_{FX}$; in other words, η_X is the map whose linear extension is the identity on FX. Now, the (unique) map whose linear extension is the identity is the map $X \to UFX$ which assigns to each $x \in X$ the "trivial" linear combination $1 \cdot x \in UFX$. That is, $\eta_X : X \to UFX$ is the canonical *embedding of* X into the set underlying FX.

The universal property now reads as follows: For every vector space V and for every function $f : X \to UV$, i.e. function from X to

V seen as a set, there is a unique linear extension $f^\flat : FX \to V$ of f such that the following diagram commutes:

$$\begin{array}{ccccc} X & \xrightarrow{\eta_X} & UFX & & FX \\ & \searrow^{f} & \downarrow{\scriptstyle Uf^\flat} & \xleftarrow{\quad U \quad} & \downarrow{\scriptstyle f^\flat} \\ & & UV & & V. \end{array}$$

The triangle commuting on the left says that if we compose $Uf^\flat$ with η_X, that is, if we restrict $f^\flat$ to the "trivial" linear combinations (the elements of X), then it agrees with the original f. In other words, $f^\flat$ is *really extending* f. This is precisely the universal property of a basis; it states that X is a basis of FX.

Exercise 4.2.3 (linear algebra). Why is the embedding $\eta_X : X \to UFX$ *canonical*? Is anything natural here? (Spoiler: For the answer, you can also read further down).

Exercise 4.2.4 (group theory, topology). What are the units of the adjunctions between sets and groups (Exercise 4.1.9) and between sets and topological spaces (Example 4.1.3)?

Example 4.2.5 (sets and relations, analysis). Consider now the Galois connection of Example 4.1.11 between the subsets and convex subsets of $\mathbb{R}^2$. The unit of the adjunction $c \dashv i$ amounts to the relation of containment $S \subseteq c(S)$, for all subsets $S \subseteq \mathbb{R}^2$ (why?). This inequality indeed holds since every set is contained in its convex hull.

So far, we have kept the object C fixed. If we now let C vary, naturality in C implies that the map η_C is actually natural in C.

We first start with the following very useful technical lemma.

Lemma 4.2.6 ([Rie16, Lemma 4.1.3]). *The naturality condition of the adjunction implies that for every $f^\flat : FC \to D$, $g^\flat : FC' \to D'$, $h : C \to C'$ of $\mathbf{C}$ and $k : D \to D'$ of $\mathbf{D}$, the diagram on the left commutes if and only* *if the diagram on the right commutes:*

$$\begin{array}{ccc} FC & \xrightarrow{f^\flat} & D \\ \big\downarrow{\scriptstyle Fh} & & \big\downarrow{\scriptstyle k} \\ FC' & \xrightarrow{g^\flat} & D' \end{array} \qquad\qquad \begin{array}{ccc} C & \xrightarrow{f^\sharp} & GD \\ \big\downarrow{\scriptstyle h} & & \big\downarrow{\scriptstyle Gk} \\ C' & \xrightarrow{g^\sharp} & GD'. \end{array}$$

Note that the diagram on the left is in **D**, while the diagram on the right is in **C**. One way to interpret the lemma is that the "sharp" and "flat" isomorphisms permit to pass "rigidly" from $\mathrm{Hom}_{\mathbf{D}}(FC, D)$ to $\mathrm{Hom}_{\mathbf{C}}(C, GD)$, i.e. in a way which plays well with the morphisms of **C** and **D**.

Proof of Lemma 4.2.6. Naturality in D means that for every $k : D \to D'$ of **D**, the following diagram must commute:

$$\begin{array}{ccc} \mathrm{Hom}_{\mathbf{D}}(FC, D) & \xrightarrow[\cong]{\sharp} & \mathrm{Hom}_{\mathbf{C}}(C, GD) \\ \big\downarrow{\scriptstyle k\circ-} & & \big\downarrow{\scriptstyle Gk\circ-} \\ \mathrm{Hom}_{\mathbf{D}}(FC, D') & \xrightarrow[\cong]{\sharp} & \mathrm{Hom}_{\mathbf{C}}(C, GD'), \end{array}$$

which means that for every $f^\flat : FC \to D$, we must have

$$(k \circ f^\flat)^\sharp = Gk \circ f^\sharp. \tag{4.2.2}$$

Similarly, naturality in C means that for every $h : C \to C'$ of **C**, the following diagram must commute:

$$\begin{array}{ccc} \mathrm{Hom}_{\mathbf{D}}(FC', D') & \xrightarrow[\cong]{\sharp} & \mathrm{Hom}_{\mathbf{C}}(C', GD') \\ \big\downarrow{\scriptstyle -\circ Fh} & & \big\downarrow{\scriptstyle -\circ h} \\ \mathrm{Hom}_{\mathbf{D}}(FC, D') & \xrightarrow[\cong]{\sharp} & \mathrm{Hom}_{\mathbf{C}}(C, GD'), \end{array}$$

which means that for every $g^\flat : FC' \to D'$, we must have

$$(g^\flat \circ Fh)^\sharp = g^\sharp \circ h. \tag{4.2.3}$$

Now, by Equations (4.2.2) and (4.2.3) and since the maps $\sharp$ and $\flat$ are bijections, we have that

$$k \circ f^\flat = g^\flat \circ Fh$$

if and only if

$$Gk \circ f^\sharp = g^\sharp \circ h;$$

that is, the diagram on the left commutes if and only if the diagram on the right does. □

> **Exercise 4.2.7.** Show that Lemma 4.2.6 has a converse, namely, that the equivalence between the commutativity of the two diagrams (for all the arrows involved) is actually *equivalent* to the naturality of the bijections.

Now, let's see in detail why and how η is natural.

Lemma 4.2.8. *The maps* $\eta_C : C \to GFC$ *assemble to a natural transformation* $\eta : \mathrm{id}_\mathbf{C} \Rightarrow G \circ F$ *of endofunctors* $\mathbf{C} \to \mathbf{C}$.

Definition 4.2.9. The natural transformation $\eta : \mathrm{id}_\mathbf{C} \Rightarrow G \circ F$ with components $\eta_C : C \to GFC$ is called the **unit** of the adjunction.

Proof of Lemma 4.2.8. We have to prove that for all $h : C \to C'$ in $\mathbf{C}$, this diagram in $\mathbf{C}$ commutes:

$$\begin{array}{ccc} C & \xrightarrow{\eta_C} & GFC \\ \downarrow{\scriptstyle h} & & \downarrow{\scriptstyle GFh} \\ C' & \xrightarrow{\eta_{C'}} & GFC'. \end{array}$$

Now, we can use Lemma 4.2.6 to "rigidly replace G on the right of the diagram by F on the left of the diagram," replacing also the horizontal morphisms with their "sharp" counterpart. We obtain, therefore, the equivalent diagram in **D**:

$$\begin{array}{ccc} FC & \xrightarrow{\mathrm{id}_{FC}} & FC \\ \Big\downarrow{\scriptstyle Fh} & & \Big\downarrow{\scriptstyle Fh} \\ FC' & \xrightarrow{\mathrm{id}_{FC'}} & FC', \end{array}$$

which clearly commutes. □

So, the adjunction $F \dashv G$ defines a natural transformation $\eta : \mathrm{id}_{\mathbf{C}} \Rightarrow G \circ F$ whose components satisfy a universal property. Dually, we can also obtain a natural transformation on the other side of the adjunction. Namely, let's now keep D fixed in the natural bijection:

$$\mathrm{Hom}_{\mathbf{D}}(FC, D) \xrightarrow{\cong} \mathrm{Hom}_{\mathbf{C}}(C, GD).$$

We have then a natural isomorphism between the presheaves:

$$\mathrm{Hom}_{\mathbf{D}}(F-, D) \quad \text{and} \quad \mathrm{Hom}_{\mathbf{C}}(-, GD) : \mathbf{C}^{\mathrm{op}} \longrightarrow \mathbf{Set};$$

in other words, we are saying that the presheaf $\mathrm{Hom}_{\mathbf{D}}(F-, D) : \mathbf{C}^{\mathrm{op}} \longrightarrow \mathbf{Set}$ is representable, and it is represented by the object GD of **C**.

> **Exercise 4.2.10.** What is the interpretation of this condition in terms of "representable presheaves as observations"?

Dually to the case of η, we get a universal arrow $FGD \longrightarrow D$ of **D** for each object D, which is given by $(\mathrm{id}_{GD})^{\flat}$. Again, this resulting universal map is natural in D, giving a natural transformation $F \circ G \Rightarrow \mathrm{id}_{\mathbf{D}}$ of endofunctors $\mathbf{D} \longrightarrow \mathbf{D}$.

Definition 4.2.11. The natural transformation $F \circ G \Rightarrow \mathrm{id}_{\mathbf{D}}$ induced by the adjunction $F \dashv G$ is called the **counit** of the adjunction, and it is denoted by $\varepsilon : F \circ G \Rightarrow \mathrm{id}_{\mathbf{D}}$.

So, for each object D of **D**, the component of the counit at D satisfies $\varepsilon_D = (\mathrm{id}_{GD})^\flat$ and $(\varepsilon_D)^\sharp = \mathrm{id}_{GD}$.

> **Exercise 4.2.12 (important!).** Derive in detail the natural transformation ε, as we did above for η, and show that it is natural. (*Hint*: Dualize what was done above.)

Let's see how the universal property of ε_D looks in practice. For every object C of **C**, apply F to it to get the object FC of **D**. Now, the bijection $\mathrm{Hom}_\mathbf{D}(FC, D) \to \mathrm{Hom}_\mathbf{C}(C, GD)$ states that, given any morphism $g^\flat : FC \to D$ of **D**, there is a unique morphism $g^\sharp : C \to GD$ of **C** such that the following triangle of **D** commutes:

$$
\begin{array}{ccc}
C & & FC \\
\Big\downarrow{\scriptstyle g^\sharp} & \xmapsto{\ \ F\ \ } & {\scriptstyle Fg^\sharp}\Big\downarrow \searrow^{g^\flat} \\
GD & & FGD \xrightarrow[\varepsilon_D]{} D.
\end{array}
\qquad (4.2.4)
$$

Example 4.2.13 (linear algebra). Let's look at the counit of the adjunction between sets and vector spaces (Example 4.1.6). Let V be a vector space. The counit $\varepsilon_V : FUV \to V$ is a map from the vector space of *formal linear combinations of the vectors of V* to V. Which map in particular? We know that this map must be equal to $(\mathrm{id}_{UV})^\flat$; that is, it must be the linear extension of the identity of V (seen as a map between sets). In other words, $\varepsilon_V : FUV \to V$ must be the unique linear map such that for each $v \in V$, $\varepsilon_V(1v) = v$ and for all nontrivial linear combinations, the value of ε_V is uniquely specified by linearity. Therefore, applying ε_D to a generic expression,

$$a_1 v_1 + \cdots + a_n v_n,$$

of FUV, we get the *result in V* of the expression above. Now, note the difference between a *formal linear combination* and the *actual result of the linear combination*: For example, if $V = \mathbb{R}$, we have

$$\varepsilon_\mathbb{R}(2 \cdot 3 + 2 \cdot 1) = 8.$$

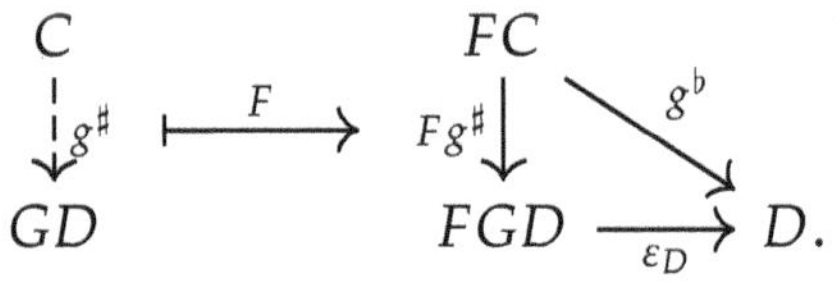

Note that, *as formal expressions*, "$2 \cdot 3 + 2 \cdot 1$" and "8" are not the same, (but they have the same result). The map $\varepsilon_{\mathbb{R}}$ *actually performs* the operations prescribed by the formal expression.

The universal property of ε_V says now that for every set X, we can form the free vector space FX, and given any linear map $g : FX \to V$, there is a unique function $g^\sharp$ from X to V (seen as a set) such that the following diagram commutes:

$$\begin{array}{ccccc} X & & FX & & \\ {\scriptstyle g^\sharp}\downarrow & \overset{F}{\longmapsto} & {\scriptstyle Fg^\sharp}\downarrow & \overset{g}{\searrow} & \\ UV & & FUV & \xrightarrow[\varepsilon_V]{} & V. \end{array}$$

Keeping Example 4.2.2 in mind, this states that any linear map is uniquely specified by its action on a basis.

As we have seen above, the counit map $\varepsilon_V : FUV \to V$ turns a *formal linear combination of the elements of* V into their *actual* linear combination. In other words, the map ε_V *is* that operation (akin to how addition is a map $+ : \mathbb{R} \times \mathbb{R} \to \mathbb{R}$). Put differently again, this map is exactly what makes V a vector space: it is its vector space structure (a vector space is precisely a set equipped with an operation of linear combination). The fact that the "extra structure" is captured by the counit is a very common phenomenon, which will be made precise by the concept of *Eilenberg–Moore* or *monadic adjunction* (see Sections 5.2.3 and 5.5). Not every adjunction has this property. For example, the one between sets and topological spaces (Example 4.1.3) does not (and we will shortly see why).

Exercise 4.2.14 (group theory). Show that the counit of the adjunction between sets and groups (Exercise 4.1.9), analogously to the case of vector spaces, can be seen as the map giving the group structure.

Can you think of other forgetful functors giving rise to adjunctions that have a similar property?

4.2.1 Alternative definition of adjunctions

We have seen how to obtain the unit and counit natural transformations from an adjunction, and we know that by the Yoneda lemma, these natural transformations (or better, their components) satisfy a universal property and thus determine the natural bijections $\mathrm{Hom}_{\mathbf{D}}(FC, D) \to \mathrm{Hom}_{\mathbf{C}}(C, GD)$ and $\mathrm{Hom}_{\mathbf{C}}(C, GD) \to \mathrm{Hom}_{\mathbf{D}}(FC, D)$ uniquely. We can now ask the converse question: Given functors $F : \mathbf{C} \to \mathbf{D}$ and $G : \mathbf{D} \to \mathbf{C}$, does a pair of natural transformations $\eta : \mathrm{id}_{\mathbf{C}} \Rightarrow G \circ F$ and $\varepsilon : F \circ G \Rightarrow \mathrm{id}_{\mathbf{D}}$ induce an adjunction $F \dashv G$? The answer is: almost, but not quite. The reason is that, thanks to the Yoneda lemma, the universal maps η_C and ε_D specify natural *transformations* but not necessarily natural *isomorphisms* between $\mathrm{Hom}_{\mathbf{D}}(FC, D)$ and $\mathrm{Hom}_{\mathbf{C}}(C, GD)$. In order for η_C and ε_D to specify isomorphisms, we need a couple of additional conditions to be satisfied, known as the **triangle identities**.

Lemma 4.2.15 (triangle identities). *Let $\eta : \mathrm{id}_{\mathbf{C}} \Rightarrow G \circ F$ and $\varepsilon : F \circ G \Rightarrow \mathrm{id}_{\mathbf{D}}$ be the unit and counit of an adjunction $F \dashv G$. Then, the following diagrams of natural transformations commute:*

$$\begin{array}{ccc} F & \overset{F\eta}{\Longrightarrow} & FGF \\ & \underset{\mathrm{id}_F}{\searrow\!\!\searrow} & \Downarrow{\scriptstyle \varepsilon F} \\ & & F \end{array} \qquad \begin{array}{ccc} G & \overset{\eta G}{\Longrightarrow} & GFG \\ & \underset{\mathrm{id}_G}{\searrow\!\!\searrow} & \Downarrow{\scriptstyle G\varepsilon} \\ & & G. \end{array} \tag{4.2.5}$$

In components, the triangle identities read as follows (why?): For every object C of $\mathbf{C}$ and D of $\mathbf{D}$,

$$\begin{array}{ccc} FC & \xrightarrow{F\eta_C} & FGFC \\ & \underset{\mathrm{id}_{FC}}{\searrow} & \downarrow{\scriptstyle \varepsilon_{FC}} \\ & & FC \end{array} \qquad \begin{array}{ccc} GD & \xrightarrow{\eta_{GD}} & GFGD \\ & \underset{\mathrm{id}_{GD}}{\searrow} & \downarrow{\scriptstyle G\varepsilon_D} \\ & & GD. \end{array} \tag{4.2.6}$$

Note that the first diagram is in $\mathbf{D}$, and the second diagram is in $\mathbf{C}$.

Proof of Lemma 4.2.15. We can rewrite the diagrams of (4.2.6) as the following squares:

$$\begin{array}{ccc} FC & \xrightarrow{F\eta_C} & FGFC \\ {\scriptstyle \mathrm{id}_{FC}}\big\downarrow & & \big\downarrow{\scriptstyle \varepsilon_{FC}} \\ FC & \xrightarrow[\mathrm{id}_{FC}]{} & FC \end{array} \qquad\qquad \begin{array}{ccc} GD & \xrightarrow{\eta_{GD}} & GFGD \\ {\scriptstyle \mathrm{id}_{GD}}\big\downarrow & & \big\downarrow{\scriptstyle G\varepsilon_D} \\ GD & \xrightarrow[\mathrm{id}_{GD}]{} & GD, \end{array}$$

and flip the first one (for convenience),

$$\begin{array}{ccc} FC & \xrightarrow{\mathrm{id}_{FC}} & FC \\ {\scriptstyle F\eta_C}\big\downarrow & & \big\downarrow{\scriptstyle \mathrm{id}_{FC}} \\ FGFC & \xrightarrow[\varepsilon_{FC}]{} & FC \end{array} \qquad\qquad \begin{array}{ccc} GD & \xrightarrow{\eta_{GD}} & GFGD \\ {\scriptstyle \mathrm{id}_{GD}}\big\downarrow & & \big\downarrow{\scriptstyle G\varepsilon_D} \\ GD & \xrightarrow[\mathrm{id}_{GD}]{} & GD. \end{array}$$

Now, using Lemma 4.2.6, we can apply "flat" to the first diagram, replacing F on the left by G on the right and replacing the horizontal arrows with their "flat" versions. Similarly, we can apply "sharp" to the second diagram, replacing G on the right of the diagram with F on the left and replacing the horizontal morphisms by their "sharp" versions. We get the following two diagrams, equivalent to the previous ones:

$$\begin{array}{ccc} C & \xrightarrow{\eta_C} & GFC \\ {\scriptstyle \eta_C}\big\downarrow & & \big\downarrow{\scriptstyle \mathrm{id}_{GFC}} \\ GFC & \xrightarrow[\mathrm{id}_{GFC}]{} & GFC \end{array} \qquad\qquad \begin{array}{ccc} FGD & \xrightarrow{\mathrm{id}_{FGD}} & FGD \\ {\scriptstyle \mathrm{id}_{FGD}}\big\downarrow & & \big\downarrow{\scriptstyle \varepsilon_D} \\ FGD & \xrightarrow[\varepsilon_D]{} & D, \end{array}$$

and these clearly commute. □

So, the unit and counit of an adjunction must always satisfy the triangle identities (4.2.5). This condition is actually sufficient: If a pair of natural transformations $\eta : \mathrm{id}_{\mathbf{C}} \Rightarrow G \circ F$ and $\varepsilon : F \circ G \Rightarrow \mathrm{id}_{\mathbf{D}}$ satisfies the triangle identities, then the natural transformations $\mathrm{Hom}_{\mathbf{D}}(FC, D) \to \mathrm{Hom}_{\mathbf{C}}(C, GD)$ and $\mathrm{Hom}_{\mathbf{C}}(C, GD) \to \mathrm{Hom}_{\mathbf{D}}(FC, D)$ that they induce (which we will shortly see how) are bijections. Indeed, recall that by the Yoneda lemma, as we have seen in Section 2.3, the map $\eta_C : C \to GFC$ should correspond to the natural transformation with components (on objects D of $\mathbf{D}$)

$$\begin{aligned} \mathrm{Hom}_{\mathbf{D}}(FC, D) &\xrightarrow{\ \sharp\ } \mathrm{Hom}_{\mathbf{C}}(C, GD) \\ f^\flat &\longmapsto f^\sharp = Gf^\flat \circ \eta_C . \end{aligned} \tag{4.2.7}$$

(To see this more explicitly, look at the diagram (4.2.1).) Similarly, the map $\varepsilon_D : D \to GFD$ corresponds to the natural transformation with components (on objects C of $\mathbf{C}$)

$$\begin{aligned} \mathrm{Hom}_{\mathbf{C}}(C, GD) &\xrightarrow{\ \flat\ } \mathrm{Hom}_{\mathbf{D}}(FC, D) \\ g^\sharp &\longmapsto g^\flat = \varepsilon_D \circ Fg^\sharp . \end{aligned} \tag{4.2.8}$$

(To see this more explicitly, look at the diagram (4.2.4).)

Lemma 4.2.16. *Let $F : \mathbf{C} \to \mathbf{D}$ and $G : \mathbf{D} \to \mathbf{C}$ be functors, and let $\eta : \mathrm{id}_{\mathbf{C}} \Rightarrow G \circ F$ and $\varepsilon : F \circ G \Rightarrow \mathrm{id}_{\mathbf{D}}$ satisfy the triangle identities* (4.2.5)*. Then, the assignments* (4.2.7) *and* (4.2.8) *are mutually inverse, giving a bijection* $\mathrm{Hom}_{\mathbf{D}}(FC, D) \to \mathrm{Hom}_{\mathbf{C}}(C, GD)$ *(and so an adjunction $F \dashv G$).*

Proof. Let's start with $f^\flat : FC \to D$. We apply (4.2.7) to get $Gf^\flat \circ \eta_C : C \to GD$, and then we apply (4.2.8) to it to get $\varepsilon_D \circ F(Gf^\flat \circ \eta_C) : FC \to D$. We have to prove that

$$\varepsilon_D \circ F(Gf^\flat \circ \eta_C) \;=\; f^\flat . \tag{4.2.9}$$

Now, consider the following diagram:

$$\begin{array}{ccccc} FC & \xrightarrow{F\eta_C} & FGFC & \xrightarrow{FGf^\flat} & FGD \\ & \searrow^{\mathrm{id}_{FC}} & \downarrow{\scriptstyle \varepsilon_{FC}} & & \downarrow{\scriptstyle \varepsilon_D} \\ & & FC & \xrightarrow{f^\flat} & D. \end{array}$$

The triangle on the left commutes, as it's one of the triangle identities (4.2.6), and the square on the right commutes by the naturality of ε. So, Equation (4.2.9) follows.

Dually, let's now take $g^\sharp : C \to GD$. We apply (4.2.8) to get $\varepsilon_D \circ F g^\sharp : FC \to D$ and then (4.2.7) to get $G(\varepsilon_D \circ F g^\sharp) \circ \eta_C$. We have to prove that

$$G(\varepsilon_D \circ F g^\sharp) \circ \eta_C \;=\; g^\sharp. \tag{4.2.10}$$

So, consider the diagram

$$\begin{array}{ccccc}
C & \xrightarrow{\;g^\sharp\;} & GD & & \\
{\scriptstyle \eta_C}\big\downarrow & & {\scriptstyle \eta_{GD}}\big\downarrow & \searrow^{\mathrm{id}_{GD}} & \\
GFC & \xrightarrow[GFg^\sharp]{} & GFGD & \xrightarrow[G\varepsilon]{} & GD.
\end{array}$$

The square on the left commutes by the naturality of η, and the triangle on the right is the other triangle identity of (4.2.6). Therefore, Equation (4.2.10) follows, which means that (4.2.7) and (4.2.8) are mutually inverse. □

Moreover, one can see that the unit and counit resulting from the adjunction above are again η and ε.

We have therefore proven the following.

Theorem 4.2.17 (alternative definition of adjunction). *An adjunction* $F \dashv G$ *is equivalently given by a pair of natural transformations* $\eta : \mathrm{id}_{\mathsf{C}} \Rightarrow G \circ F$ *and* $\varepsilon : F \circ G \Rightarrow \mathrm{id}_{\mathsf{D}}$ *satisfying the triangle identities* (4.2.5).

This result is very convenient since the unit and counit are more explicit and, sometimes, easier to work with in practice.

> **Exercise 4.2.18 (linear algebra).** Show that the triangle identities hold for the adjunction between the sets and vector spaces of Example 4.1.6.

> **Exercise 4.2.19 (group theory).** Show that the triangle identities hold for the adjunction between the sets and groups of Exercise 4.1.9.

4.2.2 Example: Adjunction between categories and multigraphs

Here is an example of how to describe an adjunction in terms of its units and counits, in this case between categories and multigraphs.

Let $\mathbf{C}$ be a small category. We can associate to $\mathbf{C}$ its **underlying multigraph** $U(\mathbf{C})$, where the vertices of $U(\mathbf{C})$ are the objects of $\mathbf{C}$ and the edges of $U(\mathbf{C})$ are the morphisms of $\mathbf{C}$ (all of them, including the identities). Take now another small category $\mathbf{D}$ and a functor $F : \mathbf{C} \to \mathbf{D}$. There is an induced morphism of multigraphs $U(\mathbf{C}) \to U(\mathbf{D})$, which we denote by $U(F)$, which maps the vertex of $U(\mathbf{C})$ corresponding to an object X of $\mathbf{C}$ to the vertex of $U(\mathbf{D})$ corresponding to the object FX of $\mathbf{D}$. Regarding edges, $U(F)$ maps the edge of $U(\mathbf{C})$ corresponding to a morphism $f : X \to Y$ of $\mathbf{C}$ to the edge of $U(\mathbf{D})$ corresponding to the morphism $Ff : FX \to FY$ of $\mathbf{D}$. The assignment $\mathbf{C} \mapsto U(\mathbf{C}), F \mapsto U(F)$ preserves identities and composition, and so we have a functor $U : \mathbf{Cat} \to \mathbf{MGraph}$. We can view U as a forgetful functor, which takes a (small) category and forgets its identity and composition structure, giving only its underlying graph.

We now want to construct a left-adjoint to the functor U. So, consider a directed multigraph G, and let's try to obtain a (small) category from it in a canonical way, which we call $\mathbf{P}(G)$. The objects of $\mathbf{P}(G)$ are the vertices of G. As morphisms, we would like to have all possible chains in G, including the trivial chain at each vertex (which will be the identity). Now, recall from Example 2.1.2 that an n-chain of G is an n-tuple of edges

$$(e_1, \dots, e_n)$$

of G which are head-to-tail, i.e. the target of e_i is the source of e_{i+1}, for all i. Let's also, for convenience, call 1-chains simply the edges of G and 0-chains the vertices, which are seen as "trivial chains." Denoting by $Chain_n(G)$ the set of n-chains of G, we now define the set of morphisms of $\mathbf{P}(G)$ to be the disjoint union

$$\coprod_{n\geq 0} \mathrm{Chain}_n(G),$$

i.e. the morphisms of $\mathbf{P}(G)$ are all the chains of arbitrary (but finite) length in G. We define the source of $(e_1, \ldots, e_n)$ to be the source of e_1 and the target to be the target of e_n. The identities will be the chains of length zero, and the composition is given by the concatenation of chains.

Exercise 4.2.20. Show that the composition of chains, defined as above, is associative and unital so that $\mathbf{P}(G)$ is indeed a category. (*Hint*: There is not that much to prove.)

Definition 4.2.21. The category $\mathbf{P}(G)$ associated to the multigraph G goes under the names of **free category over** G, **fundamental category of** G, or also **path category of** G.

Now, let's define the action of $\mathbf{P}$ on the morphisms. Given the multigraphs G, H and a morphism of multigraphs $f : G \to H$, we define the functor $\mathbf{P}f : \mathbf{P}(G) \to \mathbf{P}(H)$ to be such that:

- on objects of $\mathbf{P}(G)$, it maps the object corresponding to the vertex x of G to the object corresponding to the vertex $f(x)$ of H;
- on morphisms of $\mathbf{P}(G)$, it maps the morphism corresponding to the chain $(e_1, \ldots, e_n)$ of G to the morphism corresponding to the chain $(f(e_1), \ldots, f(e_n))$ of H. This is again a chain since f preserves incidence (it is a morphism of multigraphs).

This makes the assignment $G \mapsto \mathbf{P}G, f \mapsto \mathbf{P}f$ a functor $\mathbf{P} : \mathbf{MGraph} \to \mathbf{Cat}$.

Exercise 4.2.22. Prove that this assignment preserves identities and composition so that $\mathbf{P}$ is indeed a functor. (*Hint*: There is not that much to prove.)

Note the different levels, which are possibly confusing: **P** is a functor **MGraph** $\longrightarrow$ **Cat**. Moreover, given a morphism $f : G \longrightarrow H$ of **MGraph**, its image $\mathbf{P}f : \mathbf{P}G \longrightarrow \mathbf{P}H$ is a functor too in the sense of a morphism of **Cat**.

We now show that **P** is left-adjoint to U and describe the adjunction in terms of the unit and counit. Let's start with the unit. Consider a multigraph G. We can form the category $\mathbf{P}G$, and then look at the underlying multigraph $U\mathbf{P}G$. This is in general not the same as the multigraph we started with: There will be additional edges, corresponding to the identities and compositions that we have added. For example, if we start with the graph G given by

$$x \xrightarrow{e_1} y \xrightarrow{e_2} z,$$

the multigraph $U\mathbf{P}G$ will be given by

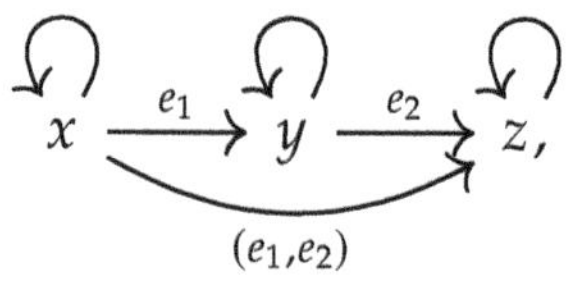

where the loops depict the identities in $\mathbf{P}G$.

Remark 4.2.23. This corresponds to the fact that the category *generated* by this diagram,

$$X \xrightarrow{f} Y \xrightarrow{g} Z,$$

is implicitly the one that *actually* looks like this:

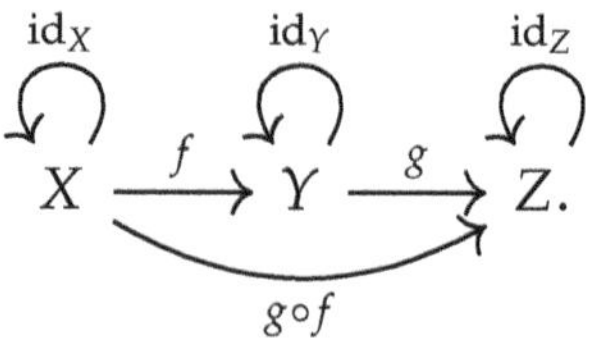

The functor **P** is actually the mathematically rigorous way of generating a category by "adding in identities and composites." We have implicitly used it all the times we have used diagrams to represent pieces of a category.

So, in general, G and $U\mathbf{P}G$ are different. However, there is a canonical inclusion of G into $U\mathbf{P}G$ (see also the example above):

- G and $U\mathbf{P}G$ have the same vertices.
- All the edges of G are also edges of $U\mathbf{P}G$ (but the latter may have more edges). This is because all the edges of G appear in the morphisms of $\mathbf{P}G$ (they are the 1-chains), which in turn are the edges of $U\mathbf{P}G$.

We denote the inclusion of G into $U\mathbf{P}G$ by $\eta_G : G \to U\mathbf{P}G$. It respects incidence, and so it is a morphism of **MGraph**.

Exercise 4.2.24. Show that the map $\eta_G : G \to U\mathbf{P}G$ is natural in G. That is, show that for each morphism of multigraphs $f : G \to H$, the following diagram commutes:

$$\begin{array}{ccc} G & \xrightarrow{\eta_G} & U\mathbf{P}G \\ \downarrow{\scriptstyle f} & & \downarrow{\scriptstyle U\mathbf{P}f} \\ H & \xrightarrow{\eta_H} & U\mathbf{P}H. \end{array}$$

(*Hint*: This becomes almost trivial once you see how the morphism $U\mathbf{P}f$ acts on those edges of $U\mathbf{P}G$ which lie in the image of η_G.)

By the exercise above, we have a natural transformation $\eta : \mathrm{id}_{\mathbf{MGraph}} \Rightarrow \mathbf{P} \circ U$. This will be the unit of our adjunction. Note the analogy with the unit of the adjunction between sets and vector spaces (Example 4.2.2). Here, just like there, the unit is the embedding of the "trivial cases." Here, it selects the "primitive morphisms," the ones not obtained by the composition of other morphisms, whereas in Example 4.2.2, it selected the "primitive vectors," the ones obtained by the trivial linear combinations of the form $1x$ for $x \in X$.

Let's now turn to the counit. Let $\mathbf{D}$ be a small category. We can look at its underlying multigraph $U\mathbf{D}$ and then at the category $\mathbf{P}U\mathbf{D}$

generated by it. Again, this will in general be different from the category we started with. In particular:

- The objects of $\mathbf{D}$ and $\mathbf{P}U\mathbf{D}$ are the same.
- The morphisms of $\mathbf{P}U\mathbf{D}$ are *tuples* of composable morphisms of $\mathbf{D}$, including empty tuples, one for each object of $\mathbf{D}$. This is because each morphism of $\mathbf{D}$ gives rise to an edge of the multigraph $U\mathbf{D}$, and by applying $\mathbf{P}$, the morphisms of $\mathbf{P}U\mathbf{D}$ will be the chains of edges of $U\mathbf{D}$ (including the 0-chains), i.e. the chains of composable morphisms.

Now, consider the difference between a *chain of composable morphisms* $(f_1, \dots, f_n)$ and their *actual composition* $f_n \circ \dots \circ f_1$. The morphisms of $\mathbf{P}U\mathbf{D}$ are in the form $(f_1, \dots, f_n)$, where all the f_i are morphisms of $\mathbf{D}$. This should already give a feeling of what the counit map will be: It will be the *map actually performing the composition*. More rigorously, $\varepsilon_\mathbf{D} : \mathbf{P}U\mathbf{D} \to \mathbf{D}$ is the functor which:

- on objects, it maps an object of $\mathbf{P}U\mathbf{D}$ to the corresponding object in $\mathbf{D}$ (remember that the two categories have the same objects).
- on morphisms, it maps a tuple of composable morphisms of $\mathbf{D}$ $(f_1, \dots, f_n)$ (which make up a morphism of $\mathbf{P}U\mathbf{D}$) to their actual composition in $\mathbf{D}$, $f_n \circ \dots \circ f_1$:

$$A \xrightarrow{f_1} B \xrightarrow{f_2} C \xrightarrow{f_3} D \qquad \xmapsto{\quad \varepsilon_\mathbf{D} \quad} \qquad A \xrightarrow{f_3 \circ f_2 \circ f_1} D.$$

  It also maps empty tuples (there's one for each object) to the identity of the corresponding object of $\mathbf{D}$.

  This gives a functor $\varepsilon_\mathbf{D} : \mathbf{P}U\mathbf{D} \to \mathbf{D}$ for each small category $\mathbf{D}$.

Exercise 4.2.25. Show that $\varepsilon_\mathbf{D}$ is indeed a functor, i.e. it respects identity and composition. (*Hint*: We have just said what it does on the identities.)

So, $\varepsilon_\mathbf{D}$ is a morphism of the category **Cat**.

Exercise 4.2.26. Show that $\varepsilon_{\mathbf{D}}$ is natural in $\mathbf{D}$. That is, for every small category $\mathbf{C}, \mathbf{D}$ and every functor $F : \mathbf{C} \to \mathbf{D}$, the following diagram commutes:

$$\begin{array}{ccc} \mathbf{P}U\mathbf{C} & \xrightarrow{\varepsilon_{\mathbf{C}}} & \mathbf{C} \\ {\scriptstyle \mathbf{P}U(F)}\downarrow & & \downarrow{\scriptstyle F} \\ \mathbf{P}U\mathbf{D} & \xrightarrow{\varepsilon_{\mathbf{D}}} & \mathbf{D}. \end{array}$$

(*Hint*: The assertion becomes almost trivial once you see what the functor $\mathbf{P}U(F)$ does on the morphisms of $\mathbf{P}U\mathbf{C}$.)

Therefore, we get a natural transformation $\varepsilon : \mathbf{P} \circ U \Rightarrow \mathrm{id}_{\mathbf{Cat}}$. This will be the counit of our adjunction. Note the similarity with the counit of the adjunction between sets and vector spaces (Example 4.2.13). There, the counit was mapping formal linear expressions to their actual results. Here, the counit is mapping sequences of composable morphisms to their actual composition (and trivial chains to identities). Here too, the counit is exactly encoding the extra structure that categories have over graphs: identities and composition. (So, additionally, this adjunction is monadic, as we will see in Section 5.5.1.)

In order to establish the adjunction, we now need to prove the triangle identities (4.2.6). The first identity says that for each multigraph G, the following diagram of **Cat** must commute:

$$\begin{array}{ccc} \mathbf{P}G & \xrightarrow{\mathbf{P}\eta_G} & \mathbf{P}U\mathbf{P}G \\ & {\scriptstyle \mathrm{id}_{\mathbf{P}G}}\searrow & \downarrow{\scriptstyle \varepsilon_{\mathbf{P}G}} \\ & & \mathbf{P}G. \end{array}$$

Now, all the categories in the diagram above have the same objects, and all the functors are simply the identities on objects. So, let's see what happens on morphisms. A morphism of $\mathbf{P}G$ is a chain $(e_1, \ldots, e_n)$ of edges of G. The map $\mathbf{P}\eta_G$, by the definition of $\mathbf{P}$ on morphisms, acts on $(e_1, \ldots, e_n)$ by applying η_G to each entry of the tuple; that is, it gives the tuple $(\eta_G(e_1), \ldots, \eta_G(e_n))$. Each entry of the tuple is in the form $\eta_G(e_i)$. What η_G does on an edge e_i is that it includes it in the morphisms of $\mathbf{P}G$ as the 1-chain (e_i). Therefore, we get the chain

$$((e_1), \ldots, (e_n)).$$

Note that this has *two* levels of brackets: This is a *chain of* 1-*chains*. Applying the counit $\varepsilon_{\mathbf{P}G}$ of $\mathbf{P}G$ gives the composition

$$(e_n) \circ \cdots \circ (e_1).$$

But the composition in $\mathbf{P}G$ is just a concatenation of chains, in this case, of 1-chains, and so we get

$$(e_1, \ldots, e_n),$$

which gives the tuple we started with, and so it gives the same result as applying the identity of $\mathbf{P}G$. Therefore, the diagram commutes.

The second triangle identity says that for every category $\mathbf{D}$, the following diagram of multigraphs must commute:

$$\begin{array}{ccc} U\mathbf{D} & \xrightarrow{\eta_{U\mathbf{D}}} & U\mathbf{P}U\mathbf{D} \\ & \searrow^{\mathrm{id}_{U\mathbf{D}}} & \downarrow U\varepsilon_{\mathbf{D}} \\ & & U\mathbf{D}. \end{array}$$

Now, all the multigraphs in the diagram have the same vertices, and all the morphisms in the diagram are the identity on vertices. So, let's focus on edges. Since the edges of $U\mathbf{D}$ are the morphisms of $\mathbf{D}$, let $f : X \to Y$ be a morphism of $\mathbf{D}$, seen as an edge of $U\mathbf{D}$. The unit map $\eta_{U\mathbf{D}}$ embeds this edge into the morphisms of $\mathbf{P}U\mathbf{D}$ (seen as edges of $U\mathbf{P}U\mathbf{D}$) as the 1-chain (f). Now, we apply the map $U\varepsilon_{\mathbf{D}}$, which returns the edge associated to the morphism $\varepsilon_{\mathbf{D}}((f))$. The latter is a "trivial composition" of only one morphism, so it simply returns itself, i.e. f (it's a trivial chain of only one arrow). Therefore, we just get f, which again corresponds to the edge we started with, so it is the same result as applying the identity of $U\mathbf{D}$. Therefore, this diagram also commutes.

In conclusion, we have proven that the functor $\mathbf{P} : \mathbf{MGraph} \to \mathbf{Cat}$ is left-adjoint to the functor $U : \mathbf{Cat} \to \mathbf{MGraph}$. This is one of the most important connections between graph theory and category theory.

Exercise 4.2.27. Write down explicitly the universal properties of the unit and counit associated to this adjunction.

Exercise 4.2.28 (linear algebra, group theory). If you have solved Exercise 4.2.18 or Exercise 4.2.19, can you see the resemblance between the meaning of the triangle identities here and in those contexts?

Exercise 4.2.29. Construct an adjunction between sets and monoids similar to the one in Exercise 4.1.9. What is the relation between this construction and the adjunction $\mathbf{P} \dashv U$ that we have just seen?

Exercise 4.2.30 (algebraic topology, graph theory). Why is the category $\mathbf{P}G$ sometimes called the "fundamental category" of G? What is its relation with the fundamental group (Example 1.3.22)?

Exercise 4.2.31 (linear algebra). Write a formal version of Definition 2.3.12.

4.3 Adjunctions, Limits and Colimits

Adjoint functors interact very well with limits and colimits.

Theorem 4.3.1. *Right-adjoint functors are continuous.*

Let's see what this statement means. Let $\mathbf{C}$ and $\mathbf{D}$ be categories, and let $L : \mathbf{C} \to \mathbf{D}$ and $R : \mathbf{D} \to \mathbf{C}$ be functors with an adjunction

$L \dashv R$. Also, let $E : \mathbf{J} \to \mathbf{D}$ be a diagram in $\mathbf{D}$ admitting a limit. Then, the limit of the diagram

$$\mathbf{J} \xrightarrow{E} \mathbf{D} \xrightarrow{R} \mathbf{C}$$

exists in C, and we have an isomorphism

$$\lim R \circ E \;\cong\; R(\lim E)$$

compatible with the universal cones.

Reversing all the arrows of $\mathbf{C}$ and $\mathbf{D}$, we obtain immediately the following.

Corollary 4.3.2. *Left-adjoint functors are cocontinuous.*

As we did with Theorem 3.3.16, let's first try to understand the statement for the simpler case of binary products, and then prove it in general. Finally, we look at examples.

Warning. Note that, differently from what we did when we dualized Theorem 3.3.16 (obtaining Corollary 3.3.17), we are now reversing the arrows of *both* $\mathbf{C}$ and $\mathbf{D}$. Therefore, left-adjoint functors do *not* map colimits into limits (as representable presheaves do); they map colimits into *colimits*, i.e. they preserve colimits.

4.3.1 Right-adjoints and binary products

Let A and B be objects of $\mathbf{D}$ admitting a product $A \times B$, with projections $p_1 : A \times B$ and $p_2 : A \times B \to B$. We want to find an isomorphism

$$R(A \times B) \;\cong\; RA \times RB$$

compatible with the respective product projections. To do so, it suffices to check that the cone

$$\begin{array}{ccc} & R(A \times B) & \\ {\scriptstyle Rp_1} \swarrow & & \searrow {\scriptstyle Rp_2} \\ RA & & RB \end{array}$$

is a limit cone in **D**. This means that we have to check that for every object C of **C**, to each pair of arrows $C \to RA$, $C \to RB$ there corresponds a unique arrow $C \to R(A \times B)$, filling in the following diagram:

$$\begin{array}{ccccc} & \swarrow & C & \searrow & \\ & & \downarrow & & \\ RA & \xleftarrow[Rp_1]{} & R(A \times B) & \xrightarrow[Rp_2]{} & RB. \end{array}$$

In other words, we have to show that the assignment

$$\begin{aligned} \mathrm{Hom}_{\mathbf{C}}(C, R(A \times B)) &\longrightarrow \mathrm{Hom}_{\mathbf{C}}(C, RA) \times \mathrm{Hom}_{\mathbf{C}}(C, RB) \\ f &\longmapsto (Rp_1 \circ f, Rp_2 \circ f) \end{aligned}$$

is a natural bijection.

Now, we have the following chain of natural bijections:

$$\begin{array}{cc} \mathrm{Hom}_{\mathbf{C}}(C, R(A \times B)) & f \\ \downarrow \cong & \downarrow \\ \mathrm{Hom}_{\mathbf{D}}(LC, A \times B) & f^\flat \\ \downarrow \cong & \downarrow \\ \mathrm{Hom}_{\mathbf{D}}(LC, A) \times \mathrm{Hom}_{\mathbf{D}}(LC, B) & (p_1 \circ f^\flat, p_2 \circ f^\flat) \\ \downarrow \cong & \downarrow \\ \mathrm{Hom}_{\mathbf{C}}(C, RA) \times \mathrm{Hom}_{\mathbf{C}}(C, RB) & ((p_1 \circ f^\flat)^\sharp, (p_2 \circ f^\flat)^\sharp), \end{array}$$

given respectively by the adjunction, the universal property of $A \times B$ (or Theorem 3.3.16, with naturality coming from Exercise 3.4.8), and the adjunction again (twice). It suffices to prove now that $(p_1 \circ f^\flat)^\sharp = Rp_1 \circ f$ and $(p_2 \circ f^\flat)^\sharp = Rp_2 \circ f$. Instantiating now Lemma 4.2.6 with $k = p_1$ and $h = \mathrm{id}$, we get that, since the left diagram commutes, the right one does:

$$\begin{array}{ccc} & \overset{f^{\flat}}{\nearrow} & A \times B \\ LC & & \Big\downarrow p_1 \\ & \underset{p_1 \circ f^{\flat}}{\searrow} & A \end{array} \qquad \begin{array}{ccc} & \overset{f}{\nearrow} & R(A \times B) \\ C & & \Big\downarrow Rp_1 \\ & \underset{(p_1 \circ f^{\flat})^{\sharp}}{\searrow} & RA \end{array}$$

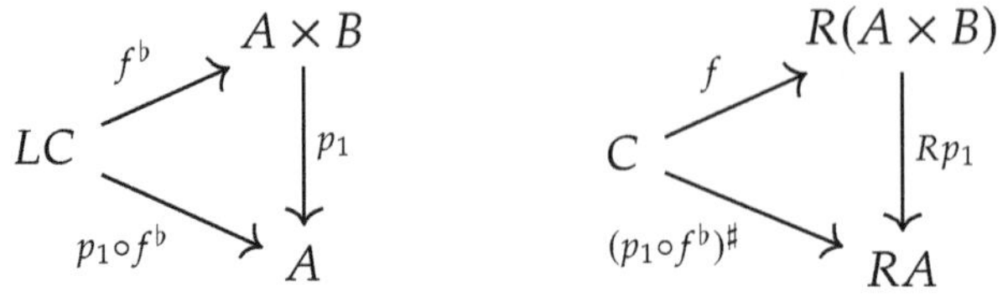

so that $(p_1 \circ f^{\flat})^{\sharp} = Rp_1 \circ f$. The same can be done for p_2.

This proves the statement for the case of binary products. But before moving on, let's try to interpret this fact intuitively. R being right-adjoint to L means the following. For each object D of **D**, the information that we can extract from D by first applying R and then probing the resulting object RD in **C** with an object C (by mapping C to RD in all possible ways) gives the same result as directly probing D with the object LC of **D**. Now, given the product $A \times B$, the features that we can extract by applying R and then probing with C are the same as those that we would obtain if we were to experiment with A and B separately. This is because applying R and then probing with C is equivalent to probing $A \times B$ directly with LC, and we know that, by the universal property of the product, probing $A \times B$ directly corresponds to probing A and B separately. (See also the remarks after Theorem 3.3.16.) So, in other words, right-adjoint functors *do not exhibit complex behavior on products* in the sense of Section 3.3.1.

> **Exercise 4.3.3.** Can you interpret the behavior of left-adjoint functors dually?

4.3.2 General proof of Theorem 4.3.1

In order to prove the theorem, we make use of the following lemma, which takes care of diagram-chasing.

Lemma 4.3.4. *Suppose that $L \dashv R$, and let $E : \mathbf{J} \to \mathbf{D}$ be a diagram in **D**. Then, the adjunction induces a bijection*

$$\mathrm{Cone}(LC, E) \cong \mathrm{Cone}(C, R \circ E),$$

*for all objects C of **C**, natural in C.*

Proof of Lemma 4.3.4. A cone α over E with tip LC consists, in particular, of a tuple of arrows $\alpha_I : LC \to EI$, for each object I of $\mathbf{J}$, in the form

$$\begin{array}{ccccc} & & LC & & \\ & {}^{\alpha_I}\swarrow & \downarrow{\scriptstyle \alpha_{I'}} & \searrow & \\ EI & & EI' & & \cdots \end{array}$$

Moreover, in order for this tuple to be a cone, we need that for every morphism $m : I \to I'$ of $\mathbf{J}$, $\alpha_{I'} = Em \circ \alpha_I$, i.e. all these triangles must commute:

$$\begin{array}{ccc} & LC & \\ {}^{\alpha_I}\swarrow & & \searrow^{\alpha_{I'}} \\ EI & \xrightarrow[Em]{} & EI'. \end{array} \tag{4.3.1}$$

The adjunction $\mathrm{Hom}_{\mathbf{D}}(LC, D) \to \mathrm{Hom}_{\mathbf{C}}(C, RD)$ now maps all these arrows $\alpha_I : LC \to EI$ bijectively to arrows $\alpha_I^\sharp : C \to REI$ in the form

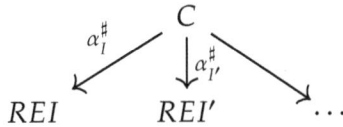

$$\begin{array}{ccccc} & & C & & \\ & {}^{\alpha_I^\sharp}\swarrow & \downarrow{\scriptstyle \alpha_{I'}^\sharp} & \searrow & \\ REI & & REI' & & \cdots \end{array}$$

and arrows in this latter form belong to a cone over $R \circ E$ if and only if, for every morphism $m : I \to I'$ of $\mathbf{J}$, we have $\alpha_{I'}^\sharp = REm \circ \alpha_I^\sharp$, i.e. all these triangles must commute:

$$\begin{array}{ccc} & C & \\ {}^{\alpha_I^\sharp}\swarrow & & \searrow^{\alpha_{I'}^\sharp} \\ REI & \xrightarrow[REm]{} & REI'. \end{array} \tag{4.3.2}$$

Moreover, every cone over $R \circ E$ with tip C is in this form. But now, by naturality of the bijection $\sharp : \mathrm{Hom}_{\mathbf{D}}(LC, D) \to \mathrm{Hom}_{\mathbf{C}}(C, RD)$ in the argument D (or by Lemma 4.2.6), the diagram (4.3.1) commutes if and only if the diagram (4.3.2) does. Therefore, the cones in $\mathrm{Cone}(LC, E)$ are mapped bijectively to the cones in $\mathrm{Cone}(C, R \circ E)$.

The naturality of the bijection is guaranteed by the naturality of the bijection $\sharp : \mathrm{Hom}_{\mathbf{D}}(LC, D) \to \mathrm{Hom}_{\mathbf{C}}(C, RD)$ in the argument C. □

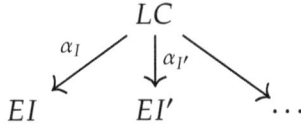

We are now ready to prove the theorem.

Proof of Theorem 4.3.1. We have to show that the object $R(\lim E)$ of **C** satisfies the universal property of the limit of $R \circ E$. That is, we have to give a bijection

$$\mathrm{Hom}_{\mathbf{C}}(C, R(\lim E)) \;\cong\; \mathrm{Cone}(C, R \circ E),$$

for all objects C of **C**, natural in C, and compatible with the universal cones.

Denote by $\lambda_I : \lim E \to EI$ (for I an object of **J**) the arrows of the limiting cone of E in **D**. Similarly to the case of products, we have the following chain of natural bijections:

$$\begin{array}{ccccccc} \mathrm{Hom}_{\mathbf{C}}(C, R(\lim E)) & \xrightarrow{\cong} & \mathrm{Hom}_{\mathbf{D}}(LC, \lim E) & \xrightarrow{\cong} & \mathrm{Cone}(LC, E) & \xrightarrow{\cong} & \mathrm{Cone}(C, R \circ E) \\ f & \longmapsto & f^{\flat} & \longmapsto & (\lambda_I \circ f^{\flat})_{I \in \mathbf{J}_0} & \longmapsto & ((\lambda_I \circ f^{\flat})^{\sharp})_{I \in \mathbf{J}_0} \end{array}$$

obtained, in order, by the adjunction, the universal property of $\lim E$, and the bijection of Lemma 4.3.4. Therefore, $R(\lim E)$ is a limit of $R \circ E$. Note also that, as in the case of products, we can use Lemma 4.2.6 for the components of the cone to get $(\lambda_I \circ f^{\flat})^{\sharp} = R\lambda_I \circ f$. Therefore, by setting $C = R(\lim E)$ and $f = \mathrm{id}$, one can see that the universal arrows of the limiting cone of $R(\lim E)$ are exactly the images $R\lambda_I$ of the limiting cone of $\lim E$. □

4.3.3 Examples

Example 4.3.5 (topology). The forgetful functor **Top** $\to$ **Set** has a left-adjoint (Example 4.1.3) as well as a right-adjoint (Exercise 4.1.5). Therefore, it is continuous and cocontinuous. In particular, the underlying set of the coproduct of topological spaces is their disjoint union, and the underlying set of the product of topological spaces is their cartesian product. Compare this also with the other results of Example 3.3.2.

Example 4.3.6 (linear algebra). The forgetful functor **Vect** $\longrightarrow$ **Set** has a left-adjoint (Example 4.1.6); therefore, it preserves limits. In particular:

- the product of two vector spaces, isomorphic to their direct sum, has as the underlying set the cartesian product of the respective sets;
- the equalizer E of a parallel pair of linear maps $f, g : V \longrightarrow W$ is mapped to the equalizer of the underlying functions, which gives a subset $U(E) \subseteq U(V)$. In other words, E is a vector space whose underlying set is a subset of the set underlying V, i.e. a vector subspace of V.

Compare this also with the other results of Example 3.3.4. Note in particular that colimits are not preserved in general.

Dually, the free functor F : **Set** $\longrightarrow$ **Vect** preserves colimits. Therefore:

- the vector space freely generated by a disjoint union $A \sqcup B$ is the direct sum of the spaces generated by A and B, respectively;
- the quotient set of an equivalence relation gives the basis of a quotient vector space (why is this a colimit?);
- the empty set generates the zero vector space.

Exercise 4.3.7 (group theory). Make statements similar to the example above for the adjunction between sets and groups (Exercise 4.1.9).

Exercise 4.3.8. Redefine the product of **Cat** in terms of the product of multigraphs. (*Hint*: By the adjunction of Section 4.2.2 between categories and multigraphs, the underlying multigraph to the product of two categories is necessarily the product of the two respective multigraphs.)

Exercise 4.3.9. Define the coproduct of two categories in **Cat**. Show that the fundamental category of a disjoint union of graphs is the coproduct of the respective fundamental categories.

4.4 The Adjoint Functor Theorem for Preorders

It turns out that, under some conditions, Theorem 4.3.1 admits a converse. Converse-like statements to Theorem 4.3.1 are known under the name of **adjoint functor theorems** and state that under some conditions, if a functor $G : \mathbf{D} \to \mathbf{C}$ preserves all limits, then it has a left-adjoint. There are several adjoint functor theorems in the literature, see for example Section 4.6 in [Rie16], or the nLab page on adjoint functor theorems.[4]

An intuitive interpretation of the phenomenon underlying the adjoint functor theorems is the following. In some cases, we can *reconstruct* what the left-adjoint to G would do, thanks to the fact that G preserves limits; that is, it does not lose information on how more complex objects are created from simpler ones (for example by forming products). Of course, the condition that G has to preserve all limits is necessary by Theorem 4.3.1. The adjoint functor theorems state that, in some cases, this is also sufficient.

Here, we look at the simplest of the adjoint functor theorems, the **adjoint functor theorem for preorders**, following the approach [FS19].

Theorem 4.4.1 (Adjoint functor theorem for preorders). *Let $(X, \leq)$ and $(Y, \leq)$ be partial orders. Suppose that Y has all infima (or meets), and let $g : Y \to X$ be a monotone function preserving all infima. Then, g has a lower adjoint $f : X \to Y$, given by*

$$f(x) = \inf\{y \in Y \mid x \leq g(y)\}.$$

[4] https://ncatlab.org/nlab/show/adjoint+functor+theorem.

In particular, a monotone map $g : Y \to X$ is the upper (or right-) adjoint in a Galois connection if and only if it preserves all infima.

> **Exercise 4.4.2 (sets and relations).** What is, explicitly, the corresponding statement for preorders?

Immediately we also get the dual statement, as follows.

Corollary 4.4.3. *Suppose that X has all suprema (or joins). A monotone map $f : X \to Y$ is lower (or left-) adjoint if and only if it preserves all suprema.*

> **Exercise 4.4.4 (sets and relations).** In this case, what is the explicit formula for the lower adjoint to f?

Before proving the theorem, let's look at a practical example of how the proof of the theorem works.

4.4.1 The case of convex subsets

Consider again the case of Example 4.1.11. The set X is the set of all subsets of $\mathbb{R}^2$. The set Y is the set of convex subsets of $\mathbb{R}^2$. Both sets are ordered by inclusion; and the "forgetful" inclusion map $g : Y \to X$ is monotone.

The infima in both sets are given by the intersection of sets (why?), and the map $g : Y \to X$ preserves those infima since the intersection of convex sets is again convex (why?). By Theorem 4.4.1, this is enough to establish that g has a left-adjoint f. Plugging in the explicit construction of the theorem, the map f takes a subset $S \subseteq \mathbb{R}^2$ and gives the convex subset

$$f(S) = \bigcap \{C \subseteq \mathbb{R}^2 \text{ convex} \mid S \subseteq C\};$$

that is, $f(S)$ must be the intersection of all the convex subsets of $\mathbb{R}^2$ which contain S. We know from Example 4.1.11 that is exactly the convex hull.

The actual proof of the theorem follows this same line of reasoning, which, it turns out, is purely about order theory and not really about convexity.

4.4.2 Proof of Theorem 4.4.1

We prove the statement for partial orders, and the statement for more general preorders is analogous (why?).

Define for each x,

$$f(x) := \inf\{y \in Y \mid x \leq g(y)\}.$$

First of all, we have to show that this assignment $x \mapsto f(x)$ is a monotone map (or a functor) $X \to Y$. So, suppose that $x \leq x'$. Then, for all y such that $x' \leq g(y)$, we also have $x \leq x' \leq g(y)$, which means that

$$\{y \in Y \mid x \leq g(y)\} \supseteq \{y \in Y \mid x' \leq g(y)\}.$$

Taking an infimum over a *larger or equal* subset gives us a *smaller or equal* infimum, so

$$f(x) = \inf\{y \in Y \mid x \leq g(y)\} \leq \inf\{y \in Y \mid x' \leq g(y)\} = f(x').$$

So, f is monotone.

Now, we need to prove that f is indeed lower adjoint to g. We use the unit and counit characterization of Section 4.2. Now, since g preserves infima, we have that for each $x \in X$,

$$g(f(x)) = g\big(\inf\{y \in Y \mid x \leq g(y)\}\big) = \inf\{g(y) \mid y \in Y,\, x \leq g(y)\},$$

and in particular, the infimum on the right-hand side exists. Since clearly x is less than or equal to every element of the set $\{g(y) \mid y \in Y, x \leq g(y)\}$, by the universal property of the infimum x is also less than or equal to the infimum of that set, which means

$$x \leq \inf\{g(y) \mid y \in Y, x \leq g(y)\} = g(f(x)).$$

This inequality is the unit of the adjunction. For the counit, let $z \in Y$. Then,

$$f(g(z)) \;=\; \inf\{y \in Y \mid g(z) \le g(y)\}.$$

Now, since $g(z) \le g(z)$, the element z belongs to the set $\{y \in Y \mid g(z) \le g(y)\}$, and so z must be larger than or equal to the infimum over that set, i.e.

$$z \;\ge\; \inf\{y \in Y \mid g(z) \le g(y)\} \;=\; f(g(z)).$$

This inequality is the counit of the adjunction. The naturality and triangle diagrams, since we are in a preorder, commute trivially. Therefore, f is left-adjoint to g.

4.4.3 Further considerations and examples

Theorem 4.4.1 states that a monotone map is the upper adjoint of a Galois connection if and only if it preserves infima, but this says nothing about suprema. That is, an upper adjoint in general does *not* preserve suprema. (*Lower* adjoints preserve suprema, of course.)

This is why the intersection of convex subsets is convex, but the union of convex subsets, in general, is not. This is a very general phenomenon: Many special properties of subsets are preserved by intersections but not by unions — this is a sign that there is a Galois connection involved.

Example 4.4.5 (linear algebra, group theory). Consider the following statements:

- The intersection of vector subspaces of $\mathbb{R}^3$ is always a vector subspace. The union is in general not (take the union of two lines, for example).
- The intersection of subgroups of a group G is again a subgroup. Their union is in general not (can you give a counterexample?).

Exercise 4.4.6 (sets and relations, linear algebra, group theory). Show that, for the two examples above, there is a Galois connection involved. (*Hint*: Use Theorem 4.4.1.)

Example 4.4.7 (topology). Let T be a topological space. Let X be the poset of subsets of T, ordered by inclusion, and let Y be the subposet of *closed* subsets of X. There is a "forgetful inclusion" $g : Y \to X$ which maps closed subsets to only subsets. This map preserves infima because the intersection of an arbitrary number of closed sets is closed. Therefore, it has a lower adjoint $f : X \to Y$. Following the construction of Theorem 4.4.1, this lower adjoint takes a set $S \subseteq T$ and gives the intersection of all the closed subsets of T containing S. This is known as the **closure of** S in topology.

Exercise 4.4.8 (topology, tricky!). The *union* of two closed sets is again closed. So, does the map $g : Y \to X$ above also admit an *upper* adjoint?

Example 4.4.9 (sets and relations). The inclusion $\mathbb{Z} \to \mathbb{R}$ clearly preserves both infima and suprema. Define now $\bar{\mathbb{Z}}$ and $\bar{\mathbb{R}}$ to be $\mathbb{Z}$ and $\mathbb{R}$, respectively, extended to include $+\infty$ and $-\infty$ too, so that we have all suprema and infima. The inclusion $\bar{\mathbb{Z}} \to \bar{\mathbb{R}}$ still preserves infima and suprema. Therefore, it has an upper and a lower adjoint. This gives a slick solution to Exercise 4.1.15.

5

Monads and Comonads

This chapter is dedicated to monads and comonads, which sit at the intersection of mathematics and computer science. Here are the main definitions, which we interpret in several ways in the following sections.

Definition 5.0.1. Let $\mathbf{C}$ be a category. A **monad on C** consists of:

- a functor $T : \mathbf{C} \to \mathbf{C}$,
- a natural transformation $\eta : \mathrm{id}_{\mathbf{C}} \Rightarrow T$ called **unit**,
- a natural transformation $\mu : TT \Rightarrow T$ called **composition** or **multiplication**,

such that the following diagrams commute, called **left and right unitality** and **associativity**, respectively:

$$\begin{tikzcd} T \arrow[r, Rightarrow, "\eta T"] \arrow[dr, Rightarrow, "\mathrm{id}"'] & TT \arrow[d, Rightarrow, "\mu"] \\ & T \end{tikzcd} \qquad \begin{tikzcd} T \arrow[r, Rightarrow, "T\eta"] \arrow[dr, Rightarrow, "\mathrm{id}"'] & TT \arrow[d, Rightarrow, "\mu"] \\ & T \end{tikzcd} \qquad \begin{tikzcd} TTT \arrow[r, Rightarrow, "T\mu"] \arrow[d, Rightarrow, "\mu T"] & TT \arrow[d, Rightarrow, "\mu"] \\ TT \arrow[r, Rightarrow, "\mu"] & T. \end{tikzcd}$$

We can also write down the natural transformations in terms of their components. For each object X of $\mathbf{C}$, the unit is a morphism $\eta_X : X \to TX$ and the multiplication is a morphism $\mu_X : TTX \to TX$, such that the following diagrams commute:

$$\begin{tikzcd} TX \arrow[r, "\eta_{TX}"] \arrow[dr, "\mathrm{id}"'] & TTX \arrow[d, "\mu_X"] \\ & TX \end{tikzcd} \qquad \begin{tikzcd} TX \arrow[r, "T\eta_X"] \arrow[dr, "\mathrm{id}"'] & TTX \arrow[d, "\mu_X"] \\ & TX \end{tikzcd} \qquad \begin{tikzcd} TTTX \arrow[r, "T\mu_X"] \arrow[d, "\mu_{TX}"] & TTX \arrow[d, "\mu_X"] \\ TTX \arrow[r, "\mu_X"] & TX. \end{tikzcd} \tag{5.0.1}$$

Definition 5.0.2. A **comonad** on $\mathbf{C}$ is a monad on $\mathbf{C}^{\mathrm{op}}$.

Explicitly, a comonad on $\mathbf{C}$ consists of:

- a functor $C : \mathbf{C} \to \mathbf{C}$,
- a natural transformation $\varepsilon : C \Rightarrow \mathrm{id}_{\mathbf{C}}$ called **counit**,
- a natural transformation $\nu : C \Rightarrow CC$ called **comultiplication**,

such that the following diagrams, in components, commute for each object X, called **left and right counitality** and **coassociativity**, respectively:

$$\begin{array}{ccc} CX & \xrightarrow{\nu_X} & CCX \\ & {\scriptstyle \mathrm{id}}\searrow & \downarrow{\scriptstyle \varepsilon_{CX}} \\ & & CX, \end{array} \qquad \begin{array}{ccc} CX & \xrightarrow{\nu_X} & CCX \\ & {\scriptstyle \mathrm{id}}\searrow & \downarrow{\scriptstyle C\varepsilon_X} \\ & & CX, \end{array} \qquad \begin{array}{ccc} CX & \xrightarrow{\nu_X} & CCX \\ \downarrow{\scriptstyle \nu_X} & & \downarrow{\scriptstyle \nu_{CX}} \\ CCX & \xrightarrow{C\nu_X} & CCCX. \end{array} \tag{5.0.2}$$

5.1 Monads as Extensions of Spaces

Here is the first intuitive idea of what a monad can look like.

Idea. *A monad can look like a consistent way of extending spaces to include generalized elements and generalized functions of a specific kind.*

Let's illustrate this aspect with some examples.

Example 5.1.1 (sets and relations). Consider the power set functor (Example 1.3.14) on the category **Set** of sets and functions. Given a set X, its power set PX can be considered an extension of X, where "subsets generalize elements." Given sets X and Y and a function $f : X \to Y$, we get the function $Pf : PX \to PY$ given by taking the image, it maps each $A \in PX$, which is a subset of X, to the subset of Y given by the image of A under f. Equivalently, it gives the subset of Y obtained by applying f to all the elements of A. One can consider the subsets of X as "generalized elements" and Pf as the "obvious" extension of f to the generalized elements of X (mapping them to generalized elements of Y).

In which way can we see subsets as generalized elements? In strict terms, the elements of X are not particular subsets since an element is not technically a subset. However, each element $x \in X$ defines a subset canonically: the singleton $\{x\}$. As we saw in Example 1.4.9, the embedding $X \to PX$ given by singletons is part of a natural transformation: $\sigma : \mathrm{id}_{\mathsf{C}} \Rightarrow P$. Explicitly:

(a) for each X, we consider the map given by singletons $\sigma_X : X \to PX$, with the interpretation being that PX, the extension, "includes" the old space X, or that PX "extends" X.

(b) for each $f : X \to Y$, the extended function Pf must agree with f on the "old elements," i.e. the elements coming from X via σ. In other words, this diagram has to commute:

$$\begin{array}{ccc} X & \xrightarrow{f} & Y \\ \downarrow{\scriptstyle \sigma_X} & & \downarrow{\scriptstyle \sigma_Y} \\ PX & \xrightarrow{Pf} & PY. \end{array}$$

Let's now turn to the multiplication map. The set PPX contains *subsets of subsets* of X. Given a subset of subsets of X, there is a canonical way of obtaining a subset of X: via the union. For example, if $x, y, z \in X$, a subset of subsets is for instance in the form

$$\mathcal{A} = \{\{x, y\}, \{y, z\}, \{\,\}\} \in PPX.$$

From the element above, we can take the union of the subsets contained in it, which is

$$\bigcup_{A \in \mathcal{A}} A = \{x, y, z\} \in PX.$$

We can view $\cup$ as the map which "removes the inner brackets." This gives an assignment $\cup : PPX \to PX$. Given a function $f : X \to Y$, the following diagram commutes:

$$\begin{array}{ccc} PPX & \xrightarrow{PPf} & PPY \\ \downarrow{\scriptstyle \cup_X} & & \downarrow{\scriptstyle \cup_Y} \\ PX & \xrightarrow{Pf} & PY, \end{array}$$

which means that the union of the images is the image of the union. In symbols,

$$f\left(\bigcup_{A \in \mathcal{A}} A\right) = \bigcup_{A \in \mathcal{A}} f(A).$$

Therefore, the map $\cup$ gives a natural transformation $PP \Rightarrow P$. Additional motivation for these natural transformations will be given in Section 5.1.1.

Let's now see how $(P, \sigma, \cup)$ is a monad. We have to show that the diagrams (5.0.1) commute. Now, the left unitality diagram

$$\begin{array}{ccc} PX & \xrightarrow{\sigma_{PX}} & PPX \\ & \searrow^{\text{id}} & \downarrow{\scriptstyle \cup_X} \\ & & PX \end{array}$$

states the following. Take a subset of X, say $\{x, y, z\}$ with $x, y, z \in X$, and form the singleton subset of PX containing just this set, i.e. $\{\{x, y, z\}\}$ (mind the double brackets: now we have a set of subsets of X, containing just one subset). Now, take the union of all the sets in this set. There is just one set in there, so we get back $\{x, y, z\}$. In symbols,

$$\begin{array}{ccc} \{x, y, z\} & \overset{\sigma}{\longmapsto} & \{\{x, y, z\}\} \\ & \searrow^{\text{id}} & \downarrow{\scriptstyle \cup} \\ & & \{x, y, z\}. \end{array}$$

The right unitality diagram

$$\begin{array}{ccc} PX & \xrightarrow{P\sigma_X} & PPX \\ & \searrow^{\text{id}} & \downarrow{\scriptstyle \cup_X} \\ & & PX \end{array}$$

states something similar. Namely, take again a subset of X, such as $\{x, y, z\}$, as above. Now, take the *image* of the map σ_X; that is, apply σ to each element of $\{x, y, z\}$. This replaces every element with its corresponding singleton set, giving the set of subsets $\{\{x\}, \{y\}, \{z\}\}$. Taking the union of the sets in this set gives again the original set $\{x, y, z\}$. In symbols,

$$\begin{array}{ccc} \{x, y, z\} & \xmapsto{P\sigma} & \{\{x\}, \{y\}, \{z\}\} \\ & \searrow^{\text{id}} & \downarrow{\scriptstyle \cup} \\ & & \{x, y, z\}. \end{array}$$

(This might remind the reader of the concatenation of edges in a graph in Section 4.2.2.) The associativity diagram

$$\begin{array}{ccc} PPPX & \xrightarrow{P\cup_X} & PPX \\ \downarrow{\scriptstyle \cup_{PX}} & & \downarrow{\scriptstyle \cup_X} \\ PPX & \xrightarrow{\cup_X} & PX \end{array}$$

states the following. Take a subset of subsets of subsets of X (three times), for example

$$\{\{\{x, y\}, \{x, z\}\}, \{\{a, b\}\}\}.$$

Then, removing the innermost brackets and then the (remaining) innermost brackets has the same result as removing the mid-level brackets and then the (remaining) innermost brackets. In symbols,

$$\begin{array}{ccc} \{\{\{x, y\}, \{x, z\}\}, \{\{a, b\}\}\} & \xmapsto{P\cup_X} & \{\{x, y, z\}, \{a, b\}\} \\ \downarrow{\scriptstyle \cup_{PX}} & & \downarrow{\scriptstyle \cup_X} \\ \{\{x, y\}, \{x, z\}, \{a, b\}\} & \xmapsto{\cup_X} & \{x, y, z, a, b\}. \end{array}$$

Since diagrams (5.0.1) commute, the triple $(P, \sigma, \cup)$ is a monad on **Set**.

Example 5.1.2 (basic probability). The probability functors given in Examples 1.3.16–1.3.18 can be also given a monad structure, in a way that is similar to the power set case. Let's do it for the set case; the other cases are left as exercises (see the following).

Recall that the probability functor $\mathcal{P}$ of Example 1.3.16 takes a set X and returns the set $\mathcal{P}X$ of finitely supported probability measures on X. Every element $x \in X$ gives a "deterministic" probability measure, namely δ_x (see Exercise 1.4.10). There is a natural embedding $\delta : X \to \mathcal{P}X$ which we can interpret as follows. Considering a probability measure $p \in \mathcal{P}X$ as a "random point" of X, then $\mathcal{P}X$ is the extension of X to account for "random points" too, with the old deterministic points included via the unit map $\delta : X \to \mathcal{P}X$ (compare with the power set case).

The multiplication map $E : \mathcal{P}\mathcal{P}X \to \mathcal{P}X$ is given as follows. For $\pi \in PPX$,

$$E(\pi)(x) := \sum_{p \in \mathcal{P}X} \pi(p)\, p(x).$$

Here is a way to interpret this map. Suppose that you have two coins in your pocket: One coin is fair, with "heads" on one face and "tails" on the other face; the second coin has "heads" on both sides. Suppose now that you draw a coin randomly and flip it. We can sketch the probabilities in the following way:

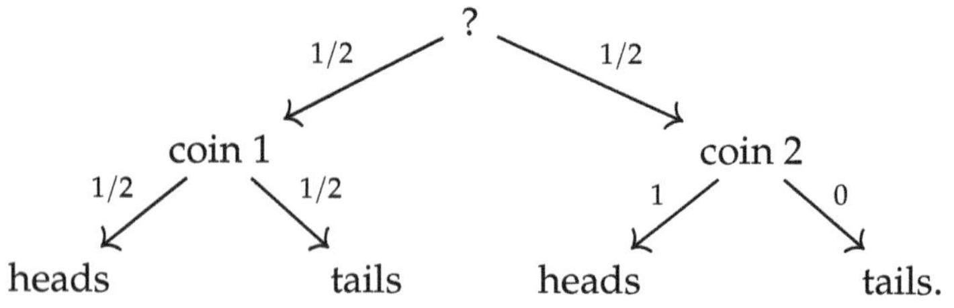

Let X be the set {"heads", "tails"}. A coin gives a **law** according to which we will obtain "heads" or "tails," so it determines an element of $\mathcal{P}X$. Since the choice of coin is also random (we also have a *law on the coins*), the law on the coins determines an element of $\mathcal{P}\mathcal{P}X$. By averaging, the resulting overall probabilities are

$$\begin{array}{ccc} & ? & \\ {}^{3/4}\swarrow & & \searrow^{1/4} \\ \text{heads} & & \text{tails.} \end{array}$$

In other words, the "average" or "composition" can be considered an assignment $E : \mathcal{P}\mathcal{P}X \to \mathcal{P}X$ from laws of "random random variables" to the laws of ordinary random variables.

To show that the map E is natural, let $f : X \to Y$ be a function. Since the map $\mathcal{P}f : \mathcal{P}X \to \mathcal{P}Y$ is given by the pushforward of measures, which is usually written as f_*, denoted for brevity as $\mathcal{P}f$ by f_*. Then, the following diagram commutes:

$$\begin{array}{ccc} \mathcal{P}\mathcal{P}X & \xrightarrow{(f_*)_*} & \mathcal{P}\mathcal{P}Y \\ \downarrow{\scriptstyle E} & & \downarrow{\scriptstyle E} \\ \mathcal{P}X & \xrightarrow{f_*} & \mathcal{P}Y. \end{array}$$

Indeed, for each $\pi \in \mathcal{P}\mathcal{P}X$ and $y \in Y$,

$$\begin{aligned} E((f_*)_*(\pi))(y) &= \sum_{q \in \mathcal{P}Y} (f_*)_*(\pi)(q)\, q(y) \\ &= \sum_{q \in \mathcal{P}Y} \sum_{p \in (f_*)^{-1}(q)} \pi(p)\, q(y) \\ &= \sum_{p \in \mathcal{P}X} \pi(p)\, f_*(p)(y) \\ &= \sum_{x \in f^{-1}(y)} \sum_{p \in \mathcal{P}X} \pi(p)\, p(x) \\ &= \sum_{x \in f^{-1}(y)} E\pi(x) \\ &= f_*(E\pi)(y). \end{aligned}$$

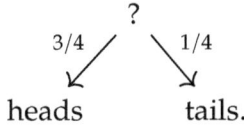

The following exercise shows that δ and E satisfy the monad axioms. Therefore, $(\mathcal{P}, \delta, E)$ is a monad on **Set**. It is known in the literature under many names, such as **probability monad, distribution monad,** or **convex combination monad.**

Exercise 5.1.3 (basic probability). Show that $(\mathcal{P}, \delta, E)$ is a monad, that is, the following diagrams commute:

$$
\begin{array}{ccc}
\mathcal{P}X & \xrightarrow{\delta_{\mathcal{P}X}} & \mathcal{P}^2X \\
& \searrow^{\mathrm{id}} & \downarrow{\scriptstyle E_X} \\
& & \mathcal{P}X
\end{array}
\qquad
\begin{array}{ccc}
\mathcal{P}X & \xrightarrow{\mathcal{P}\delta_X} & \mathcal{P}^2X \\
& \searrow^{\mathrm{id}} & \downarrow{\scriptstyle E_X} \\
& & \mathcal{P}X
\end{array}
\qquad
\begin{array}{ccc}
\mathcal{P}^3X & \xrightarrow{\mathcal{P}E_X} & \mathcal{P}^2X \\
\downarrow{\scriptstyle E_{\mathcal{P}X}} & & \downarrow{\scriptstyle E_X} \\
\mathcal{P}^2X & \xrightarrow{E_X} & \mathcal{P}X.
\end{array}
$$

The following exercises give the analogous construction for more general probability measures (using measure theory).

Exercise 5.1.4 (measure theory, probability). Consider the Giry functor defined on **Meas** in Exercise 1.3.17. Define a monad structure analogous to the one in the exercise above, with unit $\delta : X \to \mathcal{P}X$ given by the Dirac measures (see Exercise 1.4.11) and multiplication $E : \mathcal{P}\mathcal{P}X \to \mathcal{P}X$ given by integration as follows:

$$
E(\pi)(A) := \int_{\mathcal{P}X} p(A)\, d\pi(p),
$$

for each measurable $A \subseteq X$. Note that you have to prove that δ and E are measurable and natural and that they satisfy diagrams (5.0.1). (*Hint*: To prove that E is measurable, first prove that for every measurable function $f : X \to [0,1]$, the "integration" map $i_f : \mathcal{P}X \to [0,1]$ given by

$$
p \longmapsto \int f\, dp
$$

is measurable as a function of p, for the σ-algebra defined on $\mathcal{P}X$ in Exercise 1.3.17.)

The resulting monad is known as the **Giry monad.**

Exercise 5.1.5 (measure theory, probability). Consider the Radon functor on **CHaus** defined in Exercise 1.3.18. Define a monad structure in a way analogous to the exercises above.

The resulting monad is known as the **Radon monad**.

Monads are widely used in computer science. Here is a basic example, given by the list construction.

Exercise 5.1.6 (basic computer science, combinatorics). Consider the list functor of Example 1.3.19. Equip it with a monad structure, with the unit given by the map in Exercise 1.4.12 and the multiplication given by flattening a double list (or, equivalently, concatenating lists).

The "extension" interpretation is that the set of lists LX extends the old set X, which we can view as having as elements only lists of length one.

The following monad, in different forms, appears almost everywhere in mathematics. It is used in applied mathematics and computer science to model processes that have "extra costs," or produce "additional output," and in pure mathematics to describe fiber bundles and group actions.

Example 5.1.7 (several fields). Let M be a monoid or a group, and write it additively (denote its neutral element by 0 and its binary operation by $+$).[1] Given any set X, forming the set $X \times M$ is a functorial operation, where a function $f : X \to Y$ is mapped to $f \times \mathrm{id}_M : X \times M \to Y \times M$. That is, the pair $(x, m) \in X \times M$ is mapped to $(f(x), m) \in Y \times M$ (see Exercise 3.2.6).

[1] This does not necessarily imply that the monoid is commutative. In mathematics, it is customary to denote commutative operations by $+$, but not, for example, in computer science, where the concatenation of strings is also often denoted by $+$.

This functor admits a canonical monad structure, inherited by the monoid structure of M. The unit is given by the map

$$\begin{aligned} X &\xrightarrow{\eta} X \times M \\ x &\longmapsto (x, 0), \end{aligned}$$

so it is defined using the neutral element of M. The multiplication is given by the map

$$\begin{aligned} (X \times M) \times M &\xrightarrow{\mu} X \times M \\ ((x, m), n) &\longmapsto (x, m + n), \end{aligned}$$

so it is defined using the multiplication (i.e. addition) of M. (Why are these maps natural?)

We can interpret the elements of M as being "costs" or "side effects" of some kind. If we have two of them, we can multiply them (or sum them) using the multiplication (addition) map of the monoid. The neutral element 0 corresponds to zero cost. The "extension interpretation" is as follows: The elements (x, m) can be seen as more general than x if we see the latter as $(x, 0)$ (via the unit map). We extend X to account for elements which have a "cost" or "impact" or "side effect."

Note that, since M is a monoid (i.e. associative and unital), the monad axioms (5.0.1) hold automatically, as they correspond to associativity and unitality of M (this example is one of the reasons why they are called that way). The left unitality diagram says that for each $(x, m) \in X \times M$, $\mu \circ \eta(x, m) = (x, m)$, which explicitly states that $(x, m + 0) = (x, m)$. This is true since $m + 0 = m$ (as M is a monoid, it is unital). Analogously, right unitality boils down to saying $0 + m = m$, and associativity corresponds exactly to the associativity of M. Therefore, we have a monad on **Set**. Let's denote this monad by (T_M, η, μ).

This monad is known in computer science as the **writer monad**, for reasons that will be explained in Example 5.1.14. In pure mathematics, it is either known as the **trivial bundle monad** (since $X \times M$ is the trivial M-bundle over X) or as the **M-action monad**, for reasons that will be explained in Example 5.2.10.

5.1.1 Kleisli morphisms

Given a monad T, we can not only talk about generalized elements (of an extended space) but also of generalized *functions*. Given spaces X and Y, we can form functions which intuitively have as output the generalized elements of Y, i.e. functions $k : X \to TY$.

Definition 5.1.8. Let (T, η, μ) be a monad on a category $\mathbf{C}$. A **Kleisli morphism** of T from X to Y is a morphism $k : X \to TY$ of $\mathbf{C}$.

In mathematics, it often happens that one would like to obtain a function from X to Y from some construction (for example, a limit), but the result is not always well defined or unique, or not always existing. Maybe the output of a process is subject to fluctuations, so that a probabilistic output models the situation better. Or, there could be extra side effects ... and so on. Allowing more general results or outputs provides sometimes a better description of the system, i.e. replacing Y with the extension TY. Generalized functions include ordinary functions in the same way as generalized elements include ordinary elements: via the map η. A function $f : X \to Y$ canonically defines a map $X \to TY$ given by $\eta \circ f$. Note that this is *different* from extending an existing $f : X \to Y$ to TX. We are not extending an existing function to generalized elements, we are allowing more general functions on X which *take values* in elements of TY that may not come from Y. In particular, a generalized element can be seen as a constant generalized function.

Example 5.1.9 (sets and relations). In the case of the power set, Kleisli morphisms, or generalized maps, are precisely *relations*: Given sets X and Y, a map $k : X \to PY$ assigns to each element of X a *subset* of Y (possibly empty), i.e. a multi-valued (possibly no-valued) function.

Example 5.1.10 (probability). A Kleisli morphism for the distribution monad $\mathcal{P}$ of Example 5.1.2 is a function $X \to \mathcal{P}Y$, which we can view as a **stochastic map** $X \to Y$.

Analogously, a Kleisli morphism for the Giry monad of Exercise 5.1.4 is a *Markov kernel*. It consists precisely of a measurable assignment $X \to \mathcal{P}Y$.

Kleisli morphisms, according to the definition above, are just ordinary morphisms of a particular form. The reason why they are important enough to deserve their own name is because of how they *compose*. As we will see in the examples, this gives a very fruitful construction.

Definition 5.1.11. Let (T, η, μ) be a monad on a category **C**. Let $k : X \to TY$ and $h : Y \to TZ$. We define the **Kleisli composition** of k and h to be the morphism $(h \circ_{kl} k) : X \to TZ$ given by

$$X \xrightarrow{k} TY \xrightarrow{Th} TTZ \xrightarrow{\mu} TZ. \tag{5.1.1}$$

In other words, the Kleisli composition permits the composition of generalized functions from X to Y with generalized functions from Y to Z to give generalized functions from X to Z.

Example 5.1.12 (sets and relations). Let's see what happens for the power set monad. Relations can be composed as follows. Given relations $k : X \to PY$ and $h : Y \to PZ$, as in the picture

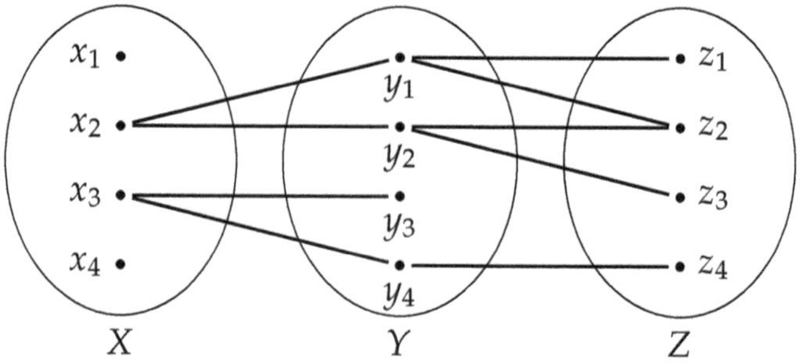

we can compose the two and forget about Y, obtaining a relation $X \to PZ$:

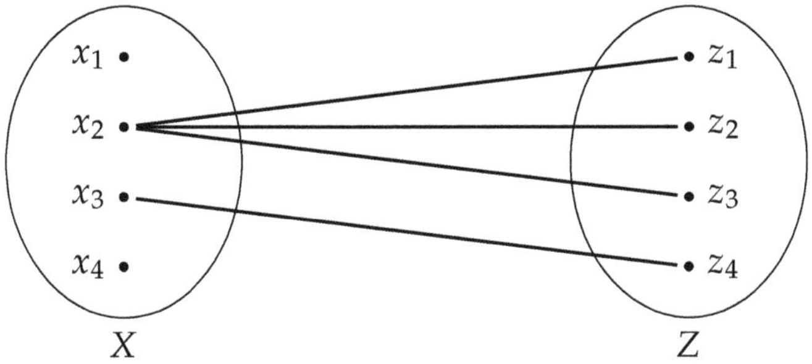

The idea is that we can go from $x \in X$ to $z \in Z$ if and only if there is a $y \in Y$ such that we can go from x to y and from y to z.

Formally, though, what occurred is that we have first applied $k : X \to PY$, which assigns to each $x \in X$ a subset of Y:

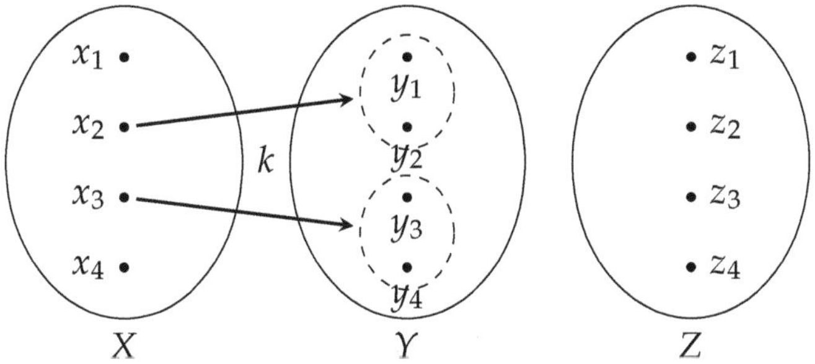

Then, we have applied h to *elementwise to each subset in the image of* k:

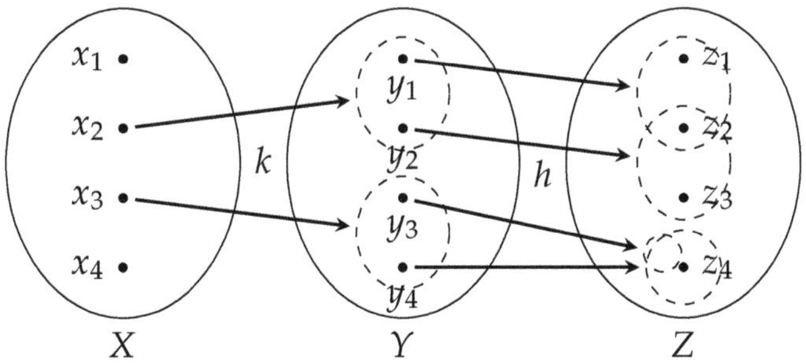

In other words, we have taken the *image* of the subsets of Y $h : Y \to PZ$, which we know is given by the map $Ph : PY \to PPZ$. Technically, to each subset of Y, we have a *subset of subsets* of Z, which contains the images of h:

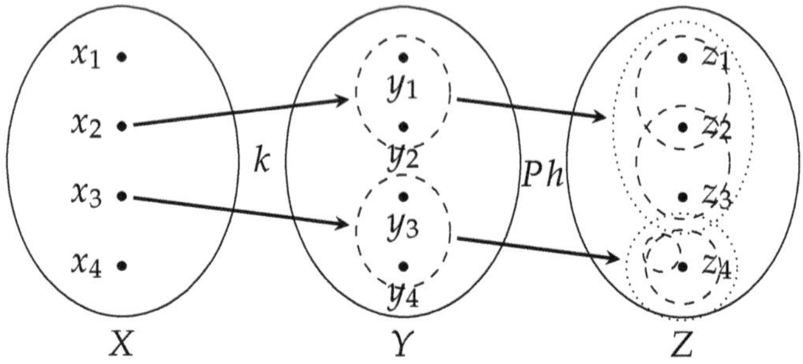

Now, for each subset of Y, we take the *union* of the subsets in its image:

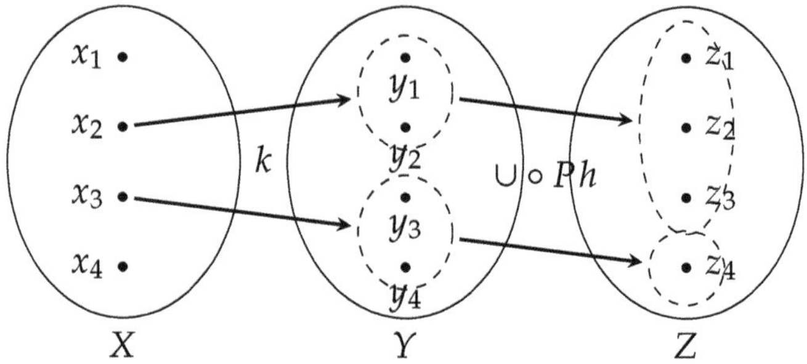

thereby obtaining the composite relation $X \to PZ$:

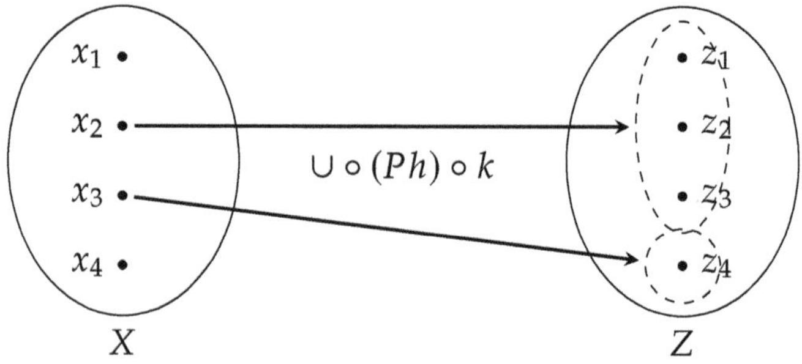

Example 5.1.13 (probability). The Kleisli composition for the distribution monad works as follows. Given sets X, Y, and Z and maps $k : X \to \mathcal{P}Y$ and $h : Y \to \mathcal{P}Z$, the composition (5.1.1) gives us

$$h \circ_{kl} k \;=\; E \circ \mathcal{P}h \circ k,$$

which for each $x \in X$ maps $z \in Z$ to

$$\begin{aligned}(h \circ_{kl} k)(x)(z) \;&=\; \sum_{q \in \mathcal{P}Z} q(z)\, h_* k(x)(q) = \sum_{q \in \mathcal{P}Z} \sum_{y \in h^{-1}(q)} q(z)\, k(x)(y) \\ &=\; \sum_{y \in Y} h(y)(z)\, k(x)(y).\end{aligned}$$

In other words, interpreting the stochastic maps as conditionals (i.e. writing $k(x)(y)$ as $p(y|x)$), we are composing conditionals by summing over all the intermediate states:

$$p(z|x) \;=\; \sum_{y \in Y} p(z|y)\, p(y|x). \tag{5.1.2}$$

This is the famous **Chapman–Kolmogorov formula.**

Let's see now what happens more generally with the Giry monad. Given the Markov kernels $k : X \to \mathcal{P}Y$ and $h : Y \to \mathcal{P}Z$, the composition formula gives us for each $x \in X$ and measurable set $A \subseteq Z$,

$$(h \circ_{kl} k)(x)(A) = \int_{\mathcal{P}Z} q(A)\, h_* k(x)(dq) = \int_Y h(y)(A)\, k(x)(dy).$$

Again, in conditional notation,

$$p(A|x) \;=\; \int_Y p(A|y)\, p(dy|x),$$

which is analogous to Equation (5.1.2) if one replaces sums by integrals.

The Kleisli composition for probability monads is exactly the usual composition of stochastic maps and Markov kernels; compare Example 1.3.29.

As the previous two examples show, Kleisli morphisms capture known structures in mathematics (in this case, relations and stochastic maps) not just in how they look but also in how they behave.

Let's now look at the Kleisli morphisms of the *writer monad* of Example 5.1.7 and see why it is called that way.

Example 5.1.14 (several fields). Let M be a monoid, which we again write additively. A Kleisli morphism of the "writer monad" T_M is a map $k : X \to Y \times M$. We can interpret it as a process which, when given an input $x \in X$, does not just produce an output $y \in Y$, but also an element of M. For example, it could be energy released by a chemical reaction, waste generated, or transaction cost. In computer science, this is the behavior of a function that computes a certain value but that also writes into a log file (or to the standard output) that something has happened (the monoid operation being the concatenation of strings). For example, when you compile a LaTeX document, a log file is produced alongside your output file, hence the name "writer monad."

Let's now look at the Kleisli composition. If we have processes $k : X \to Y \times M$ and $h : Y \to Z \times M$, then $h \circ_{kl} k : X \to Z \times M$ is given by

$$\begin{array}{ccccccc} X & \xrightarrow{k} & Y \times M & \xrightarrow{h \times \mathrm{id}_M} & Z \times M \times M & \xrightarrow{\mathrm{id}_Z \times +} & Z \times M \\ x & \longmapsto & (y, m) & \longmapsto & (z, n, m) & \longmapsto & (z, n + m). \end{array}$$

What it does is as follows:

(a) It executes the process k with an input $x \in X$, giving as output an element of $y \in Y$ as well as a cost (or extra output) $m \in M$.

(b) It executes the process h, taking as input the $y \in Y$ produced by k and giving an element $z \in Z$ as well as an extra cost $n \in M$ (all of this while keeping track of the first cost m).

(c) The two costs m and n are summed (or the extra outputs are concatenated).

So, for example, the cost of executing two processes one after another is the sum of the costs. The same is true for the release of energy in a chemical reaction and for waste generated. Similarly, executing two programs one after another will produce a concatenation of text in a log file (or two log files).

When you compile a LaTeX file, after compilation your compiler tells you how many errors (as well as warnings and bad boxes) happened during the process. This can be modeled in terms of Kleisli morphisms. Each step in the compilation consists of a Kleisli morphism taking an input (for example a keyword) and giving an output (for example a mathematical symbol) together with a number, the number of errors encountered. The compilation consists of several steps that pass information to one another, and the final number of errors will be the sum of the errors encountered during all the steps (hopefully zero).

Again, the Kleisli composition is the "right" notion of composition for these functions with generalized output.

The example of the writer monad (Example 5.1.7) motivated the usage of the terms "unit," "multiplication," "associativity," and "unitality" for monads. Here is another, almost independent, reason for the last two.

Proposition 5.1.15. *Let (T, η, μ) be a monad on a category* **C**. *The Kleisli morphisms of T form themselves a category, where*

- *the objects are the objects of* **C**;
- *the morphisms are the Kleisli morphisms of T;*
- *the identities are given by the units $\eta : X \to TX$, for each object X;*
- *the composition is given by the Kleisli composition.*

In other words, the Kleisli morphisms form themselves a category, which we can think of as "having as morphisms the generalized maps."

Definition 5.1.16. The category defined above is called the **Kleisli category** of T, and it is denoted by $\mathbf{C}_T$.

Proof. In order for $\mathbf{C}_T$ to be a category, we need the identities (i.e. the unit maps) to behave indeed like identities and the composition (i.e. the Kleisli composition) to be associative:

- The *right* unitality condition for T, for each $k : X \to TY$, gives a commutative diagram:

$$X \xrightarrow{\ k\ } TY \xrightarrow{\ T\eta\ } TTY$$
$$\text{id} \searrow \quad \downarrow \mu$$
$$TY,$$

  which means that $\eta \circ_{kl} k = k$. That is, η behaves like an identity on the *left* side for the Kleisli composition.

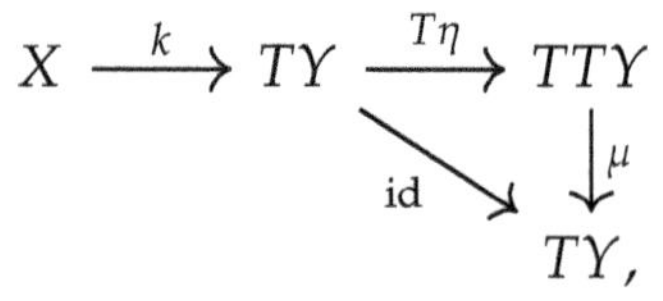

- The *left* unitality condition for T, together with the naturality of η, for each $k : X \to TY$, gives a commutative diagram:

$$TX \xrightarrow{\ Tk\ } TTY$$
$$\eta \nearrow \quad \eta \nearrow \quad \downarrow \mu$$
$$X \xrightarrow{\ k\ } TY \xrightarrow{\ \text{id}\ } TY,$$

  which means that $k \circ_{kl} \eta = k$. That is, η behaves like an identity on the *right* side for the Kleisli composition. So, the maps η are indeed the identities of $\mathbf{C}_T$.

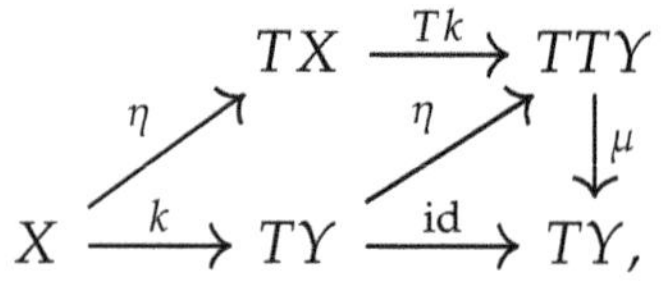

- The associativity square, together with the naturality of μ, gives for each $\ell : W \to TX$, $k : X \to TY$, and $h : Y \to TZ$ a commutative diagram:

$$W \xrightarrow{\ \ell\ } TX \xrightarrow{\ Tk\ } TTY \xrightarrow{\ TTh\ } TTTZ \xrightarrow{\ T\mu\ } TTZ$$
$$\downarrow \mu \qquad\qquad \downarrow \mu \qquad\qquad \downarrow \mu$$
$$TY \xrightarrow{\ Th\ } TTZ \xrightarrow{\ \mu\ } TZ,$$

  which means that $h \circ_{kl} (k \circ_{kl} \ell) = (h \circ_{kl} k) \circ_{kl} \ell$ (why?). That is, the Kleisli composition is associative. $\square$

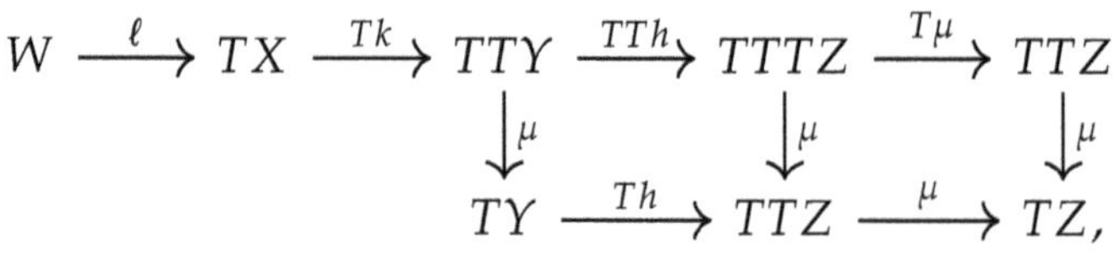

Example 5.1.17 (sets and relations). The Kleisli category of the power set monad is the category of sets and *relations*. The identity of X is given by the singleton map $\sigma : X \to PX$. As said earlier, the interpretation is that the power set "forms spaces of generalized elements in a consistent way," and that the associated "consistent generalization" of functions is relations.

Example 5.1.18 (probability). The Kleisli category of the distribution monad is the category of sets and *stochastic maps*. The identity is given by the delta map $\delta : X \to \mathcal{P}X$. Once again, the interpretation is that the probability construction "forms spaces of generalized elements in a consistent way" (which we can see as "random elements") and that the associated "consistent generalization" of functions is stochastic maps.

For the Giry monad, the Kleisli category is the category of measurable spaces and Markov kernels.

Exercise 5.1.19 (basic computer science). What is the Kleisli category of the list monad?

Exercise 5.1.20 (several fields). How does the Kleisli category of the writer monad look explicitly?

5.1.2 The Kleisli adjunction

Let (T, η, μ) be a monad on a category $\mathbf{C}$. As we said already, every ordinary morphism $f : X \to Y$ defines "trivially" a Kleisli morphism via

$$X \xrightarrow{f} Y \xrightarrow{\eta} TY.$$

For example, every function defines in particular a (very special) relation, given by "x is related to y if and only if $y = f(x)$." Analogously, every function defines a *deterministic* stochastic map, given by "$y = f(x)$ with probability one."

This assignment is actually a functor $\mathbf{C} \to \mathbf{C}_T$, which we denote by L_T; see the following exercise.

Exercise 5.1.21 (important!). Prove that L_T is indeed a functor, where on objects, $L_T(X) = X$, and on morphisms, $L_T(f) = \eta \circ f$.

It has to be proven that it preserves identities (i.e. $L_T(\mathrm{id}_X) = \eta_X$ for each object X) and composition (i.e. $L_T(f \circ g) = L_T(f) \circ_{kl} L_T(g)$).

In our power set example, this says in particular that if we consider functions as special relations, then composing them as functions or as relations gives the same result. The same is true for stochastic maps.

Conversely, suppose we have a Kleisli morphism $k : X \to TY$. In general, this does not come from a map $X \to Y$ and cannot be used to define such a map. However, we can obtain from k a map between the *extended* spaces, i.e. $TX \to TY$, as follows:

$$TX \xrightarrow{Tk} TTY \xrightarrow{\mu} TY.$$

Denote this map by $R_T(k)$. This is again a functorial assignment. That is, the assignment $X \mapsto TX$, $k \mapsto \mu \circ Tk$ is a functor $\mathbf{C}_T \to \mathbf{C}$.

Exercise 5.1.22 (important!). Prove that R_T is indeed a functor. This means that it preserves identities (i.e. $R_T(\eta_X) = \mathrm{id}_{TX}$ for each object X) and composition (i.e. $R_T(k \circ_{kl} h) = R_T(k) \circ R_T(h)$).

Example 5.1.23 (sets and relations). Let $k : X \to PY$ be a Kleisli morphism of the power set monad, i.e. a relation between X and Y. We can assign to a subset $S \subseteq X$ the subset of all the points of Y which are related to at least one element of S. This gives the set

$$\bigcup_{x \in S} k(x) \quad = \quad \cup \circ Pk(S).$$

If we do this for each $S \in PX$, we have then the map $PX \to PY$.

Example 5.1.24 (probability). Let $k : X \to \mathcal{P}Y$ be a Markov kernel (or a stochastic map, for readers who are unfamiliar with measure theory). Given $p \in \mathcal{P}X$ we can form the measure on Y as

$$A \longmapsto \int_X k(x)(A)\, dp(x).$$

Doing this for each $p \in \mathcal{P}X$ gives the desired assignment $\mathcal{P}X \to \mathcal{P}Y$. This is sometimes called the **pushforward of measures along a Markov kernel**; however, for our formalism, this name is slightly inaccurate: It corresponds precisely to the pushforward along the map $k : X \to \mathcal{P}Y$, *followed by integration*. In symbols,

$$\mathcal{P}X \xrightarrow{\mathcal{P}k} \mathcal{P}\mathcal{P}Y \xrightarrow{\mu} \mathcal{P}Y.$$

Proposition 5.1.25. *Let (T, η, μ) be a monad on* $\mathbf{C}$*:*

(a) *The composite functor $R_T \circ L_T : \mathbf{C} \to \mathbf{C}$ is naturally isomorphic to T.*

(b) *The functor L_T is left-adjoint to R_T.*

(c) *The unit of the adjunction is given by the unit of the monad η.*

We call the adjunction above the **Kleisli adjunction**. This is one of the many ways in which monads and adjunctions are related; more will come in the following sections.

Proof. First of all, let X and Y be objects of $\mathbf{C}$, but consider Y as an object of $\mathbf{C}_T$. Note that $L_T(X) = X$ and $R_T(Y) = TY$ from the way we constructed the functors:

(a) Set now $Y = L_T(X)$. We get $R_T(L_T(X)) = R_T(X) = TX$. Let now $f : X \to X'$ be a morphism of $\mathbf{C}$. We have that

$$R_T(L_T(f)) \;=\; R_T(\eta \circ f) \;=\; \mu \circ T(\eta \circ f) \;=\; \mu \circ T\eta \circ Tf.$$

By the right unitality triangle of (5.0.1), $\mu \circ T\eta$ is the identity, and so we are left with Tf. Therefore, $R_T \circ L_T = T$.

(b) Let's now turn to the adjunction. There is a natural bijection between functions $X \to TY$ and the Kleisli morphisms from X to Y, given by

$$\begin{aligned} \mathrm{Hom}_{\mathbf{C}}(X, TY) &\xrightarrow{\cong} \mathrm{Hom}_{\mathbf{C}_T}(X, Y) \\ k &\longmapsto k. \end{aligned}$$

The fact that this is a bijection follows directly from the definition of the Kleisli morphism: Kleisli morphisms from X to Y are *by definition* the morphisms $X \to TY$ of $\mathbf{C}$.
Naturality in X states that for $g : W \to X$, the following diagram commutes:

$$\begin{array}{ccc} \mathrm{Hom}_{\mathbf{C}}(X,TY) & \xrightarrow{\cong} & \mathrm{Hom}_{\mathbf{C}_T}(X,Y) \\ \big\downarrow{\scriptstyle -\circ g} & & \big\downarrow{\scriptstyle -\circ_{kl} L_T(g)} \\ \mathrm{Hom}_{\mathbf{C}}(W,TY) & \xrightarrow{\cong} & \mathrm{Hom}_{\mathbf{C}_T}(W,Y). \end{array}$$

In other words, we need to show that for all $k : X \to TY$, $k \circ g = k \circ_{kl} L_T(g)$, or, more explicitly, that $k \circ g = \mu \circ Tk \circ \eta \circ g$. This can be shown by forming the following diagram:

$$\begin{array}{ccccccc} & & TX & \xrightarrow{Tk} & TTY & & \\ & & \big\uparrow{\scriptstyle \eta} & & \big\uparrow{\scriptstyle \eta} & \overset{\mu}{\searrow} & \\ W & \xrightarrow{g} & X & \xrightarrow{k} & TY & \xrightarrow[\mathrm{id}]{} & TY, \end{array}$$

which commutes by the naturality of η and by the left unitality diagram of (5.0.1).
Naturality in Y states that for every $h : Y \to TZ$, the following diagram commutes:

$$\begin{array}{ccc} \mathrm{Hom}_{\mathbf{C}}(X,TY) & \xrightarrow{\cong} & \mathrm{Hom}_{\mathbf{C}_T}(X,Y) \\ \big\downarrow{\scriptstyle R_T(h)\circ -} & & \big\downarrow{\scriptstyle h\circ_{kl} -} \\ \mathrm{Hom}_{\mathbf{C}}(X,TZ) & \xrightarrow{\cong} & \mathrm{Hom}_{\mathbf{C}_T}(X,Z). \end{array}$$

In other words, we need to show that for all $k : X \to TY$, $R_T(h) \circ k = h \circ_{kl} k$. But both sides of the equation translate explicitly to $\mu \circ Th \circ h$, and so they coincide.

(c) To see what the unit of the adjunction is, set $X = Y$, and consider the Kleisli identity $Y \to Y$. This corresponds, under the bijection above, with the map $\eta : Y \to TY$. Therefore, η is the unit of the adjunction. $\square$

Exercise 5.1.26. Write down explicitly the universal properties associated to this adjunction.

5.1.3 Closure operators and idempotent monads

The interpretation of monads as extensions is helpful also for the case of monads on posets. These are simpler constructions which go under the name of *closure operators*. Let's see this in detail. Let $(X, \leq)$ be a poset (or a preorder). Plugging in the definition, a monad on X amounts to a monotone map $t : X \to X$ (the "endofunctor"), and with pointwise inequalities ("natural transformations") $\mathrm{id}_X \leq t$ and $t^2 \leq t$, which means that for every $x \in X$, $x \leq t(x)$ and $t(t(x)) \leq t(x)$. Since t is monotone, the first inequality implies that $t(x) \leq t(t(x))$ too; therefore, equivalently, we have the following.

Definition 5.1.27. A **closure operator** on a poset $(X, \leq)$ is a map $t : X \to X$ satisfying the following properties:

(a) **Monotonicity**: for each $x \leq y \in X$, $t(x) \leq t(y)$;

(b) **Extensivity**: for each $x \in X$, $x \leq t(x)$ (i.e. "applying the map lets us step up");

(c) **Idempotency**: for each $x \in X$, $t(t(x)) = t(x)$ (i.e. once we apply the map, applying it again has no effect).

Warning. In topology, one calls "closure operator" a function satisfying, in addition to the properties above, the property of preserving finite joins. This is a distinct notion. In order to avoid confusion, the operators in the form we have given above are also called "Moore closure operators," while those in topology are called "Kuratowski closure operators."

The following exercises show why closure operators can be interpreted as "extensions" or "completions" of some kind. Before reading the following exercises, make sure you understand the examples given in Section 4.4.3.

Exercise 5.1.28 (sets and relations, analysis). Let S be a subset of $\mathbb{R}^2$, and denote by $t(S)$ its convex hull, see Example 4.1.11. (In the notation of Example 4.1.11, $t = i \circ c$.) Prove that t is a closure operator on $P(\mathbb{R}^2)$.

Exercise 5.1.29 (topology). Prove that the topological closure (see Example 4.4.7) is indeed a closure operator on the subsets of a topological space.

Exercise 5.1.30 (linear algebra). Let V be a vector space. Recall that the *span* of a subset $S \subseteq V$ is the smallest vector subspace of V containing S, or, equivalently, the set of all vectors of V which can be expressed as linear combinations of the elements of S.

Show that the span gives a closure operator on the subsets of V.

You may have noted a pattern in these exercises. It is as follows.

Exercise 5.1.31 (important!). Let X and Y be posets, and let $f : X \to Y$ and $g : Y \to X$ form a Galois connection $f \dashv g$. Prove that $g \circ f$ is a closure operator.

This statement will be generalized in Section 5.5.

Monads can be "idempotent" even when they are defined on a category which is not a poset — in that case, one has to allow for "idempotency up to isomorphism."

Definition 5.1.32. A monad (T, η, μ) on a category $\mathbf{C}$ is called **idempotent** if the multiplication $\mu : TT \Rightarrow T$ is a natural isomorphism.

Idempotent monads model, for example, situations in which, after "extending" our spaces, we have "completed" our space in a way that cannot be further extended, "everything is there already." Here is a standard example.

Exercise 5.1.33 (analysis). Let X be a metric space. The **Cauchy completion** of X is a space CX which we can think of as "containing all the limit points of X too" and is constructed as follows. Let SX be the set of Cauchy sequences in X. Given sequences $\{x_n\}$ and $\{y_n\}$, define

$$d\big(\{x_n\},\{y_n\}\big) \;:=\; \lim_{n\to\infty} d(x_n, y_n),$$

where the distance on the right is the one of X. Let now CX be the quotient of SX given by identifying all the elements of SX which have a distance of zero (i.e. sequences which "would have the same limit," even if the limit does not exist in X). The space CX is complete (why?). Moreover, there is a canonical dense isometric embedding $\eta : X \to CX$ that maps the point x to the constant sequence at x, which obviously has x as limit. Therefore, CX extends X by allowing for "limits of points of X," which may not have been in X before. If X is complete to begin with, then CX is isometric to X via the map η: Indeed, an inverse to η is given by the map

$$\{x_n\} \longmapsto \lim_{n\to\infty} x_n, \tag{5.1.3}$$

which always exists in X if (and only if) X is complete. (Why is this an inverse?)

The Cauchy completion is functorial in the following sense. Let X and Y be metric spaces and $f : X \to Y$ be Lipschitz (or even just uniformly continuous). We can extend f "by continuity" to a map $Cf : CX \to CY$ by setting

$$Cf(\{x_n\}) \;:=\; \{f(x_n)\}.$$

This is well defined, as it does not depend on the choice of the representative sequence (why?), and so C is a functor on the category **Lip** of metric spaces and Lipschitz maps. The embedding $\eta : X \to CX$ is natural (why?).

We can make C an idempotent monad as follows. First of all, since CX is complete, we have that $\eta : CX \to CCX$ is an isometry.

Then, set $\mu : CCX \to CX$ to be the inverse of η, i.e. the "lim" map of Equation (5.1.3).

Show that (C, η, μ) is a monad on **Lip**. Note that by construction, the monad is idempotent.

We conclude this section by showing that *not all monads can be thought of as extensions*. The interpretation as extensions is often helpful, but not always accurate, and the reason is that the unit $\eta : X \to TX$ of the monad is not always a monomorphism. This is particularly true for some idempotent monads. Sometimes, for example, the map η is even a quotient so that instead of "extending" X, it is "compressing" it by identifying different elements. The basic example is the following.

Exercise 5.1.34 (sets and relations). Let **Eq** be the category whose objects are sets equipped with an equivalence relation $(X, \sim)$, and morphisms are functions respecting the equivalence, i.e. $x \sim x'$ implies $f(x) \sim f(x')$ (recall Example 1.3.3). Assign now to each $(X, \sim)$ the quotient set $X/\sim$ obtained by identifying all elements of X which are related by $\sim$ (recall Exercise 1.3.4). Take $X/\sim$ equipped with the identity relation. Denote by q the quotient map $(X, \sim) \to X/\sim$. Show that given a map $f : (X, \sim) \to (Y, \sim)$ respecting the equivalence, there exists a unique map $\tilde{f} : X/\sim \to Y/\sim$ that makes the following diagram commute:

$$\begin{array}{ccc} (X,\sim) & \xrightarrow{f} & (Y,\sim) \\ \downarrow{\scriptstyle q} & & \downarrow{\scriptstyle q} \\ X/\sim & \overset{\tilde{f}}{\dashrightarrow} & Y/\sim . \end{array}$$

Why is this map well defined? (Reading Section 3.2.4 again may help you.) The assignment $(X, \sim) \mapsto X/\sim$, $f \mapsto \tilde{f}$ is therefore an endofunctor on **Eq**.

Show that this functor induces an idempotent monad on **Eq**, with unit given by the quotient maps q.

Here are similar examples from different fields.

Exercise 5.1.35 (topology). Show that the Kolmogorov quotient of topological spaces gives an idempotent monad on **Top**.

Exercise 5.1.36 (group theory). Show that the abelianization of a group gives an idempotent monad on **Grp**.

In the most general case, the unit of a monad is neither monic nor epi. For example, the unit of the Giry monad is in general neither (to see why it is not always monic, take as X a codiscrete topological space with its Borel σ-algebra).

5.2 Monads as Theories of Operations

Here is another way to look at monads, which is very useful in algebra.

Idea. *A monad can look like a consistent choice of formal expressions of a specific kind, together with ways to evaluate them.*

Before looking at the examples, recall the discussion about formal expressions as opposed to their result given in Example 4.2.13. A formal expression is something that "looks like an operation" but may have no result defined in any sense. For example, if $x, y \in X$, we can write the expression $x + y$, even if there is no sum defined on X. The expression $x + y$ does not *actually* mean that we are summing them, it is only written there formally. Whenever we work with variables instead of numbers, we do this all the time: We know that $2 + 1 = 3$, but what's $a + b$? The expression $a + b$ cannot be further evaluated unless we know the values of a and b (or, in computer science, unless each variable has been assigned a value and we have access to it). Even if $a + b$ has no result, we can still use it in mathematical manipulations because there are some facts which will be true regardless of the values of a and b, for example, that $a + b = b + a$.[2]

Example 5.2.1 (algebra). Let X be a set. We can form the set FX of *formal sums of elements of* X. That is, the formal expressions

$$x_1 + \cdots + x_n$$

of finite length (but arbitrarily large), including the empty expression, satisfying the usual commutativity law for addition.

We can make F functorial as follows. Let $f : X \to Y$ be a function. We define the function $Ff : FX \to FY$ to "apply f elementwise in the expression." That is,

$$Ff(x_1 + \cdots + x_n) \;:=\; f(x_1) + \cdots + f(x_n),$$

and Ff applied to the empty expression on X gives the empty expression on Y. This way, F is a functor **Set** $\to$ **Set**.

Let's now equip F with a monad structure. Given $x \in X$, we can form the trivial formal expression

$$x$$

in which x is the only addendum, and there is "nothing else to add." Doing this for every $x \in X$, we get then an assignment $\eta : X \to FX$. This is natural in X. The naturality diagram

$$\begin{array}{ccc} X & \xrightarrow{f} & Y \\ \downarrow{\scriptstyle \eta} & & \downarrow{\scriptstyle \eta} \\ FX & \xrightarrow{Ff} & FY \end{array}$$

commutes, as for each $x \in X$, both paths in the diagram give the trivial formal expression $f(x) \in FY$.

---

[2] By "addition," here we mean an operation satisfying the axioms of a commutative monoid, such as the addition of natural numbers. We will always denote such an operation, formal or not, by "+."

Consider now the space FFX, whose elements are *formal expressions of formal expressions*. A way to represent its elements is to add brackets, as in

$$(x_1 + x_2) + (x_1 + x_3),$$

and so on. The terms in the sum above are themselves formal sums. Given such a nested formal expression, we can always remove the brackets, obtaining a plain formal expression

$$x_1 + x_2 + x_1 + x_3.$$

This gives an assignment $\mu : FFX \to FX$, which will be the multiplication of the monad. (Why is this natural?)

The maps η and μ satisfy the monad axioms (5.0.1). Let's see what the axioms mean in this case.

- The left unitality diagram says that if we take a formal expression, such as

  $$x_1 + x_2 + x_3,$$

  and we embed it into FFX via η, forming the formal expression

  $$(x_1 + x_2 + x_3)$$

  (which is a nested formal expression with only one addendum), then removing the brackets via the map μ gives the original expression

  $$x_1 + x_2 + x_3.$$

- The right unitality diagram says that if we again take a formal expression, as above, and now embed it into FFX via the map $F\eta$, that is, embedding each term as its own trivial formal expression,

  $$(x_1) + (x_2) + (x_3)$$

  (this is a nested formal expression whose addenda are themselves formal expressions of one addendum), then removing the brackets via the map μ gives again the original expression.

- The associativity square states the following. Take a formal expression of formal expressions of formal expressions (three times, so two levels of nesting), such as

$$((x_1 + x_2) + (x_3)) + ((x_4)).$$

We can then either remove the (outer) brackets via the map μ,

$$(x_1 + x_2) + (x_3) + (x_4),$$

or instead apply the map $F\mu$, which removes the brackets in each term, i.e. the inner brackets,

$$(x_1 + x_2 + x_3) + (x_4).$$

These two expressions differ, but applying μ one more time to either of them removes the remaining brackets and makes the two expressions coincide:

$$x_1 + x_2 + x_3 + x_4.$$

Therefore, (F, η, μ) is a monad. It is called the **(free) commutative monoid monad**. "Commutative monoid" because the formal operations that it encodes, additions, are exactly those giving the algebraic structure of a commutative monoid. Additional motivation are given in Example 5.2.6. The word "free" is explained in Section 5.2.2.

Exercise 5.2.2 (several fields). Consider the list monad of Exercise 5.1.6. Given a set X, we can view the elements of LX, instead of as "lists on X," as "formal products of elements of X," writing

$$[x_1, \ldots, x_n]$$

as

$$x_1 \ldots x_n.$$

The unit and multiplication have a similar interpretation, as in Example 5.2.1, in terms of "one-term expression" and "removing brackets." Note that this time, differently from Example 5.2.1, the order of the terms matters. This encodes the structure of a monoid, but not necessarily commutative. Because of this, the list monad is also known under the name of **free monoid monad.**

Exercise 5.2.3 (analysis, probability). Just as the list monad can be reinterpreted in terms of formal expressions, the same can be done for the probability monad of Example 5.1.2. Given a set X, we can view an element $p \in \mathcal{P}X$ as a *formal average* or *formal convex combination* of elements of x. Here's an example.

Consider a coin flip, where "heads" and "tails" both have a probability of $1/2$. Then, *in some sense,* this is a formal convex combination of "heads" and "tails." The word *"formal"* here is the key: The set {"heads", "tails"} is not a convex set, so one can't really take actual averages of its elements. There is no "intermediate state between heads and tails." How would the coin land? It is a *formal* convex combination without a result. Every probability measure on X can be interpreted in a similar way. The unit and multiplication, once again, can be interpreted in terms of "one-term convex combination" and "removing the brackets" (how exactly?).

Exercise 5.2.4 (algebra, group theory). Given a set X, consider the set GX of words whose letter are either elements of X or formal inverses thereof, such as

$$x_1 \, x_2{}^{-1} \, x_3,$$

reduced according to the group inverse law (i.e. that expressions in the form $x\,x^{-1}$ and $x^{-1}\,x$, within a word, are equal to the empty expression). Can you equip GX with a monad structure?

5.2.1 Algebras of a monad

Given a monad, we don't only have a way to form formal expressions but also a setting in which those expressions have a result.

Definition 5.2.5. Let (T, η, μ) be a monad on a category $\mathbf{C}$. An **algebra of** T, or **algebra over** T, or T**-algebra**, consists of

- an object A of $\mathbf{C}$,
- a morphism $e : TA \to A$ of $\mathbf{C}$,

such that the following diagrams commute, called **unit** and **multiplication** diagrams, respectively:

$$\begin{array}{ccc} A & \xrightarrow{\eta} & TA \\ & \searrow^{\mathrm{id}} & \downarrow e \\ & & A \end{array} \qquad \begin{array}{ccc} TTA & \xrightarrow{Te} & TA \\ \downarrow \mu & & \downarrow e \\ TA & \xrightarrow{e} & A. \end{array} \tag{5.2.1}$$

Warning. The word "algebra" has other meanings in mathematics. In order to avoid confusion, the algebras of a monad are often called "Eilenberg–Moore algebras."

The interpretation of a T-algebra in terms of formal expressions is that an algebra is a place which is *equipped* with the operations specified by the monad, i.e. those expressions have an actual result. The map $e : TA \to A$ maps a formal expression to its actual result.

Example 5.2.6 (algebra). Let's show that the algebras of the "free commutative monoid monad" given in Example 5.2.1 are exactly commutative monoids.

First of all, let A be a commutative monoid (for example, natural numbers with addition). Take as map $e : FA \to A$ the function that maps each formal sum in A to its actual result (and the empty expression to zero). For example, $2 + 2 + 1 \mapsto 5$. This satisfies the algebra axioms (5.2.1):

- The unit diagram of the algebra states that if we evaluate a one-element sum, the result is just that element. For example, given $a \in A$, the evaluation of the trivial formal sum containing only a gives as result again a.
- If we have a formal expression of formal expressions, we can either first remove the brackets and then evaluate the result, or first evaluate the content of the brackets and then evaluate the resulting expression. The composition diagram states that the result will be the same. For example, the expression in the top-left corner of the following diagram can be evaluated in these two equivalent ways:

$$\begin{array}{ccc} (2+3)+(1+2) & \overset{Fe}{\longmapsto} & 5+3 \\ \downarrow{\scriptstyle\mu} & & \downarrow{\scriptstyle e} \\ 2+3+1+2 & \overset{e}{\longmapsto} & 8. \end{array}$$

Therefore, every commutative monoid is an F-algebra.

Conversely, let A be an F-algebra with structure map $e : FA \to A$. Then, A has a canonical monoid structure given as follows:

- The neutral element is the element of A obtained by applying e to the empty expression.
- Given $a, b \in A$, their (actual) sum is defined as the result (via e) of the formal expression $a + b$, i.e. $e(a + b)$.

This operation is associative, unital, and commutative (why?), and so A is canonically a commutative monoid.

Exercise 5.2.7 (several fields). Prove that the algebras of the free monoid monad (or list monad) of Exercises 5.1.6 and 5.2.2 are exactly monoids.

Exercise 5.2.8 (algebra, group theory). If you have solved Exercise 5.2.4, what are the algebras of the resulting monad?

Example 5.2.9 (analysis, probability). The algebras of the probability monad (see Example 5.1.2 and Exercise 5.2.3) are called **convex spaces**. These are spaces A where one can (actually) take weighted averages, as specified by the elements of $\mathcal{P}A$ (the nonzero values of $p \in \mathcal{P}A$ are the weights).

For example, the unit interval $[0,1]$ is a $\mathcal{P}$-algebra with the usual convex structure: The average

$$\frac{1}{2}0 + \frac{1}{2}1$$

gives as result $1/2$. Compare with the coin example of Exercise 5.2.3, where the average of "heads" and "tails" had no result. More generally, every convex subset of a vector space is a $\mathcal{P}$-algebra (the converse is not true: there are $\mathcal{P}$-algebras which cannot be embedded into vector spaces).

In probability theory, the operation of *expectation value* is one of the most important — this is captured by the notion of an algebra of the probability monad. (For non-finitely supported measures, take for example the Giry monad.)

In the following example, we see that the writer monad is related to group and monoid actions on sets (and a similar monad can be constructed for topological or other types of spaces).

Example 5.2.10 (several fields). Let's study the algebras of the writer monad (or action monad) T_M, where M is a monoid (Example 5.1.7). In order for this interpretation to be as suggestive as possible, it is helpful to change the notation slightly. We write $M \times A$ instead of $A \times M$, we write the monoid M multiplicatively instead of additively (with the neutral element as 1), and we denote the map $e : M \times A \to A$ simply by a dot, i.e. we write $e(m, a)$ by $m \cdot a$.

Plugging in the definition, a T_M-algebra is then a set A together with a map $M \times A \to A$ such that the following diagrams commute:

$$
\begin{array}{ccc}
A & \xrightarrow{\eta} & M \times A \\
& {\scriptstyle \mathrm{id}} \searrow & \downarrow {\scriptstyle e} \\
& & A
\end{array}
\qquad
\begin{array}{ccc}
M \times M \times A & \xrightarrow{\mathrm{id}_M \times e} & M \times A \\
\downarrow {\scriptstyle \mu} & & \downarrow {\scriptstyle e} \\
M \times A & \xrightarrow{\quad e \quad} & A.
\end{array}
$$

Explicitly, the diagram states that for all $a \in A$ and $m, n \in M$,

$$1 \cdot a = a \qquad \text{and} \qquad (mn) \cdot a = m \cdot (n \cdot a).$$

In other words, a T_M-algebra is exactly a set equipped with an M-action (also called an M-set). This is why the writer monad is also called the **action monad**, or **M-action monad**.

We have already seen a way to talk about monoids and groups acting on sets (and vector spaces, and so on): namely, a monoid M acting on a set is the same as a functor $\mathbf{B}M \to \mathbf{Set}$ (see Examples 1.3.8, 1.3.10, 1.4.6, and 1.4.7). This is a different (but related) way to talk categorically about the same structure.

This also says that we can interpret the monad T_M in terms of formal expressions: An element $(m, a) \in M \times A$ is a "formal action" or "formal move": m is "ready to act on a," but the operation is not evaluated (yet). (As usual, this can also be done on sets which are not necessarily algebras, in which case formal actions are never evaluated.)

From the point of view of computer science, remember that we were interpreting M, for example as a "cost" or "side effect" of some kind. Algebras are settings in which the side effect can be reincorporated by possibly "acting" on A, i.e. it has an effect on the output data. Using again the LaTeX example, if a reference in the document is wrong or missing, the compiler not only reports an error but also adds question marks in the compiled file, in place of the reference. The error has an effect on the main output data.

The algebras of a given monad form a category whose morphisms we can think of as "preserving the specified operations." Let's see the general definition.

Definition 5.2.11. Let (A, e) and (B, e) be algebras of a monad T on $\mathbf{C}$. A **morphism of T-algebras**, or **T-morphism**, is a morphism $f : A \to B$ of $\mathbf{C}$ such that the following diagram commutes:

$$\begin{array}{ccc} TA & \xrightarrow{Tf} & TB \\ \downarrow{\scriptstyle e} & & \downarrow{\scriptstyle e} \\ A & \xrightarrow{f} & B. \end{array} \tag{5.2.2}$$

The category of T-algebras and T-morphisms is called the **Eilenberg–Moore category**, or the **category of algebras of T**, and it is denoted by $\mathbf{C}^T$.

Example 5.2.12 (algebra). Consider two algebras of the free commutative monad (Example 5.2.1), i.e. two commutative monoids, A and B. Not every function between them respects addition. The function f preserves the addition and neutral elements (a property called *additivity*) if and only if evaluating expressions before or after applying f does not change the result. That is, $f(a+b) = f(a) + f(b)$ and $f(0) = 0$. In other words, f is additive if and only if it makes the diagram (5.2.2) commute.

Therefore, the Eilenberg–Moore category of the free commutative monoid monad is the category of commutative monoids and their monoid homomorphisms.

Exercise 5.2.13 (algebra). Prove that the Eilenberg–Moore category of the list monad (or free monoid monad, Exercise 5.1.6) is the category of monoids and their morphisms.

Exercise 5.2.14 (algebra, group theory). If you have solved Exercise 5.2.4, what is the Eilenberg–Moore category of the resulting monad?

Example 5.2.15 (several fields). Consider the writer monad (or action monad) T_M, with M a monoid (Example 5.2.10). For T_M-algebras (i.e. M-sets) A and B, a function $f : A \to B$ is a morphism of algebras if and only if the following commutes:

$$\begin{array}{ccc} M \times A & \xrightarrow{\mathrm{id}_M \times f} & M \times B \\ \downarrow{\scriptstyle e} & & \downarrow{\scriptstyle e} \\ A & \xrightarrow{f} & B. \end{array}$$

Equivalently, if and only if for every $m \in M$ and $a \in A$,

$$f(m \cdot a) = m \cdot f(a).$$

In other words, the morphisms of algebras are precisely the equivariant maps, i.e. those preserving "symmetries" or "specified operations" (see Section 1.4.2).

Therefore, the Eilenberg–Moore category of T_M is equivalent to the functor category $[\mathbf{B}M, \mathbf{Set}]$ (see Example 1.4.20).

Exercise 5.2.16 (linear algebra). Construct a monad on **Set** whose Eilenberg–Moore category is the category **Vect** of vector spaces and linear maps. (*Hint*: Adapt to the vector space case the construction of the free commutative monad. Reading Example 4.2.13 again could help.)

Exercise 5.2.17 (probability, analysis). What are the morphisms of algebras of the probability monad of Example 5.1.2?

Exercise 5.2.18 (sets and relations, basic computer science). The **maybe monad** is a monad on **Set** whose functor part adds an extra point to each set; that is it maps X to $X \sqcup 1$ (where 1 is the singleton). What can its unit and multiplication be?

Show that the Eilenberg–Moore category of the maybe monad is (equivalent to) the category of pointed sets and base point-preserving functions.

Exercise 5.2.19 (sets and relations; difficult!). Show that the algebras of the power set monad of Example 5.1.1 are complete semilattices, with the algebra structure map given by the join (or the meet). What are the morphisms of algebras?

Exercise 5.2.20 (difficult!). Show that for every T-algebra (A, e), the diagram

$$TTA \underset{Te}{\overset{\mu}{\rightrightarrows}} TA \xrightarrow{e} A$$

is a coequalizer diagram. (*Hint*: The unique maps induced by the universal property of the coequalizer can be obtained using the unit η.)

Exercise 5.2.21. Show that a monad is idempotent if and only if all its algebra structure maps are isomorphisms. (*Hint*: Use the exercise above, after proving that for every algebra (A, e) of an idempotent monad T, μ and Te must coincide.)

The exercise above implies that, for example,

- the algebras of the Cauchy completion monad of Exercise 5.1.33 are the complete metric spaces;
- the algebras of the abelianization monad of Exercise 5.1.36 are the abelian groups;
- the algebras of the Kolmogorov quotient monad of Exercise 5.1.35 are the T_0 (or *Kolmogorov*) topological spaces.

What are the morphisms of algebras in each case? If you can't answer this question, see the following exercise.

Exercise 5.2.22. Using Exercise 5.2.21, show that if T is an idempotent monad on $\mathbf{C}$, given any two T-algebras A and B, any morphism $f : A \to B$ of $\mathbf{C}$ is automatically a morphism of algebras.

5.2.2 Free algebras

Let (T, η, μ) be a monad on $\mathbf{C}$. Given an object X of $\mathbf{C}$, the object TX is canonically a T-algebra with structure morphism $\mu : TTX \to TX$. The diagrams (5.2.1) in this setting become

$$\begin{array}{ccc} TX & \xrightarrow{\eta} & TTX \\ & \underset{\text{id}}{\searrow} & \downarrow{\scriptstyle\mu} \\ & & TX \end{array} \qquad \begin{array}{ccc} TTTX & \xrightarrow{T\mu} & TTX \\ \downarrow{\scriptstyle\mu} & & \downarrow{\scriptstyle\mu} \\ TTX & \xrightarrow{\mu} & TX, \end{array}$$

which are just the left unitality and associativity diagram of the monad, respectively, as in (5.0.1), and so they commute.

Definition 5.2.23. We call a T-algebra in the form (TX, μ), for some X of $\mathbf{C}$, a **free T-algebra**.

Moreover, let $f : X \to Y$ be a morphism of $\mathbf{C}$. The induced map $Tf : TX \to TY$ is a morphism of algebras since the diagram

$$\begin{array}{ccc} TTX & \xrightarrow{TTf} & TTY \\ \downarrow{\scriptstyle\mu} & & \downarrow{\scriptstyle\mu} \\ TX & \xrightarrow{Tf} & TY \end{array}$$

commutes by the naturality of μ.

An interpretation of free algebras, in terms of formal expressions, is that they contain formal expressions and that the structure map μ is just the simplification of nested formal expressions (or the juxtaposition of them). Let's see some examples. The word "free," as we said before, will be motivated in Section 5.2.3.

Example 5.2.24 (several fields). Consider the list monad L of Exercise 5.1.6. We know its algebras are monoids. Now, given a set X, the set LX of *lists* or *words* in X is canonically a monoid. As a neutral element, it has the empty list, and as multiplication the concatenation of strings

$$[x_1, \ldots, x_n] \cdot [y_1, \ldots, y_m] := [x_1, \ldots, x_n, y_1, \ldots, y_m].$$

Example 5.2.25 (algebra). Consider the free commutative monoid monad F of Example 5.2.1. Given a set X, the free algebra FX contains formal sums such as

$$x_1 + \cdots + x_n,$$

and the structure map, which is the multiplication of the monad, sums formal expressions by just removing the brackets:

$$(x_1 + \cdots + x_n) + (y_1 + \cdots + y_m) \ := \ x_1 + \cdots + x_n + y_1 + \cdots + y_m.$$

Example 5.2.26 (probability). Consider the probability monad $\mathcal{P}$ of Example 5.1.2, or any of its variants in the categories of measurable or topological spaces. We have seen that its algebras can be interpreted as "convex spaces" of some kind (Example 5.2.9). The free algebras are now the convex spaces in the form $\mathcal{P}X$ for some set (or space) X, with the convex combination operation given by the mixture of probability measures, such as

$$\frac{1}{2}\left(\frac{1}{2}\text{ heads} + \frac{1}{2}\text{ tails}\right) + \frac{1}{2}(1\text{ heads} + 0\text{ tails}) \ := \ \frac{3}{4}\text{ heads} + \frac{1}{4}\text{ tails}.$$

Note that we are not mixing "heads" and "tails" (we know that's impossible); we are mixing *probability measures over them.*

Spaces in the form $\mathcal{P}X$ are also known as **simplices**, and the points in the image of $\delta : X \to \mathcal{P}X$ are called the **extreme points** of the simplex. Can you see why?

Exercise 5.2.27 (linear algebra). Consider the vector space monad of Exercise 5.2.16. Using the fact that *every vector space has a basis,* prove that each algebra is free.

Exercise 5.2.28 (group theory). What are the free algebras of the monad of Exercise 5.2.4?

5.2.3 The Eilenberg–Moore adjunction

Let (T, η, μ) be a monad on a category $\mathbf{C}$, and let A and B be T-algebras. By construction, any morphism of algebras $f : A \to B$ is first of all a morphism of $\mathbf{C}$. For example, a monoid homomorphism is first of all a function. Therefore, there is a faithful "forgetful" functor $\mathbf{C}^T \to \mathbf{C}$. Let's denote this functor by R^T. We have seen some of those forgetful functors already:

- **Vect** $\to$ **Set**,
- **Mon** $\to$ **Set**,
- **Grp** $\to$ **Set**.

(Which monads have the categories above as Eilenberg–Moore categories?)

We have also seen that, in many cases, those functors have left-adjoints. For example, the forgetful functor **Vect** $\to$ **Set** has a left-adjoint (see Example 4.1.6), with the intuition of "forming formal linear combinations of elements out of a set." This idea of "forming formal expressions," which we now know can be formalized by the ideas of a monad and of free algebras, works in general. Let's see how.

Definition 5.2.29. Let (T, η, μ) be a monad on a category $\mathbf{C}$. We define the functor $L^T : \mathbf{C} \to \mathbf{C}^T$ as follows:

- It maps an object X to the free T-algebra (TX, μ).
- It maps a morphism $f : X \to Y$ to the morphism of algebras $Tf : TX \to TY$.

The interpretation of this construction is as follows: Given X, form spaces of formal expressions out of X, and given $f : X \to Y$, apply f to each term in a formal expression to get a formal expression in Y.

We now have a result which is similar, in some sense even dual, to Proposition 5.1.25.

Proposition 5.2.30. *Let (T, η, μ) be a monad on $\mathbf{C}$:*

(a) *The composite functor $R^T \circ L^T : \mathbf{C} \to \mathbf{C}$ is naturally isomorphic to T.*

(b) *The functor L^T is left-adjoint to R^T.*

(c) *The unit of the adjunction is given by the unit of the monad η.*

(d) *The counit of the adjunction is given by the structure maps e of the algebras.*

Proof. (a) The composite functor $R^T \circ L^T$ acts as follows:

$$\begin{array}{ccccc} X & & (TX, \mu) & & TX \\ \downarrow f & \overset{L^T}{\mapsto} & \downarrow Tf & \overset{R^T}{\mapsto} & \downarrow Tf \\ Y & & (TY, \mu) & & TY; \end{array}$$

Therefore, by definition, it has almost the same action on objects and morphisms as T.

(b) We define the adjunction in terms of the unit and counit, which will show (c) and (d) as well. So, first of all, the unit of the monad $\eta : \mathrm{id}_{\mathbf{C}} \Rightarrow T = R^T \circ L^T$ is already in the desired form for a unit of the adjunction. The counit $\varepsilon : L^T \circ R^T \Rightarrow \mathrm{id}_{\mathbf{C}^T}$ is defined as follows. Given a T-algebra (A, e),

$$L^T \circ R^T(A, e) \;=\; L^T(A) \;=\; (TA, \mu).$$

The very structure map $e : TA \to A$ defines a morphism of algebras $(TA, \mu) \to (A, e)$ since the following diagram commutes:

$$\begin{array}{ccc} TTA & \xrightarrow{Te} & TA \\ \downarrow \mu & & \downarrow e \\ TA & \xrightarrow{e} & A, \end{array}$$

which is simply the multiplication diagram of (5.2.1). Moreover, every morphism of $\mathbf{C}^T$ commutes with the structure maps e

of the algebras, so that $e : (TA, \mu) \to (A, e)$ is actually natural in the algebra A, and so it induces a natural transformation $\varepsilon : L^T \circ R^T \Rightarrow \mathrm{id}_{\mathbf{C}^T}$.

In order to have an adjunction, we have to show that η and ε satisfy the triangle identities (4.2.5). The first one states (why?) that the following diagram of $\mathbf{C}^T$ has to commute for all objects X of $\mathbf{C}$:

$$\begin{array}{ccc} (TX,\mu) & \xrightarrow{T\eta} & (TTX,\mu) \\ & {\scriptstyle \mathrm{id}}\searrow & \downarrow{\scriptstyle \mu} \\ & & (TX,\mu). \end{array}$$

(Note that the map μ appearing on the right is the structure map of the free algebra (TX, μ).) This diagram commutes since it corresponds to the right unitality diagram of the monad (5.0.1). The second triangle identity states (why?) that the following diagram of $\mathbf{C}$ has to commute for each T-algebra (A, e):

$$\begin{array}{ccc} A & \xrightarrow{\eta} & TA \\ & {\scriptstyle \mathrm{id}}\searrow & \downarrow{\scriptstyle e} \\ & & A. \end{array}$$

This commutes since it is the unit condition for algebras (5.2.1). □

Corollary 5.2.31. *The adjunction amounts to a natural bijection*

$$\mathrm{Hom}_{\mathbf{C}}(X, A) \xrightarrow{\cong} \mathrm{Hom}_{\mathbf{C}^T}\big((TX, \mu), (A, e)\big)$$

for each object X of $\mathbf{C}$ and each T-algebra (A, e). In other words, it says that given a morphism $f : X \to A$, where A is (the underlying object of) a T-algebra, there is a unique morphism of T-algebras $TX \to A$ such that the following diagram of $\mathbf{C}$ commutes:

$$\begin{array}{ccc} X & & \\ {\scriptstyle \eta}\downarrow & \searrow{\scriptstyle f} & \\ TX & \dashrightarrow & A. \end{array}$$

Example 5.2.32 (linear algebra). We have already seen the adjunction between sets and vector spaces of Example 4.1.6. The morphism $TX \to A$, in that case, is the linear extension of a map from the basis X of TX to the vector space A.

Example 5.2.33 (algebra). If F is the free commutative monoid monad of Example 5.2.1, this says that given a set X and a commutative monoid M, there is a unique morphism of monoids $\tilde{f} : FX \to M$ such that $f = \tilde{f} \circ \eta$. This is the universal property of free commutative monoids. It states that FX is "freely generated" by X (in our interpretation, by forming formal sums of elements of X).

Example 5.2.34 (probability, analysis). For the probability monad of Example 5.1.2, the statement above means that given a convex space A (for example, $\mathbb{R}$ or $[0, 1]$) and a set (or space) X, there is a bijection between functions $f : X \to A$ and *affine maps* from the simplex, $\mathcal{P}X \to A$, which agrees with f on the extrema of the simplex. (A map is called **affine** if it preserves convex combinations.)

Similar statements hold true for the different variants of probability monads in their respective categories.

Exercise 5.2.35 (group theory). Write down the adjunction explicitly for the case of groups, given in Exercise 5.2.4. Compare with the adjunction in Exercise 4.1.9.

5.3 Comonads as Extra Information

We now turn to comonads. Let's look at a way in which comonads can be motivated and which helps in understanding many comonads arising in practice.

Idea. *A comonad can look like a consistent way to equip spaces with extra information of a specific kind, and let some morphisms access that information.*

As usual, let's show this by giving some examples.

Example 5.3.1 (several fields). Let E be a set. Given any (other) set X, the assignment $X \mapsto X \times E$ is functorial (see Exercise 3.2.6 and Example 5.1.7). Denote this functor by C_E. So, $C_E(X) = X \times E$, and given $f : X \to Y$, we have $C_E(f) = f \times \mathrm{id}_E : X \times E \to Y \times E$, which maps (x, e) to $(f(x), e)$.

If we view E as a set of "extra data," then an element (x, e) of $C_E(X) = X \times E$ has more information than just x. We can equip the functor C_E with a comonad structure following this intuition:

- The counit $\varepsilon : C_E(X) \to X$ is given by the projection $(x, e) \mapsto x$. In other words, this is *forgetting or discarding the extra information.*
- Information cannot only be discarded, it can also be copied.[3] The comultiplication $\nu : C_E(X) \to C_E(C_E(X))$ then *copies the extra information,* which is given by $(x, e) \mapsto (x, e, e)$.

It is easy to see that these maps are natural. The comonad axioms (5.0.2) now state the following:

- The left counitality diagram says that $\varepsilon \circ \nu = \mathrm{id}_{C_E(X)}$. Explicitly, if we start with $(x, e) \in X \times E$, then copying the extra information (getting (x, e, e)) and then discarding the last part (getting (x, e)) is the same as doing nothing.
- The right unitality diagram says that $C\varepsilon \circ \nu = \mathrm{id}_{C_E(X)}$. Again, if we start with (x, e), copying the extra information (getting (x, e, e)) and then discarding the first extra datum (the first of the "e," giving (x, e)) is again the same as doing nothing.
- The coassociativity diagram says that $\nu \circ \nu = C\nu \circ \nu$. Explicitly, starting again with (x, e), if we again copy the extra information (obtaining (x, e, e)), then copying either the first piece or the second piece of the extra information gives the same result, (x, e, e, e).

[3]This is not true for quantum information, see Example 1.4.17.

Therefore, (C_E, ε, ν) is a comonad on **Set**. We can call it the **reader comonad**. The name will be motivated in Example 5.3.6. For now, note that it is somewhat dual to the writer monad of Example 5.1.7: There we had "extra stuff" that can be put together (sum), whereas here we have "extra information" that can be copied.

Example 5.3.2 (basic computer science, dynamical systems). Let X be a set. Denote by SX the space of **infinite sequences**, or **streams**, in X, i.e. functions $\mathbb{N} \to X$. (Equivalently, we can view SX as the cartesian product of $\mathbb{N}$-many copies of X with itself.) We can denote the elements by $\{x_n\}_{n\in\mathbb{N}}$, or, more briefly, $\{x_n\}$, or sometimes more explicitly by

$$\{x_0, x_1, x_2, \dots\}.$$

This construction is canonically functorial: Given $f : X \to Y$, we can define $Sf : SX \to SY$ as the function applying f elementwise, i.e.

$$\{x_0, x_1, x_2, \dots\} \longmapsto \{f(x_0), f(x_1), f(x_2), \dots\}.$$

Clearly, a stream on X has more information than just an element of X. We can now equip S with a comonad structure as follows:

- First of all, define the counit $\varepsilon : SX \to X$ as the map

  $$\{x_0, x_1, x_2, \dots\} \longmapsto x_0,$$

  which keeps only the first (well, zeroth) value and discards the rest of the stream.

- The comultiplication map $\nu : SX \to SSX$ has to map a stream to a *stream of streams* (imagine an infinite two-dimensional matrix). Given a stream

  $$\{x_0, x_1, x_2, \dots\},$$

we form the stream of streams

$$\begin{aligned}\{\{x_0, x_1, x_2, \dots\}\\ \{x_1, x_2, x_3, \dots\}\\ \{x_2, x_3, x_4, \dots\}\\ \dots \qquad\qquad \}.\end{aligned} \tag{5.3.1}$$

In other words, we form a stream of stream which has as the first element the original stream, as the second element the original stream but starting at x_1 (in computer science, one calls it "popping," while in dynamical systems theory, one calls it "shifting"). As the third element, we have a stream that starts at x_2, and so on. In symbols,

$$\nu(\{x_n\}_{n\in\mathbb{N}}) = \{\{x_{n+m}\}_{n\in\mathbb{N}}\}_{m\in\mathbb{N}}.$$

One way to interpret SX is that the elements of SX are the elements x_0 of X together with their *history*: x_1 is where it was a second ago, x_2 is where it was two seconds ago, and so on. The unit forgets the past and keeps only the present state. The multiplication looks at the *history of the history*: One second ago, the history was only until x_1; two seconds ago, the history was only until x_2; and so on.

Let's check that the comonad axioms (5.0.2) hold:

- The left counitality says that if we take the stream $\{x_0, x_1, x_2, \dots\}$ and look at its history (5.3.1), the first element in there is the original stream.
- The right counitality diagram says that if we again take the stream $\{x_0, x_1, x_2, \dots\}$ and look at its history (5.3.1), then the stream formed by *taking the first element of each stream in the history* gives the original stream.
- The coassociativity diagram says that if we once again take the stream $\{x_0, x_1, x_2, \dots\}$ and look at its history (5.3.1), then we can look once again at its history (a three-dimensional stream).

We can also form the three-dimensional stream by looking at the history of each stream in the two-dimensional stream (5.3.1). Both constructions give the same three-dimensional stream, namely

$$\{\{\{x_{n+m+k}\}_{n\in\mathbb{N}}\}_{m\in\mathbb{N}}\}_{k\in\mathbb{N}}.$$

Therefore, (S, ε, μ) is a comonad. It is known as the **stream comonad.**

Exercise 5.3.3 (several fields). Let M be a monoid. Construct a comonad on **Set** analogous to the stream comonad, but with M in place of $\mathbb{N}$.

If $M = \mathbb{Z}$, this means that we are looking not just at the past but also at the future. If $M = \mathbb{R}$, we are looking at continuous time instead of discrete time . . . and so on.

A fairly general phenomenon is that, for many monads, the unit is an embedding of X into its "extension" TX, while for many comonads, the counit is a quotient map of CX onto X, which "forgets the extra data." (In the words of Tobias Fritz, while monads tend to have *generalized* points, comonads tend to have *refined* points.)

In the following exercise, for readers who have a background in geometry and topology, you can show that the universal covering is an idempotent comonad. The "extra information" that it encodes can be interpreted as the "winding number," or "number of times we have gone around a hole." (If you find the following example too technical but still find these concepts interesting, you can try reading Example 5.4.3, which requires less prerequisite knowledge.)

Exercise 5.3.4 (algebraic topology; difficult!). Let $\mathbf{PCLC}_*$ be the category whose:

- objects are path-connected, locally contractible,[4] pointed topological spaces;
- morphisms are continuous maps preserving the base point.

[4]The condition of local contractibility can be weakened, see [Hat02, Section 1.3].

Given a space (X, x) in the category above, denote by $p : RX \to X$ its universal covering (where RX is the space of homotopy classes of paths in X starting at x, with its usual topology, and the map p takes the endpoint of each path).

Show that R is a functor as follows. Given $f : (X, x) \to (Y, y)$ and given a path γ starting at x, $f(\gamma)$ is a path starting at y. Since f is continuous, this assignment respects homotopy, and so we have a well-defined map $RX \to RY$. Show that this map is continuous. (*Hint*: Prove that the preimage of a basic open in RY is open in RX using the fact that X is locally contractible.) Assume RX is based at the constant path at x.

Show moreover that the covering map $p : RX \to X$ is natural in X. This will form the counit of the comonad. (This does discard the extra information: It literally forgets the path and keeps track only of the endpoint.)

Since RX is simply connected, the map $p : RRX \to RX$ obtained by taking again the universal covering is a homeomorphism. Let ν be the inverse of this map. Show that (R, p, ν) is an idempotent comonad on $\mathbf{PCLC}_*$.

5.3.1 Co-Kleisli morphisms

Definition 5.3.5. Let (C, ε, ν) be a comonad on $\mathbf{C}$. A **co-Kleisli morphism** of C from X to Y is a morphism $CX \to Y$ of $\mathbf{C}$.

Here is a typical situation in science. Suppose that, during an experiment, we have a process f taking an input $x \in X$ and giving an output $y \in Y$. This could be, for example, a survey in which we ask people of different age (X) what their political views are (Y). Suppose that we repeat the experiment, feeding again the *same* input x to f. It could happen that this time we get a *different* outcome from before: $y' \neq y$. For example, suppose a person of age x expresses political view y. Another person of the same age x may express a different political view y' (this may even happen by asking the same person twice on different occasions). The usual conclusion that we draw, in science, is as follows: X alone (for example, age alone) is *not enough* to determine Y. There was some "hidden," "extra" data that the process f has access to in order to determine Y. It could depend on hidden information, it could depend on past outcomes, and so on. Therefore, a better mathematical model of the situation is not quite $f : X \to Y$, but rather $f : CX \to Y$, where CX contains more information than just X, as specified by the comonad C.[5]

Note that this is somewhat dual to Kleisli morphisms: There, functions were allowed to have more possible *outcomes*. Here, they are allowed to *depend* on possibly more information.

Example 5.3.6 (several fields). A co-Kleisli morphism for the reader comonad of Example 5.3.1 is a map $k : X \times E \to Y$. We can view it as a map that, when fed $x \in X$, also needs to read an $e \in E$ to give its output. This can model the experimental situation of "hidden variables" described above.

An example in computer science is a function that needs an extra input in order to carry out the computation, either via user interface (say, keyboard), or by having access to some external environment. This motivates the name "reader comonad."

Example 5.3.7 (basic computer science, dynamical systems). A co-Kleisli morphism for the stream comonad is a map $SX \to Y$; in other words, it is a function that potentially depends on the whole stream, not just on the current value.

In dynamical systems (and probability, and so on), this corresponds to a dynamic that *depends on the history*, or *has memory*. For a concrete example, consider the "Fibonacci" function $S\mathbb{N} \to \mathbb{N}$ given by $\{x_0, x_1, x_2, \dots\} \mapsto x_0 + x_1$.

[5] This is a good model when we have access to the extra information, or at least when we have a model for its structure.

Example 5.3.8 (complex analysis). Let z be a nonzero complex number. The **complex logarithm of** z is given by the integral

$$\int_1^z \frac{1}{z'}\,dz'.$$

This does not quite depend only on z but also on the path of integration, or, more specifically, on the homotopy class of the path of integration in $\mathbb{C}\backslash\{0\}$. Therefore, the integral above, rather than a function $\mathbb{C}\backslash\{0\} \to \mathbb{C}$, is a function $R(\mathbb{C}\backslash\{0\}) \to \mathbb{C}$, where R is the universal covering comonad of Exercise 5.3.4, and the spaces $\mathbb{C}\backslash\{0\}$ and $\mathbb{C}$ are assumed to be based at 1 and 0, respectively. In other words, the complex logarithm is, rather than an ordinary function, a co-Kleisli morphism. It takes the extra information of "how many times we have gone around 0."

Example 5.3.9 (quantum physics). The wave function of an electron in a one-dimensional periodic material (such as a crystal) can be obtained by solving Schrödinger's equation with a periodic potential and periodic boundary conditions, or equivalently, by solving Schrodinger's equation on the circle. Denoting a point on the circle by x_0, we can see the circle as a pointed space (S^1, x_0). By Bloch's theorem, the wave function of our electron will have the form

$$\psi(x) = e^{ik(x-x_0)}\,u(x),$$

where k is a real number and $u(x)$ is a function with the same periodicity as the potential. Note that the probability density of the electron is equal to

$$\|\psi(x)\|^2 = \|u(x)\|^2,$$

which is a periodic function, and so it can be written as a function on the graph (S^1, x_0). However, the wave function itself is not periodic because of its phase (k is not necessarily related to the period of the potential), and so ψ is not a well-defined function on the circle.

Instead, we can model ψ as a (well-defined) *co-Kleisli morphism* for the universal covering comonad of Exercise 5.3.4. That is, $\psi : RS^1 \to \mathbb{C}$. The "extra information" to which ψ has access is "how far we are from x_0 not only on the circle but also on its universal covering, or on the actual periodic material" (we may be a few entire cells away). Note that this extra information is reflected by the phase. The value of the phase at a given point, alone, does not encode observable information, but the *difference* of phase at different points does have physical meaning (and is responsible, for example, for interference).

As is the case for Kleisli morphisms, co-Kleisli morphisms also have a meaningful notion of composition.

Definition 5.3.10. Let (C, ε, ν) be a comonad on $\mathbf{C}$. Let $k : CX \to Y$ and $h : CY \to Z$ be co-Kleisli morphisms. Their **co-Kleisli composition** is the co-Kleisli morphism $h \circ_{ck} k : CX \to Z$ given by

$$CX \xrightarrow{\nu} CCX \xrightarrow{Ck} CY \xrightarrow{h} Z.$$

Example 5.3.11 (several fields). Let $k : X \times E \to Y$ and $h : Y \times E \to Z$ be co-Kleisli morphisms of the reader comonad of Example 5.3.1. Their co-Kleisli composition proceeds as follows:

$$\begin{array}{ccccccc} X \times E & \xrightarrow{\nu} & X \times E \times E & \xrightarrow{k \times \mathrm{id}_E} & Y \times E & \xrightarrow{h} & Z \\ (x, e) & \longmapsto & (x, e, e) & \longmapsto & (y, e) & \longmapsto & z. \end{array}$$

We first start with $(x, e) \in X \times E$. We copy the extra information to get (x, e, e). We then feed the first two arguments to k, which gives us an element $y \in Y$, and we keep track of the additional e. We then feed (y, e) to h to get $z \in Z$.

Exercise 5.3.12 (basic computer science, dynamical systems). Write down explicitly the co-Kleisli composition of two composable co-Kleisli morphisms of the stream comonad of Example 5.3.2.

Note that, just as comonads are monads in the opposite category, co-Kleisli morphisms are Kleisli morphisms in the opposite category. Therefore, they form a category too. In the following exercise, you can prove this explicitly.

Exercise 5.3.13. Prove explicitly that co-Kleisli morphisms also form a category, with identities given by the counits and composition given by the co-Kleisli composition.

Definition 5.3.14. This category is called the **co-Kleisli category** and is denoted, similarly to the Kleisli category, by $\mathbf{C}_C$.

5.3.2 The co-Kleisli adjunction

Just as Kleisli morphisms "include" ordinary morphisms via the unit, co-Kleisli morphisms also include ordinary morphisms via the counit. Explicitly, let (C, ε, ν) be a comonad on $\mathbf{C}$. Then, an ordinary morphism $f : X \to Y$ of C defines canonically a co-Kleisli morphism

$$CX \xrightarrow{\varepsilon} X \xrightarrow{f} Y.$$

The interpretation is that if we start with extra information, we first discard it, and then apply f. In other words, f does not really need, nor use, the extra information.

As for the Kleisli case, this assignment is functorial and defines an adjunction, as we will see. Denote for convenience $f \circ \varepsilon$ by $R_C(f)$.

Exercise 5.3.15. Show that R_C is part of a functor $\mathbf{C} \to \mathbf{C}_C$, which is the identity on objects, with $R_C(\mathrm{id}_X) = \varepsilon_X$ and $R_C(f \circ g) = R_C(f) \circ_{ck} R_C(g)$.

Conversely, suppose we have a co-Kleisli morphism $k : CX \to Y$. This gives canonically an ordinary morphism $CX \to CY$ as

$$CX \xrightarrow{\nu} CCX \xrightarrow{Ck} CY.$$

Denote this composite $Ck \circ \nu$ by L_C.

Example 5.3.16 (several fields). For the reader comonad, this assignment takes a function $k : X \times E \to Y$ and gives a function $L_C(k) : X \times E \to Y \times E$ which maps (x, e) to $(k(x, e), e)$, copying the extra information e to the output.

Exercise 5.3.17 (basic computer science, dynamical systems). What does this give for the stream comonad? (*Hint*: One looks at all the values that k "would have assumed in the past.")

Exercise 5.3.18. Prove that L_C is part of a functor $\mathbf{C}_C \to \mathbf{C}$, which maps an object X to CX. In particular, show that $L_C(\varepsilon_X) = \mathrm{id}_X$ and that $L_C(k \circ_{ck} h) = L_C(k) \circ L_C(h)$.

Exercise 5.3.19. Prove the dual to Proposition 5.1.25, namely that:

- $L_C \circ R_C = C$;
- L_C is left-adjoint to R_C;
- the counit of the adjunction is given by the counit ε of the comonad.

This adjunction is called the **co-Kleisli adjunction**. Note that the role of the left- and right-adjoints are reversed compared to Proposition 5.1.25.

5.4 Comonads as Processes on Spaces

Idea. *A comonad can be viewed as a consistent way to construct, from spaces, processes of a specified structure and give specific strategies or trajectories.*

Example 5.4.1 (basic computer science, dynamical systems). Consider the stream comonad S on **Set** (Example 5.3.2). We have seen that a possible interpretation is that a stream in X is the "history" of a point of X, where the first element of the stream is the current state.

We can "reverse the arrow of time" and interpret instead a stream as the *future* positions of a point of x, with the first element given by the current state. This can be thought of as a process of some kind. In fact, SX is even canonically a *dynamical system*, with the map $SX \to SX$ given by shifts, i.e. mapping the stream

$$\{x_0, x_1, x_2, \dots\} \quad \text{to} \quad \{x_1, x_2, x_3, \dots\}.$$

Example 5.4.2 (several fields). The stream comonad for a different choice of monoid (Exercise 5.3.3) has the same interpretation. Note that even if we choose a group, such as $\mathbb{Z}$ or $\mathbb{R}$, which intuitively indexes both the past and the future, we still have to reverse the direction of time in the interpretation. Moreover, by choosing different indexing monoids, such as $\mathbb{Z}^2$, the process can grow with more general shapes.

The following example can be considered a discrete analogue of the universal covering comonad of Exercise 5.3.4. Before starting with the example, let's give a preliminary definition: A **rooted tree** is a directed graph with a distinguished vertex x, called the **root**, and such that for every (other) vertex y, there is a unique chain from x to y. Rooted trees are graphs that look for example like this:

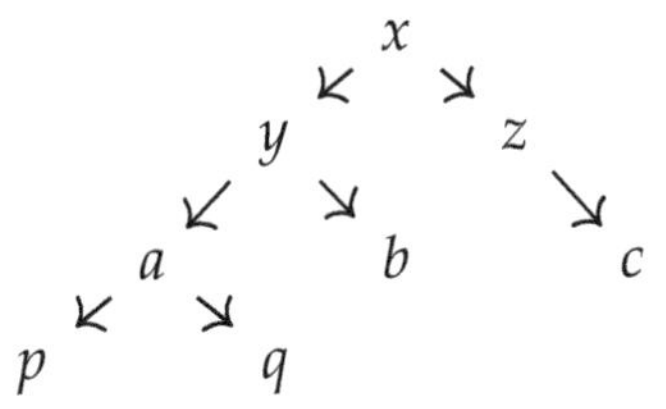

and *not* like these:

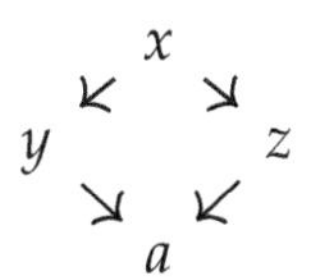

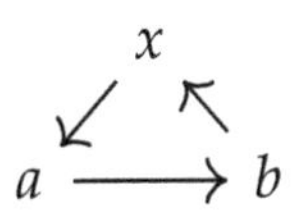

Let's now give our example.

Example 5.4.3 (graph theory). Let $\mathbf{MGraph}_*$ be the category of directed multigraphs with a distinguished vertex ("base point") and multigraph morphisms preserving the base point (and incidence).

Consider an object (G, x) which is a multigraph G with a distinguished vertex x. Now, denote by RG the graph constructed as follows:

- Vertices are *chains* in G (chains of head-to-tail edges) starting at x, of finite length, including the trivial chain 1_x at x. That is, a vertex of RG is either 1_x, or a tuple of the form
$$(e_1, \dots, e_n),$$
where, denoting source and target by s and t, $s(e_1) = x$ and $t(e_i) = s(e_{i+1})$, for all $i = 1, \dots, n-1$.
- There is a unique edge of RG from q to q' if and only if they are in the form
$$q = (e_1, \dots, e_n) \quad \text{and} \quad q' = (e_1, \dots, e_n, e_{n+1}),$$
with $s(e_{n+1}) = t(e_n)$, i.e. if q' can be obtained from q by adding a consecutive edge.

As we denote the trivial chain at x by 1_x, we have a pointed multigraph $(RX, 1_x)$.

This construction is functorial. Indeed, let $f : (G, x) \to (H, y)$ be a morphism, i.e. an incidence-preserving map with $f(x) = y$. We can construct a map $Rf : RG \to RH$ which maps:

- the trivial chain 1_x at x to the trivial chain 1_y at y;
- the chain $(e_1, \dots, e_n)$ to the tuple $(f(e_1), \dots, f(e_n))$, which is a chain in H since f preserves incidence (and so the edges are consecutive in H too), starting at y (since $f(x) = y$).

This map preserves the base point by construction, and it also preserves incidence: The unique edge between
$$(e_1, \dots, e_n) \quad \text{and} \quad (e_1, \dots, e_n, e_{n+1})$$

is mapped to the unique edge between

$$(f(e_1), \ldots, f(e_n)) \quad \text{and} \quad (f(e_1), \ldots, f(e_n), f(e_{n+1})).$$

Therefore, Rf is a morphism of pointed multigraphs. It is easy to check the functoriality axioms, so we have an endofunctor $R : \mathbf{MGraph}_* \to \mathbf{MGraph}_*$.

Let's now give to R a comonad structure. First of all, the unit $p : (RG, 1_x) \to (G, x)$ is given by the endpoints of chains. That is, we define, on vertices, $p(1_x) := x$, and

$$p(e_1, \ldots, e_n) \;:=\; t(e_n),$$

the target of the last edge e_n. On edges, we map the unique edge between

$$(e_1, \ldots, e_n) \quad \text{and} \quad (e_1, \ldots, e_n, e_{n+1})$$

to e_{n+1}. This respects incidence (why?), and by construction, it preserves the base points, so we have a morphism of pointed multigraphs. Moreover, this is natural (why?), and so we have a natural transformation $p : R \Rightarrow \mathrm{id}_{\mathbf{MGraph}_*}$.

In order to construct the comultiplication, first note that RG is always a rooted tree with root 1_x. Indeed, for any vertex $(e_1, \ldots, e_n)$ of RG, there is always a unique chain from 1_x, namely

$$1_x \to (e_1) \to (e_1, e_2) \to \cdots \to (e_1, \ldots, e_n), \tag{5.4.1}$$

where all the arrows denote the unique chains in RG. This gives a bijection $v : RG \to RRG$, which is even an isomorphism of multigraphs (why does it preserve incidence?). It is part of a natural isomorphism $v : R \Rightarrow RR$ (check naturality!).

To show that (R, p, v) is a comonad, let's check whether the diagrams (5.0.2) commute:

- The left counitality diagram says that given a chain $(e_1, \ldots, e_n)$ and applying v to form the chain of chains (5.4.1), then the endpoint of (5.4.1) is the original $(e_1, \ldots, e_n)$.

- The right counitality diagram says that, again, given a chain $(e_1, \ldots, e_n)$ and applying ν to form the chain of chains (5.4.1), if we take the tuple formed by *the endpoints of each arrow in* (5.4.1), then we get again $(e_1, \ldots, e_n)$.
- The coassociativity diagram says that if we repeat the construction (5.4.1) twice, we get the same result as if we took the construction (5.4.1) for each element of (5.4.1) itself. (Can you write down which *chain of chains of chains* you get that way?)

  Therefore, (R, p, ν) is an idempotent comonad. We call it the **comonad of rooted trees** (see Example 5.4.12 for the reason), or also **discrete universal covering comonad** (see Exercise 5.4.5 for the reason).

Also, in this example, the comonad constructs some "processes" on our spaces, in this case "chains." The counit forgets the process, and just tells us "where we are."

Exercise 5.4.4 (graph theory). Give an analogous construction for *un*directed (multi)graphs. (*Hint*: You can consider an undirected edge a pair of directed edges in opposite directions.)

Exercise 5.4.5 (graph theory, algebraic topology). If you have solved Exercises 5.3.4 and 5.4.4, establish the connection between the universal covering for graphs and the one for topological spaces given in Exercise 5.3.4.

Also, the universal covering comonad of topological spaces can be interpreted in terms of "processes": The points of the universal cover are again paths from the base point.

Exercise 5.4.6 (graph theory, dynamical systems). Can you think of a function or transition process on a graph which is best modeled by a co-Kleisli morphism of the rooted tree comonad? (*Hint*: Think of something that "depends really on the path, not just on the endpoints.")

5.4.1 Coalgebras of a comonad

We saw that a comonad can encode a "process" of some kind. In that case, its coalgebras encode settings in which the process can be started in a "default" way, or a "strategy."

Definition 5.4.7. Let (C, ε, μ) be a comonad on a category $\mathbf{C}$. A **coalgebra** of C, or **over** C, or C-coalgebra, consists of:

- an object A of $\mathbf{C}$,
- a morphism $i : A \to CA$ of $\mathbf{C}$,

such that the following diagrams commute, called **counit** and **comultiplication**, or **coalgebra square**, respectively:

$$\begin{array}{ccc} A & \xrightarrow{i} & CA \\ & \overset{\text{id}}{\searrow} & \downarrow{\scriptstyle \varepsilon} \\ & & A \end{array} \qquad \begin{array}{ccc} A & \xrightarrow{i} & CA \\ \downarrow{\scriptstyle i} & & \downarrow{\scriptstyle \nu} \\ CA & \xrightarrow{Ci} & CCA. \end{array} \tag{5.4.2}$$

The intuition for the map i is that it assigns to each $a \in A$ a distinguished process, which we can think of as being "started," or "triggered," by a. Let's see some examples.

Example 5.4.8 (several fields). The coalgebras of the stream comonad S (Example 5.3.2) are exactly dynamical systems (for the monoid $\mathbb{N}$, discrete time, for the monoid $\mathbb{R}$, continuous time, and so on). Let's see why. Plugging in the definition, a coalgebra consists of a set A together with a map $i : A \to SA$, which we can write as

$$a \longmapsto \{i_0(a), i_1(a), i_2(a), \dots\},$$

such that the diagrams (5.4.2) commute. The counit square says that for each $a \in A$, $\varepsilon(i(a)) = a$, which means precisely that $i_0(a) = a$. Therefore, i is in the form

$$a \longmapsto \{a, i_1(a), i_2(a), \dots\}. \tag{5.4.3}$$

Let's now turn to the comultiplication square. For each $a \in A$, after forming the string (5.4.3), we can either apply ν to form the "history" string of (shifted) strings,

$$\begin{aligned}\{&\{a, i_1(a), i_2(a), \dots\}\\ &\{i_1(a), i_2(a), i_3(a), \dots\}\\ &\{i_2(a), i_3(a), i_4(a), \dots\}\\ &\dots \qquad\qquad \},\end{aligned}$$

or we can apply i again to each element of (5.4.3) to form the string of strings

$$\begin{aligned}\{&\{a, i_1(a), i_2(a), \dots\}\\ &\{i_1(a), i_1(i_1(a)), i_2(i_1(a)), \dots\}\\ &\{i_2(a), i_1(i_2(a)), i_2(i_2(a)), \dots\}\\ &\dots \qquad\qquad \}.\end{aligned}$$

The comultiplication square says that these two strings of strings must agree. This implies that all the corresponding elements must agree, which means that $i_2(a) = i_1(i_1(a))$ and, more generally, $i_n(a) = {i_1}^n(a)$. Therefore, i is of the form

$$a \longmapsto \{a, f(a), f(f(a)), \dots\},$$

for some $f : A \to A$. That is, a coalgebra structure gives a dynamical system on A.

Conversely, given a function $f : A \to A$, the map assigning the orbits

$$a \longmapsto \{a, f(a), f(f(a)), \dots\}$$

gives a coalgebra structure. Therefore, the coalgebras of the stream comonad are exactly dynamical systems (indexed by $\mathbb{N}$, i.e. in discrete time).

We can view this as a "canonical way to obtain a stream from each element of A." Every element defines a "path," or "orbit." This phenomenon happens with many other comonads too.

Exercise 5.4.9 (dynamical systems). Show that if we form the stream comonad with the monoid $\mathbb{R}_{\geq 0}$ instead of $\mathbb{N}$ (as in Exercise 5.3.3), its coalgebras are the *continuous-time dynamical systems* on a set. This means that to each point of X, we associate a continuous-time trajectory $\mathbb{R} \to X$, which is x at time zero.

In the exercise above, despite having a dynamical system in continuous time, the trajectories are not continuous in any sense since we are simply working in the category of sets. Let's change that in the following exercise using continuous functions.

Exercise 5.4.10 (dynamical systems, topology; difficult!). Construct a comonad on the category **CHaus** of compact Hausdorff spaces and continuous maps, similar to the stream comonad, where as monoid you take the extended half-line $[0, \infty]$ (with $x + \infty = \infty$ for every x). As streams, consider the *continuous* maps $[0, \infty] \to X$.

Show that the coalgebras of this comonad are the continuous-time dynamical systems in the traditional sense, i.e. with continuous trajectories, with the additional constraint of having a limit at infinity.

Exercise 5.4.11 (dynamical systems). What are the coalgebras of the stream comonad for even different indexing monoids, or groups?

Here is another very important example.

Example 5.4.12 (graph theory). The coalgebras of the rooted tree comonad R (Example 5.4.3) are rooted trees (hence the name). Let's see why. Plugging in the definition, a coalgebra consists of a pointed multigraph (G, x) together with a morphism $i : G \to RG$ preserving the base point (and incidence) and making the diagrams (5.4.2) commute. The map i has to map a vertex y of G to a vertex $i(y)$ of RG, i.e. to a chain in G starting at x. The counit condition says that

$p(i(y)) = y$, that is, the chain $i(y)$ has to end at y. Therefore, the map i on vertices is of the form

$$y \longmapsto (x \to \cdots \to y).$$

In particular, $i(x) = 1_x$, the trivial chain at x (since i needs to preserve the base point). Let now e be an edge of G from y to z. In order to be incidence-preserving, the map i has to map e to an edge $i(e)$ of RG from $i(y)$ to $i(z)$. This means necessarily that $i(z)$ is exactly given by the chain which is the composite of the chain $i(y)$ and e:

$$\begin{aligned} y &\longmapsto (x \to \cdots \to y) \\ z &\longmapsto (x \to \cdots \to y \xrightarrow{e} z). \end{aligned}$$

The counit diagram now states that, on edges, $p(i(e)) = e$. Let's show that G is a tree, rooted at x. Let y be a vertex of G. Consider $i(x) = 1_x$ and $i(y)$. Since RG is a tree rooted at 1_x, there is a unique chain q from 1_x to $i(y)$. Applying p to this chain q, we get a chain in G from x to y. Therefore, there is a chain from x to y. To show that this chain is unique, suppose that we had chains r and r' from x to y. Then, $i(r)$ and $i(r')$ would both be chains in RG from 1_x to $i(y)$. But since RG is a tree, necessarily $i(r) = i(r')$. Applying p, we get

$$r = p(i(r)) = p(i(r')) = r'.$$

Therefore, G is itself a tree rooted at x. (We don't even have to look at the comultiplication square.)

Conversely, suppose that (G, x) is a tree rooted at x. Then, for each vertex y, there exists a unique chain from x to y, and this gives a morphism $i : G \to RG$ that preserves the base point and incidence, which by construction satisfies the counit diagram. Let's see that the comultiplication diagram commutes too. Let y be a vertex of G, consider the unique chain $i(y)$ from x to y, and denote its edges by $e_1, \ldots, e_n$. Then, applying the comultiplication map ν to $i(y)$, we get

$$1_x \to (e_1) \to (e_1, e_2) \to \cdots \to (e_1, \cdots, e_n),$$

and if we apply Ri to $i(y)$, i.e. i to each of intermediate vertices between x and y, we get the same result. Therefore, the comultiplication diagram commutes on vertices. We don't have to check that the diagram commutes on edges since we are in a rooted tree, and so between any two vertices, there is at most one edge. Therefore, G with this map i is a coalgebra.

Again, here, we see that each vertex of the graph defines canonically a chain leading to it (or starting from it, depending on the point of view).

A more formal way to prove the characterization above is as follows (why?).

Exercise 5.4.13. Prove the dual statement to Exercise 5.2.21, namely, that a comonad is idempotent if and only if all its coalgebra structure maps are isomorphisms.

By the same line of reasoning, we also have a similar statement for the universal covering of topological spaces.

Example 5.4.14 (algebraic topology). The exercise above implies that the coalgebra of the universal covering comonad (see Exercise 5.3.4) are the simply connected (and locally contractible) pointed topological spaces.

Exercise 5.4.15 (several fields). Prove that the coalgebras of the reader comonad C_E (Example 5.3.1) are sets A equipped with a function $e : A \to E$.

We can interpret this function as a "default value for the extra information" that every element a on A carries. (More on this in Proposition 6.2.48.)

Dually to the case of algebras over monads, coalgebras over comonads have a notion of morphisms between them.

Definition 5.4.16. Let (A, i) and (B, i) be coalgebras of a comonad C on $\mathbf{C}$. A **morphism of C-coalgebras**, or **C-morphism**, is a morphism $f : A \to B$ of $\mathbf{C}$ such that the following diagram commutes:

$$\begin{array}{ccc} A & \xrightarrow{f} & B \\ \downarrow{\scriptstyle i} & & \downarrow{\scriptstyle i} \\ CA & \xrightarrow{Cf} & CB. \end{array} \tag{5.4.4}$$

The category of C-algebras and C-morphisms is called the **category of coalgebras of** C, or, sometimes, **co-Eilenberg–Moore category**, and it is denoted, similarly to the Eilenberg–Moore category, by $\mathbf{C}^C$.

These morphisms can be interpreted as morphisms which "respect the dynamics of the processes," or "respect the default choices or strategies."

Example 5.4.17 (dynamical systems). Let (A, f) and (B, g) be dynamical systems in the category of sets, or, equivalently, coalgebras over the stream comonad. A morphism of coalgebras, considering the definition, is a map $m : A \to B$ such that for each $a \in A$, $Cm(i(a)) = i(m(a))$. Recalling that the map i is given by the orbits of a, this means that

$$\{m(a), m(f(a)), m(f(f(a)), \dots\}$$

has to be equal to

$$\{m(a), g(m(a)), g(g(m(a))), \dots\}.$$

This is equivalent to saying that $g \circ m = m \circ f$, i.e. the following diagram commutes:

$$\begin{array}{ccc} A & \xrightarrow{f} & A \\ \downarrow{\scriptstyle m} & & \downarrow{\scriptstyle m} \\ B & \xrightarrow{g} & B. \end{array}$$

In other words, the morphisms of coalgebras are precisely the morphisms of dynamical systems, as in Example 1.4.7.

Example 5.4.18 (several fields). Let (A, e) and (B, e) be coalgebras of the reader comonad of Example 5.3.1, i.e. sets equipped with functions $e : A \to E$ and $e : B \to E$ (see Exercise 5.4.15). A function $f : A \to B$ is a morphism of coalgebras if and only if the following diagram commutes (why?):

$$\begin{array}{ccc} A & \xrightarrow{f} & B \\ \downarrow{\scriptstyle e} & & \downarrow{\scriptstyle e} \\ E & \xrightarrow{\text{id}} & E, \end{array}$$

which can be interpreted as the fact that f has to preserve the "default choice of extra data."

Exercise 5.4.19 (dynamical systems; difficult!). Let (A, f) be a dynamical system in the category of sets, or, equivalently, a coalgebra over the stream comonad. Show that the coalgebra structure map $i : A \to SA$ is a morphism of dynamical systems, where the dynamics on SA is given by shifts. Show, moreover, that the map i is the unit of a *monad* (not comonad!) on the category of S-coalgebras. (*Hint*: The functor of the monad is induced by the functor S.)

5.4.2 The adjunction of coalgebras

As in the case of the Eilenberg–Moore category, every morphism of coalgebras over a given comonad C is in particular a morphism of the underlying category $\mathbf{C}$. Therefore, there is a faithful "forgetful" functor $\mathbf{C}^C \to \mathbf{C}$. Let's denote this functor by L^C.

Dually to the case of monads, this forgetful functor has a right-adjoint, which is constructed as follows. Let X be an object of C. Then, CX is canonically a C-coalgebra with the structure map ν. (Coalgebras of this form are sometimes called "cofree" since they are the dual to free algebras.) Similarly, given $f : X \to Y$, the morphism $Cf : CX \to CY$ is canonically a morphism of coalgebras (why?). This construction gives the desired functor $R^C : \mathbf{C} \to \mathbf{C}^C$.

Example 5.4.20 (dynamical systems). We know that coalgebras of the stream comonad are dynamical systems. Given a set X, we can canonically form a dynamical system from it, the one given by *streams and their shifts*. That is, we can form the set of streams SX, with the structure map given by ν. Recall from Example 5.4.8 that the structure map of a coalgebra A of the stream comonad corresponds to a function $f : A \to A$. The structure map $\nu : SX \to SSX$ corresponds to the function $SX \to SX$ given by shifts,

$$\{x_0, x_1, x_2, \dots\} \longmapsto \{x_1, x_2, x_2, \dots\}.$$

Example 5.4.21. Consider the reader comonad of Example 5.3.1. Given a set X, we can always form a coalgebra by taking the set $X \times E$, with the map $X \times E \to X \times E \times E$ given by copying the "extra information" E. Recall that, intuitively, a coalgebra of the reader comonad is a set equipped with a default choice for the extra information. Here, the "default value" of the extra information is trivially just the information that we already have.

Exercise 5.4.22. Prove that R^C is indeed right-adjoint to L^C (*Hint*: This is dual to Proposition 5.2.30.)

Corollary 5.4.23. *The adjunction above gives the following universal property. Let (C, ε, ν) be a comonad on* **C**. *Let X be an object of* **C**, *and let (A, i) be a C-coalgebra. For each morphism $f : A \to X$ of* **C**, *there exists a unique morphism of coalgebras $(A, i) \to (CX, \nu)$ such that the following diagram commutes:*

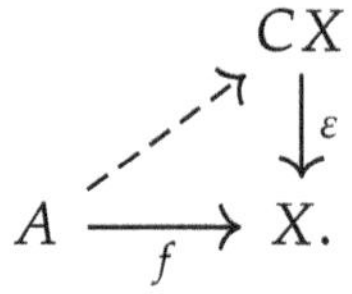

Let's see examples of this universal property. Just as the map for the monad case (Corollary 5.2.31) can be interpreted as an *extension* to arbitrary formal expressions (see the examples there), here we can interpret this unique map as a *lifting to the dynamics*.

Example 5.4.24 (dynamical systems). Let $(A, d : A \to A)$ be a dynamical system in the category of sets, and let X be a set. Given a map $g : A \to X$, we can form the morphism of dynamical systems $A \to SX$ given by

$$a \longmapsto \{g(a), g(d(a)), g(d(d(a))), \dots\}.$$

This respects the dynamics since replacing a by $d(a)$ has the same effect as shifting. This map is the unique morphism of dynamical systems $A \to SX$ such that its first component agrees with g. This map can be seen as *lifting* g to the dynamics over X, or as defining a dynamics on X derived from the dynamics on A, with g fixing the initial condition.

Example 5.4.25 (graph theory). Let (T, x) be a rooted tree, and let (G, y) be any multigraph with a distinguished vertex ("base point"). Given a map $f : T \to G$ that preserves incidence and base points, there is a unique map $T \to RG$ that preserves incidence and base points and lifts f to RG. This map is constructed as follows. A vertex v of T identifies a unique chain in T from x to v, which we had denoted by $i(v) \in RT$. Taking the image of $i(v)$ under f, we get a chain in G from y to $f(v)$. This chain can be seen as a vertex of RG. Since T and RG are trees, there is only one way of extending this assignment to edges while preserving incidence (how?), this gives the desired map $T \to RG$.

Exercise 5.4.26 (algebraic topology). What's the analogous of the example above for topological spaces and the universal covering comonad?

(Constructions of this kind are known in topology as **homotopy lifting properties.**)

5.5 Adjunctions, Monads, and Comonads

We have seen that, whenever we have a monad or a comonad, we can obtain an adjunction in two canonical ways. Conversely, whenever we have an adjunction, we can obtain a monad and a comonad canonically.

Theorem 5.5.1. *Let* **C** *and* **D** *be categories, and let* $F : \mathbf{C} \to \mathbf{D}$ *and* $G : \mathbf{D} \to \mathbf{C}$ *be adjoint functors, with* $F \dashv G$. *Denote the unit and counit by* $\eta : \mathrm{id}_{\mathbf{C}} \Rightarrow G \circ F$ *and* $\varepsilon : F \circ G \Rightarrow \mathrm{id}_{\mathbf{D}}$, *respectively. Then:*

(a) $G \circ F$ *is a monad on* **C**, *with unit* η *and multiplication* $G\varepsilon F$;

(b) $F \circ G$ *is a comonad on* **D**, *with counit* ε *and comultiplication* $F\eta G$.

Note that this theorem is a wide generalization of Exercise 5.1.31.

Since the two statements in the theorem are dual to each other, we only prove one.

Proof of (a). The specified natural transformations are by assumption natural and already in the desired form. We only have to prove that they satisfy the monad axioms (5.0.1). Explicitly, we have to show that the following diagrams commute for each object C of **C**:

$$\begin{array}{ccc} GFC & \xrightarrow{\eta} & GFGFC \\ & \searrow^{\mathrm{id}} & \downarrow G\varepsilon \\ & & GFC \end{array} \qquad \begin{array}{ccc} GFC & \xrightarrow{GF\eta} & GFGFC \\ & \searrow^{\mathrm{id}} & \downarrow G\varepsilon \\ & & GFC \end{array} \qquad \begin{array}{ccc} GFGFGFC & \xrightarrow{GFG\varepsilon} & GFGFC \\ \downarrow G\varepsilon & & \downarrow G\varepsilon \\ GFGFC & \xrightarrow{G\varepsilon} & GFC. \end{array}$$

Now:

- The first diagram commutes since it corresponds exactly to the second triangle identity (4.2.6), for $D = FC$.
- The second diagram commutes since it corresponds to the image under G of the first triangle identity (4.2.6).
- The last diagram commutes since it is the image under G of a naturality diagram for ε. □

Exercise 5.5.2. Prove (b) explicitly.

We can make further statements. Not only does every adjunction give rise to a monad and a comonad, but moreover, the adjunction will always lie "between" the Kleisli and Eilenberg–Moore adjunctions, in the way made precise and explained in the following.

Theorem 5.5.3. *Let* $\mathbf{C}$ *and* $\mathbf{D}$ *be categories, and let* $F : \mathbf{C} \to \mathbf{D}$ *and* $G : \mathbf{D} \to \mathbf{C}$ *be adjoint functors, with* $F \dashv G$ *and with unit and counit* η *and* ε*, respectively. Denote the induced monad* $G \circ F$ *by* T*:*

(a) *There is a canonical "comparison" functor* J *from the Kleisli category of* T *to* $\mathbf{D}$*, unique up to isomorphism, which makes the following diagrams commute (up to isomorphism):*

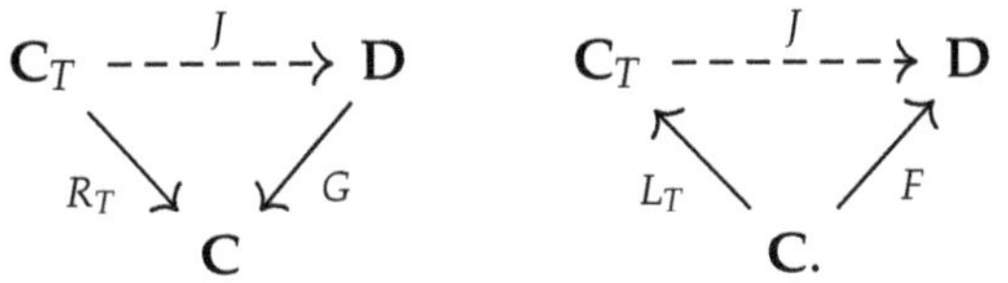

(5.5.1)

(b) *There is a canonical "comparison" functor* K *from* $\mathbf{D}$ *to the Eilenberg–Moore category of* T*, unique up to isomorphism, which makes the following diagrams commute (up to isomorphism):*

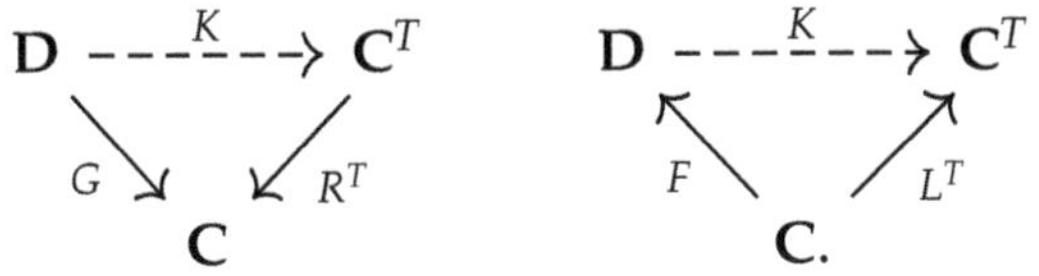

(5.5.2)

We prove the theorem by making use of the following lemma, following a similar approach to [Rie16, Proposition 5.2.12] (with extra steps).

Lemma 5.5.4. *Let* $\mathbf{A}$, $\mathbf{B}$ *and* $\mathbf{C}$ *be categories, and let* F, G, L, R, J *be functors as in the diagram*

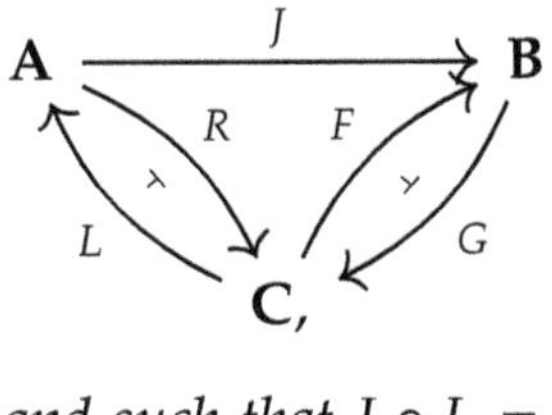

with $F \dashv G$ *and* $L \dashv R$*, and such that* $J \circ L = F$ *and* $G \circ J = R$*. Then, denoting by* $\varepsilon_{F,G}$ *and* $\varepsilon_{L,R}$ *the counits of the respective adjunctions, we have that* $J\varepsilon_{F,G} = \varepsilon_{L,R}$.

Proof of Lemma 5.5.4. First of all, let's write the counits in components. We have to prove that for every object A of $\mathbf{A}$, the following two arrows are equal:

$$\begin{array}{ccc} & JLRA = FRA = FGJA & \\ J(\varepsilon_{L,R})_A \searrow & & \swarrow (\varepsilon_{F,G})_{JA} \\ & A & \end{array}$$

Let's now show that for the "sharp" map of the adjunction (F, G), we have $\left(J(\varepsilon_{F,G})_A\right)^\sharp = \mathrm{id}_{GJA}$ (where, as usual, the map is given by $f \mapsto f^\sharp = Gf \circ \eta_{F,G}$). Since the "sharp" map is injective, this is sufficient to prove our statement. Now, denoting simply by G the action of G on hom-sets,

$$\begin{array}{ccccc} \mathrm{Hom}_{\mathbf{B}}(JLRA, JA) & \xrightarrow{G} & \mathrm{Hom}_{\mathbf{C}}(GJLRA, GJA) = \mathrm{Hom}_{\mathbf{C}}(RLRA, RA) & \xrightarrow{-\circ\eta} & \mathrm{Hom}_{\mathbf{C}}(RA, RA) \\ J(\varepsilon_{L,R})_A & \longmapsto & GJ(\varepsilon_{L,R})_A = R(\varepsilon_{L,R})_A & \longmapsto & R\varepsilon_{L,R} \circ \eta_{RA}, \end{array}$$

and by the triangle identities, $R\varepsilon_{L,R} \circ \eta_{RA} = \mathrm{id}_{RA} = \mathrm{id}_{GJA}$. Therefore, by uniqueness, $J(\varepsilon_{L,R})_A = (\varepsilon_{F,G})_{JA}$. □

Proof of Theorem 5.5.3. Let's construct the functor J explicitly. Recall that the objects of the Kleisli category are just the objects of $\mathbf{C}$ and L_T is the identity on objects. In order for the second diagram in (5.5.1) to commute, we are then forced to define, on objects, $J(X) = F(X)$. Note that this on objects makes also the diagram on the left of (5.5.1) commute: For each object X, $GJX = GFX = TX$, and recall that, on objects, $R_T(X) = TX$.

On morphisms, we want to argue that the action of J can only be given by the following composition:

$$\begin{array}{ccc} \mathrm{Hom}_{\mathbf{C}_T}(X, Y) & \xrightarrow{\quad J \quad} & \mathrm{Hom}_{\mathbf{D}}(FX, FY) \\ \sharp \searrow \cong & & \cong \nearrow \flat \\ & \mathrm{Hom}_{\mathbf{C}}(X, TY) = \mathrm{Hom}_{\mathbf{C}}(X, FGY) & \end{array}$$

where the "sharp" map is the one of the Kleisli adjunction and the "flat" map is the one of F and G. Since the "sharp" map is an isomorphism, we can equivalently prove the commutativity of the diagram with both "flat" maps, which we decompose as follows:

$$
\begin{array}{ccc}
\mathrm{Hom}_{\mathbf{C}_T}(X,Y) & \xrightarrow{\quad J \quad} & \mathrm{Hom}_{\mathbf{D}}(FX,FY) \\
\uparrow{\scriptstyle (\varepsilon_{kl})_Y \circ -} & & {\scriptstyle (\varepsilon_{F,G})_{FY} \circ -}\uparrow \\
\mathrm{Hom}_{\mathbf{C}_T}(X,TY) & \xrightarrow{\quad J \quad} & \mathrm{Hom}_{\mathbf{D}}(FX,FGFY) \\
& L_T \quad\quad F & \\
& \mathrm{Hom}_{\mathbf{C}}(X,TY) = \mathrm{Hom}_{\mathbf{C}}(X,FGY) &
\end{array}
$$

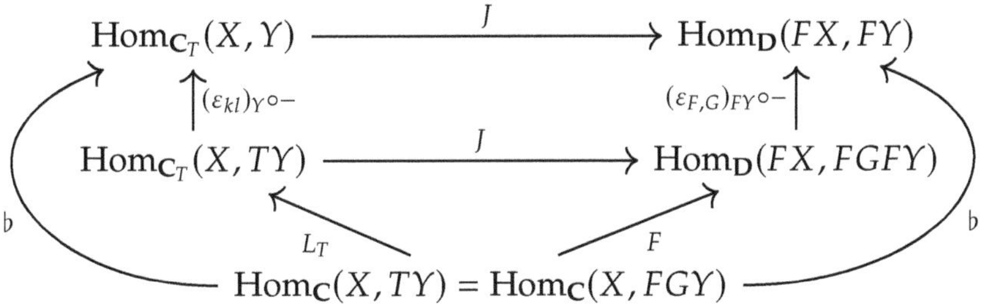

where ε_{kl} is the counit of the Kleisli adjunction and $\varepsilon_{F,G}$ is the one of F and G. Now, the lateral regions commute by definition of the "flat" maps, and the lower triangle commutes by assumption. The upper rectangle commutes since for every Kleisli morphism $f : X \to TY$ (corresponding to a morphism $X \to TTY$ of $\mathbf{C}$), we have that by functoriality and by Lemma 5.5.4,

$$J((\varepsilon_{kl})_Y \circ f) = J(\varepsilon_{kl})_Y \circ Jf = (\varepsilon_{F,G})_{FY} \circ Jf.$$

Therefore, the diagram commutes, and so the action of J on morphisms is uniquely specified. Concretely, if we write a Kleisli morphism $X \to Y$, in terms of a morphism $k : X \to TY$ of $\mathbf{C}$, then Jk is given by the "flat" of the adjunction (F, G), i.e.

$$Jk = k^{\flat} = \varepsilon_{F,G} \circ Fk,$$

which, for brevity, we simply write as $\varepsilon \circ Fk$.

It remains to be shown that J is indeed a functor. To see that it preserves identities, we have that at each object X of $\mathbf{C}$, $J(\eta) = \eta^{\flat} = \mathrm{id}_{FX}$. For composition, let $k : X \to GFY$ and $h : Y \to GFZ$. Then, by the naturality of ε, the following diagram commutes:

$$
\begin{array}{ccccc}
FGFY & \xrightarrow{FGFh} & FGFGZ & \xrightarrow{FG\varepsilon} & FGFZ \\
\downarrow{\scriptstyle \varepsilon} & & \downarrow{\scriptstyle \varepsilon} & & \downarrow{\scriptstyle \varepsilon} \\
FY & \xrightarrow{Fh} & FGFZ & \xrightarrow{\varepsilon} & FZ.
\end{array}
$$

Therefore,

$$
\begin{aligned}
J(h \circ_{kl} k) &= \varepsilon \circ F(\mu \circ Th \circ k)\\
&= \varepsilon \circ F(G\varepsilon \circ GFh \circ k)\\
&= \varepsilon \circ FG\varepsilon \circ FGFh \circ Fk\\
&= \varepsilon \circ Fh \circ \varepsilon \circ Fk\\
&= J(h) \circ J(k).
\end{aligned}
$$

Let's now turn to the functor K. Let A be an object of **D**. The functor K must map A to a T-algebra (B, b), and since $R^T(B, b) = B$, the condition $R^T \circ K = G$ forces us to have $B = GA$. Moreover, recall by Proposition 5.2.30 that the structure map of any T-algebra must be given by the counit of the Eilenberg–Moore adjunction; therefore, the map b must be exactly $(\varepsilon_{em})_{GA} : TGA \to GA$. By Lemma 5.5.4 (setting $J = K$, $(L, R) = (F, G)$ and $(F, G) = (L^T, R^T)$), we can write this map as

$$(\varepsilon_{em})_{GA} = K(\varepsilon_{F,G})_A,$$

and the condition $R^T \circ K = G$ on morphisms tells us that, as a morphism of **C**, $K(\varepsilon_{F,G})_A = G(\varepsilon_{F,G})_A$. Denoting $\varepsilon_{F,G}$ simply as ε, we must then have that, on objects,

$$KA = (GA, G\varepsilon_A).$$

To prove that this is indeed a T-algebra, note that the unit and multiplication square

$$
\begin{array}{ccc}
GA & \xrightarrow{\eta_{GA}} & GFGA \\
& \searrow^{\mathrm{id}} & \downarrow G\varepsilon_A \\
& & GA
\end{array}
\qquad\qquad
\begin{array}{ccc}
GFGFGA & \xrightarrow{GFG\varepsilon_A} & GFGA \\
\downarrow G\varepsilon_{FGA} & & \downarrow G\varepsilon_A \\
GFGA & \xrightarrow{G\varepsilon_A} & GA
\end{array}
$$

are the images under G of a unit triangle of the adjunction and of a naturality square for ε, respectively. This fixes the action of K on objects.

Let now $f : A \longrightarrow B$ be a morphism of $\mathbf{D}$. The condition $R^T \circ K = G$ on morphisms, together with the fact that R^T is faithful, imposes necessarily that $Kf = Gf$. This fixes the action of K on morphisms, and functoriality is immediately guaranteed.

Let's now check that the second diagram in (5.5.2) commutes, i.e. $K \circ F = L^T$. On objects, recalling the action of the "free" functor L^T,

$$KFX = (GFX, G\varepsilon_{FX}) = (TX, \mu_X) = L^T X.$$

On morphisms,

$$KFf = GFf = Tf = L^T f.$$

Therefore, the second diagram also commutes. □

Exercise 5.5.5 (important!). Write down the corresponding statements for comonads, dual to Theorem 5.5.3. (*Hint*: Be extra careful with the direction of the arrows.)

Definition 5.5.6. Let $\mathbf{C}$ and $\mathbf{D}$ be categories, and let $F : \mathbf{C} \longrightarrow \mathbf{D}$ and $G : \mathbf{D} \longrightarrow \mathbf{C}$ be adjoint functors, with $F \dashv G$. The adjunction $F \dashv G$ is called **monadic** if and only if the comparison functor $\mathbf{D} \longrightarrow \mathbf{C}^T$ is an equivalence of categories. In that case, we also call the right-adjoint G a **monadic functor**.

In other words, an adjunction is monadic if and only if the category $\mathbf{D}$ is, up to equivalence, the category of algebras of the induced monad.

Example 5.5.7 (several fields). Most adjunctions which we interpret as "free-forgetful" are monadic:

- the adjunction between sets and vector spaces of Exercise 4.1.9;
- the adjunction between sets and groups of Exercise 4.1.9;
- by construction, all the other examples of Section 5.2.3.

To see simple examples of non-monadic adjunctions, take the Kleisli adjunctions of most monads — usually, the Kleisli category and the Eilenberg–Moore category are not equivalent. Furthermore, by Theorem 5.5.3, given a monad T on $\mathbf{C}$, there is a canonical comparison functor $J : \mathbf{C}_T \to \mathbf{C}^T$, making the following diagrams commute (up to isomorphism):

$$\begin{array}{ccc} \mathbf{C}_T & \overset{J}{\dashrightarrow} & \mathbf{C}^T \\ & {\scriptstyle R_T}\searrow \quad \swarrow{\scriptstyle R^T} & \\ & \mathbf{C} & \end{array} \qquad \begin{array}{ccc} \mathbf{C}_T & \overset{J}{\dashrightarrow} & \mathbf{C}^T \\ & {\scriptstyle L_T}\nwarrow \quad \nearrow{\scriptstyle L^T} & \\ & \mathbf{C}. & \end{array} \tag{5.5.3}$$

Proposition 5.5.8. *The comparison functor $J : \mathbf{C}_T \to \mathbf{C}^T$ given above establishes an equivalence between the Kleisli category $\mathbf{C}_T$ and the full subcategory of $\mathbf{C}^T$ whose objects are the* free *algebras.*

Proof. First of all, let X be an object of $\mathbf{C}$ (or, equivalently, of $\mathbf{C}_T$). We have that, instantiating the proof of Theorem 5.5.3, $J(X) = L^T(X) = (TX, \mu)$. Therefore, all the objects in the image of J are free algebras. Conversely, every free algebra is in the image of J: Given a free algebra (TY, μ), the object Y is such that $J(Y) = (TY, \mu)$.

It remains to be shown that J is fully faithful. In other words, we have to prove that given the objects X and Y of $\mathbf{C}$, J induces a bijection from the Kleisli morphisms between X and Y and the morphisms of algebras between TX and TY. But this is given exactly by the universal property associated to the Eilenberg–Moore adjunction, as given in Corollary 5.2.31, by setting $(A, e) = (TY, \mu)$. □

Example 5.5.9 (probability). We have seen in Example 5.1.24 that, given a stochastic map (or a Markov kernel) $k : X \to \mathcal{P}Y$, we can canonically obtain a map $\mathcal{P}X \to \mathcal{P}Y$. We now know that this map is the unique morphism of (free) algebras, making the following diagram commute:

$$\begin{array}{ccc} X & & \\ {\scriptstyle \delta}\downarrow & \searrow^{k} & \\ \mathcal{P}X & \dashrightarrow & \mathcal{P}Y. \end{array}$$

In the finite case, such a map is an affine map, and so it can be represented by a particular matrix of nonnegative entries and whose columns sum to one, called a **stochastic matrix.**[6] Proposition 5.5.8 implies that stochastic maps and stochastic matrices encode the same information: There is an equivalence of categories between the category whose morphisms are stochastic maps between sets and the category whose morphisms are affine maps between simplices.

Here is an example where the comparison functor given above is an equivalence.

Exercise 5.5.10 (linear algebra). Prove that the Kleisli adjunction of the vector space monad (Exercise 5.2.16) *is* monadic. (*Hint*: In this case, all algebras are free.)

Why is this related to Example 1.5.19?

Exercise 5.5.11 (topology). Is the forgetful functor **Top** $\to$ **Set** of Example 4.1.3 monadic? (*Hint*: What is the induced monad on **Set**?)

A classic result of category theory, Beck's monadicity theorem, gives a necessary and sufficient condition for an adjunction to be monadic. We will not cover Beck's theorem here, but we refer the interested reader to [Rie16, Section 5.5].

We conclude this chapter with the following example, once again on the connection between graphs and categories.

5.5.1 The adjunction between categories and multigraphs is monadic

Consider the adjunction between categories and multigraphs of Section 4.2.2. Let's write explicitly the monad $T = U \circ \mathbf{P}$ associated to

[6] A stochastic matrix, in our convention, is a matrix of nonnegative entries such that the sum along each column is one.

the adjunction. Given a multigraph G, the graph TG has the same vertices as G, but, as edges, it has the *chains* of G, including the trivial chains at each vertex. Given a morphism $f : G \to H$ of multigraphs, T gives a morphism $Tf : TG \to TH$ given by taking the images of chains under f (since f preserves incidence, we get again a chain).

An algebra over T is a graph A together with a map $c : TA \to A$ that preserves incidence, which we can think of as mapping a chain to a single edge by *composing the edges of the chain*. The unit condition of (5.2.1) implies first of all that c has to be the identity on vertices (why?). Moreover, it has to map the edges in TA which come from (single) edges to A to the corresponding edges of A.

Let's now show that the category of T-algebras is equivalent to **Cat** by proving that the canonical comparison functor $K : \mathbf{Cat} \to \mathbf{MGraph}^T$ is an equivalence.

Every (small) category is canonically a T-algebra, with the map c given by identities and composition. In detail:

- for the trivial chain 1_X at X, we define $c(1_X) = \mathrm{id}_X$;
- for a one-morphism chain f, we define $c(f) = f$;
- for any two or more composable morphisms f and g, we define $c(f_1, \dots, f_n) = f_n \circ \dots \circ f_1$.

The unit condition for algebras states that the composition of a single morphism gives again that morphism, which is satisfied by construction. The multiplication condition states that if we have a composable chain of composable chains of morphisms

$$((f_{1,1}, \dots, f_{1,n_1}), \dots, (f_{m,1}, \dots, f_{m,n_m})),$$

then we can either first "flatten the array,"

$$(f_{1,1}, \dots, f_{1,n_1}, \dots, f_{m,1}, \dots, f_{m,n_m}),$$

and take the composite,

$$f_{m,n_m} \circ \dots \circ f_{m,1} \circ \dots \circ f_{1,n_1} \circ \dots \circ f_{1,1},$$

or we can first take the compositions inside the brackets,

$$(f_{1,n_1} \circ \cdots \circ f_{1,1}, \ldots, f_{m,n_m} \circ \cdots \circ f_{m,1}),$$

and then compose the resulting morphism,

$$(f_{m,n_m} \circ \cdots \circ f_{m,1}) \circ \cdots \circ (f_{1,n_1} \circ \cdots \circ f_{1,1}).$$

By associativity of composition, the two procedures produce the same result. Therefore, **C** has a canonical T-algebra structure.

Consider now two (small) categories **C** and **D** — we know these are equivalently T-algebras. Every functor $\mathbf{C} \to \mathbf{D}$ preserves incidence on the underlying graphs, and different functors give different morphisms of graphs. A morphism of graphs $f : \mathbf{C} \to \mathbf{D}$ is a morphism of T-algebras if and only if the following diagram commutes:

$$\begin{array}{ccc} T\mathbf{C} & \xrightarrow{Tf} & T\mathbf{D} \\ \downarrow{\scriptstyle c} & & \downarrow{\scriptstyle d} \\ \mathbf{C} & \xrightarrow{f} & \mathbf{D}. \end{array}$$

Now, functors preserve compositions and identities, and so they make the diagram above commute. In detail, for 1_X, the trivial chain in $T\mathbf{C}$ at X, we have that, by functoriality,

$$f(c(1_X)) \;=\; f(\mathrm{id}_X) \;=\; \mathrm{id}_{f(C)} \;=\; c(Tf(1_X)).$$

For a chain given by just one morphism g, the diagram commutes since both approaches simply return g. For a chain given by two (or more) composable morphisms $e_1, \ldots, e_n$, we have that, again by functoriality,

$$\begin{aligned} f(c(e_1, \ldots, e_n)) &= f(e_n \circ \cdots \circ e_1) = f(e_n) \circ \cdots \circ f(e_1) \\ &= c(f(e_1), \ldots, f(e_n)). \end{aligned}$$

Therefore, every functor corresponds to a morphism of T-algebras. In other words, the assignment from small categories to T-algebras is functorial. The functor commutes with the forgetful functors that take the underlying graph (of a category and of an algebra), and so this functorial assignment must be (up to natural isomorphism) the functor $J : \mathbf{Cat} \to \mathbf{MGraph}^T$ of Theorem 5.5.3.

Let's now show that J is an equivalence. First of all, not only is every category an algebra, but also every algebra is canonically a category (necessarily small, why?), with objects and morphisms given by vertices and edges and composition given indeed by the map c. In detail, let (A, c) be a T-algebra:

- For each vertex x of A, we have the trivial chain of TA at x; denote it by 1_x. Define then the identity morphism id_x as the edge $c(1_x)$ of A.
- For each pair of composable edges e_1 and e_2, define their composite $e_2 \circ e_1$ as the edge $c(e_1, e_2)$ (we write only one level of brackets).
- Given an edge e from x to y, we have

$$e \circ \mathrm{id}_x \; = \; c(1_x, e) \; = \; c(e) \; = \; e$$

(since composing with a trivial chain gives the same edge e) and

$$\mathrm{id}_y \circ e \; = \; c(e, 1_y) \; = \; c(e) \; = \; e,$$

so the composition given by c is unital.
- Given three consecutive edges e, f, g, we have by the multiplication square that

$$\begin{aligned} g \circ (f \circ e) \; &= \; c(c(e, f), c(g)) \; = \; c(e, f, g) \\ &= \; c(c(e), c(f, g)) \; = \; (g \circ f) \circ e \end{aligned}$$

(recall that μ gives the concatenation of chains), so the composition given by c is associative. Therefore, (A, c) is canonically a category.

Now, let $f : \mathbf{C} \to \mathbf{D}$ be a morphism of T-algebras, which makes the following diagram commute:

$$\begin{array}{ccc} T\mathbf{C} & \xrightarrow{Tf} & T\mathbf{D} \\ \downarrow{\scriptstyle c} & & \downarrow{\scriptstyle d} \\ \mathbf{C} & \xrightarrow{f} & \mathbf{D}. \end{array}$$

Then, for each object X of $\mathbf{C}$,

$$f(\mathrm{id}_X) \;=\; f(c(1_X)) \;=\; c(Tf(1_X)) \;=\; c(1_{f(X)}) \;=\; \mathrm{id}_{f(X)},$$

and for each composable morphisms $g : X \to Y$ and $h : Y \to Z$,

$$f(h \circ g) \;=\; f(c(g, h)) \;=\; c(f(g), f(h)) \;=\; f(h) \circ f(g).$$

Therefore, f is a functor. This means once again that the assignment from T-algebras to categories is functorial. By construction, this functor gives an inverse to the functor $J : \mathbf{Cat} \to \mathbf{MGraph}^T$ commuting with the forgetful functors. (If it's not clear why the two functorial assignments are mutually inverse, try to prove it.) Therefore, J is an equivalence, and the adjunction is monadic.

In summary, the T-algebras are precisely the (small) categories, and their morphisms are precisely the functors between them. That is, the extra structure that a graph needs to have in order to be a category is exactly encoded by a monad, whose algebras are categories.

6

Monoidal Categories

In this last chapter, we study monoidal categories. They are one of the most fruitful ideas of category theory, with applications as diverse as quantum information theory, algebraic geometry, and probability.

Monoidal categories would require an entire book of their own, or even several, just for an introduction. The material here, however, should at least be enough to get you started.

While *using* monoidal categories in mathematical practice very often saves time and effort (especially using *string diagrams*, see Section 6.1.2), *developing* the necessary theory often involves lengthy proofs and rather large diagrams. For reasons of space, we cannot include all these proofs in their entirety here — we instead provide the basic ideas and references for further learning.

6.1 General Definitions

Definition 6.1.1. A **monoidal category** or **tensor category** consists of:

- a category $\mathbf{C}$;
- a distinguished object I, called the **unit**;
- a (joint) functor[1] $\otimes : \mathbf{C} \times \mathbf{C} \to \mathbf{C}$, called the **tensor product**, **monoidal product**, or **monoidal multiplication**;

[1]Recall Section 1.3.8.

- natural isomorphisms with components

$$I \otimes A \xrightarrow[\cong]{\lambda_A} A, \quad A \otimes I \xrightarrow[\cong]{\rho_A} A, \quad (A \otimes B) \otimes C \xrightarrow[\cong]{\alpha_{A,B,C}} A \otimes (B \otimes C),$$

  called, respectively, **left unitor, right unitor,** and **associator,**

such that the following diagrams commute for all objects: $A, B, C, D \in \mathbf{C}$,

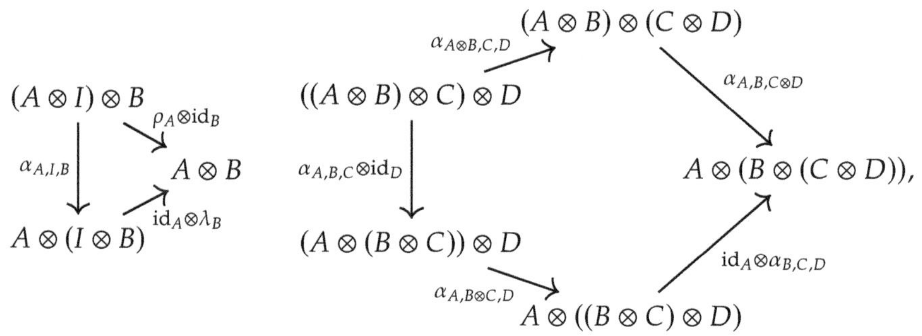

called the **triangle** and **pentagon** conditions.

Notation 6.1.2. For each object A, the action of the functor $A \otimes - : \mathbf{C} \to \mathbf{C}$, i.e.

$$B \xrightarrow{f} C \qquad \longmapsto \qquad A \otimes B \xrightarrow{A \otimes f} A \otimes C,$$

will be also denoted by

$$B \xrightarrow{f} C \qquad \longmapsto \qquad A \otimes B \xrightarrow{\mathrm{id}_A \otimes f} A \otimes C.$$

By Proposition 1.3.50, this does not lead to ambiguity. We use a similar notation for the functor $- \otimes A$.

6.1.1 Categories with multiplications and coherence

The first interpretation that we can give for monoidal categories, which also motivates the name, is that a monoidal category is a category which "looks like a monoid":

Idea. *A monoidal category is a category whose objects and whose morphisms can be multiplied together in a way that is associative and unital, as in a monoid, but only up to isomorphism. (And the isomorphisms are particularly convenient.)*

Usually, this allows us to form "more complex objects from simpler ones."

The simplest case of a monoidal category is when the associators and unitors are identities so that associativity and unitality hold strictly, just like in monoids.

Definition 6.1.3. A **strict monoidal category** consists of:

- a category $\mathbf{C}$;
- a distinguished object, I called the **unit**;
- a functor $\otimes : \mathbf{C} \times \mathbf{C} \to \mathbf{C}$, called **multiplication**,

such that the following conditions hold:

$$I \otimes A = A; \qquad A \otimes I = A; \qquad (A \otimes B) \otimes C = A \otimes (B \otimes C),$$

called, respectively, **left and right unitality**, and **associativity**.

Exercise 6.1.4. Show that in a strict monoidal categories, the unitors and associator given by identities satisfy the triangle and pentagon conditions.

Here are some examples of strict monoidal categories.

Example 6.1.5. Given a category $\mathbf{C}$, form the category $\mathbf{LC}$, where:

- objects are finite lists, $[X_1, \ldots, X_n]$, of objects of $\mathbf{C}$, including the empty list;
- morphisms are in the form $[f_1, \ldots, f_n] : [X_1, \ldots, X_n] \to [Y_1, \ldots, Y_n]$, where each $f_i : X_i \to Y_i$ is a morphism of $\mathbf{C}$, and the only morphism on the empty list is the identity;

- the tensor product, on objects and morphisms, is given by concatenation of lists, and the unit is the empty list.

The category **LC** as constructed above is strict monoidal.

Exercise 6.1.6. Show that the construction above forms a monad on **Cat**, analogous to the list monad of Exercise 5.1.6. What are its algebras?

Example 6.1.7. Let **C** be a category. Following the construction of Definition 1.4.18, we can form the functor category $[\mathbf{C}, \mathbf{C}]$, where objects are endofunctors $\mathbf{C} \to \mathbf{C}$ and morphisms are natural transformations between them. This category is strict monoidal, with the unit given by the identity functor and the tensor product given by functor composition, which is strictly associative and unital. (Why is composition functorial?)

Exercise 6.1.8. Show that a monoid can be seen as a discrete, strict monoidal category.

Exercise 6.1.9. Let M be a monoid equipped with a partial order, and suppose that the multiplication is monotone, meaning that if $a \leq a'$ and $b \leq b'$, then $ab \leq a'b'$. Show that this partial order, seen as a category, is a strict monoidal category.

One may ask, why can't we just use strict monoidal categories, with associativity and unitality holding strictly? The main reason is that for a lot of important, useful examples, associativity and unitality do *not* hold strictly. The prototypical example of this fact is given by sets and their cartesian products. Let's look at them in detail.

Example 6.1.10 (sets and relations). Consider the category **Set**, with its cartesian product and with a singleton 1 as the monoidal unit. That is, given two sets A and B, their tensor product is the categorical product $A \times B$. This construction is associative and unital only up to isomorphism. Indeed, consider three sets A, B, and C, and form the nested products

$$(A \times B) \times C \quad \text{and} \quad A \times (B \times C).$$

The first set has as elements nested ordered pairs in the form $((a, b), c)$, with $a \in A$, $b \in B$, and $c \in C$. The second set has instead elements in the form $(a, (b, c))$. The two sets are in bijection to one another:

$$\begin{aligned} (A \times B) \times C &\xrightarrow[\cong]{\alpha_{A,B,C}} A \times (B \times C) \\ ((a, b), c) &\longmapsto (a, (b, c)); \end{aligned}$$

however, they are not *the same set*. Technically, they are not *equal*. We cannot write that $(A \times B) \times C = A \times (B \times C)$, in general. Similarly, if $\bullet$ is the unique element of the singleton set 1, we have the bijections

$$\begin{aligned} 1 \times A &\xrightarrow[\cong]{\lambda_A} A \xleftarrow[\cong]{\rho_A} A \times 1 \\ (\bullet, a) &\longmapsto a \longleftarrow\!\dashv (a, \bullet), \end{aligned}$$

but once again, $A \times 1$, $1 \times A$, and A are not *equal*. Associativity and unitality hold only up to isomorphism. (As we will shortly see, this construction will make **Set** a monoidal category.)

Similar considerations hold for other categories, such as

- **Top**, with the cartesian product of topological spaces;
- **Vect**, with the cartesian product (or direct sum) of vector spaces;
- **Grp**, with the direct product of groups.

The good news is that we still get a monoidal category, just not a strict one.

Proposition 6.1.11. *The category* **Set**, *together with the cartesian product, the singleton set, and the isomorphisms given in Example* 6.1.10, *forms a monoidal category.*

Proof. First of all, let's define the action of $\times$ on morphisms. Consider two functions $f : X \to Y$ and $g : A \to B$. We define the function $f \times g : X \times A \to Y \times B$ by

$$\begin{aligned} X \times A &\longrightarrow Y \times B \\ (x, a) &\longmapsto (f(x), g(a)), \end{aligned}$$

and this makes $\times$ a (joint) functor $\mathbf{Set} \times \mathbf{Set} \to \mathbf{Set}$ (why?).

We now have to prove that the isomorphisms α, λ and ρ of Example 6.1.10 are natural and satisfy the triangle and pentagon conditions of Definition 6.1.1. Naturality is left as an exercise. Now, let A, B, C, and D be sets. If, as above, $\bullet$ is the unique element of 1, and for all $a \in A$, $b \in B$, $c \in C$, and $d \in D$, the triangle and pentagon identities give the following diagrams:

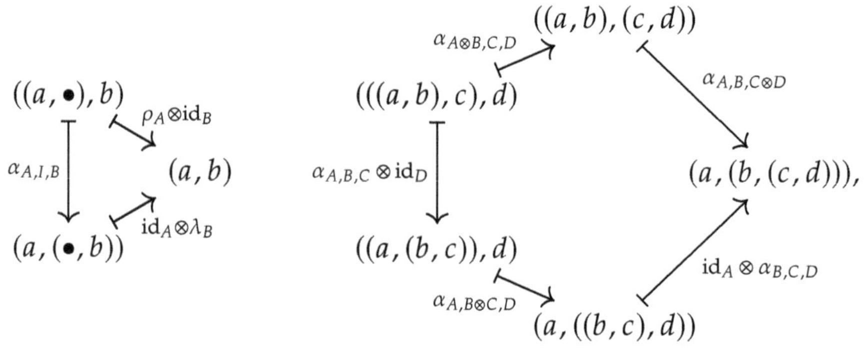

which means that the triangle and pentagon diagrams commute since in each of the diagrams here, both paths arrive at the same element. □

The same holds for other categories where we take the categorical product as tensor.

Exercise 6.1.12. Prove that if a category $\mathbf{C}$ has binary products and a terminal object, it canonically forms a monoidal category. (*Hint*: Generalize what we did in Example 6.1.10 and Proposition 6.1.11.)

When this happens, we speak of a *cartesian* monoidal category, defined as follows.

Definition 6.1.13. A **cartesian monoidal category** is a monoidal category where:

- the tensor product of any two objects is a categorical product;
- the unit is a terminal object;
- the unitors $1 \times A \to A$ and $A \times 1 \to A$ are product projections;
- the associators are the unique maps making the diagrams in the following form commute:

$$\begin{array}{ccccc} & & (A \times B) \times C & & \\ & \swarrow^{p_1 \circ p_1} & \downarrow & \searrow^{p_2 \times \mathrm{id}_C} & \\ A & \xleftarrow[p_1]{} & A \times (B \times C) & \xrightarrow[p_2]{} & B \times C, \end{array}$$

  where p_i are the product projections.

Not every monoidal category is cartesian monoidal, for example, the category constructed in Example 6.1.5, in general, is not.

A natural question to ask at this point is, why do we want the triangle and pentagon diagrams to commute, why don't we just require associativity and unitality to hold up to isomorphism, with no further conditions? This choice can be motivated by introducing the idea of *coherence*: The triangle and pentagon conditions allow every diagram of a certain kind to commute automatically.

Theorem 6.1.14 (Coherence for monoidal categories). *In a monoidal category, every formal diagram built freely out of tensor products of objects and the unit (on objects), and out of associators, unitors and their products (on morphisms) commutes.*

The proof of this statement would be too much for the purposes of this book. Here, we can give an example for some intuition, and refer the interested reader to more specialized material, in particular Refs. [Tru20] and [Yanng].

Example 6.1.15. Consider five objects $A, B, C, D,$ and E in a monoidal category. We can form tensor products nested on the left and right, as follows:

$$(((A \otimes B) \otimes C) \otimes D) \otimes E \quad A \otimes (B \otimes (C \otimes (D \otimes E))).$$

By composing associators, we can show that these two products are isomorphic. *A priori*, though, they could be isomorphic in different ways. For example, this diagram connects them in two ways, and it's not clear, *a priori*, that these operations are the same: (We omit the symbol $\otimes$ in objects for brevity.)

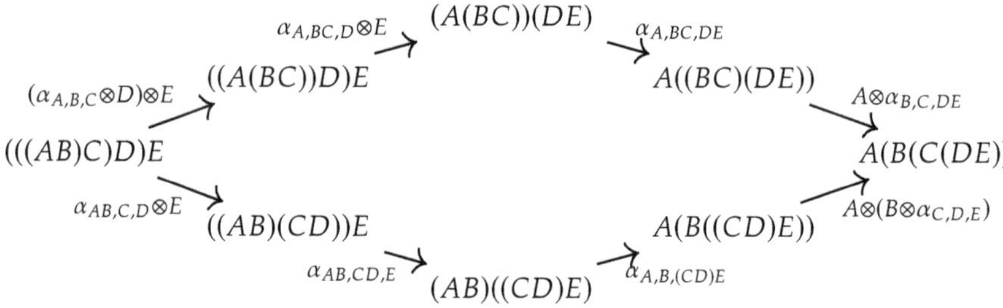

Let's now show that it commutes. We can decompose it as follows, omitting the labels for brevity:

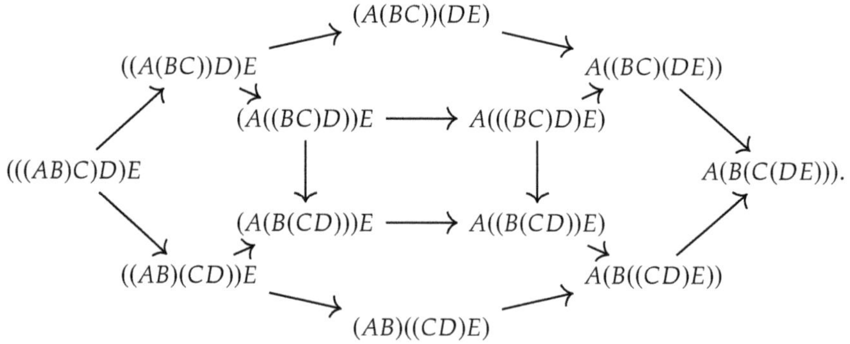

The central square commutes by naturality of the associator, and the other pentagons are instances of the pentagon condition:

- On the left, it's the pentagon condition for the objects A, B, C, D (tensored with E at the end).
- Above, it's the pentagon condition for the objects A, BC, D, E.

- Below, it's the pentagon condition for the objects A, B, CD, E.
- On the right, it's the pentagon condition for the objects B, C, D, E (tensored with A).

This way, the whole diagram commutes.

What the coherence theorem says is that every composition and product of associators and unitors, just as in the case above, is equal, no matter how complex the composition is.

Exercise 6.1.16. Add the labels to the arrows of the diagram in the previous page.

The idea and importance of coherence, similar to those of naturality, are not usually clear at the beginning, but they will hopefully be clear as we proceed further into the chapter. Very roughly, coherence can be interpreted as the fact that we can "forget" about associators and unitors, and treat them as if they were equalities, as in a strict monoidal category. (But they are not equalities, in general.) More on this in Section 6.3.4.

Because of coherence, we omit the symbols α, λ, and ρ, as well as their images under functors, whenever we have a morphism built out of these which is unique by coherence. For example, we can denote the map in Example 6.1.15 simply as

$$(((A \otimes B) \otimes C) \otimes D) \otimes E \xrightarrow{\cong} A \otimes (B \otimes (C \otimes (D \otimes E))).$$

Indeed, there is really only one choice of isomorphism built out of associators. (We retain the symbols whenever it is necessary to avoid ambiguity.) Similarly, we denote a monoidal category by $(\mathbf{C}, \otimes, I)$ instead of $(\mathbf{C}, \otimes, I, \alpha, \lambda, \rho)$.

Before we move on, let's also look at an example of a diagram that does *not* commute, where the coherence theorem does not apply, to explain what we mean by a *formal* diagram. Suppose that in a certain monoidal category $\mathbf{C}$, we have an object A such that $A \otimes A = A$.

(A famous example, given by Isbell, is to take a category equivalent to **Set** where we have one object for each isomorphism class, i.e. a *skeleton* of **Set**, and A to be a countably infinite set.) Then, we have $(A \otimes A) \otimes A = A \otimes A = A \otimes (A \otimes A)$, and we can form the following diagram:

$$(A \otimes A) \otimes A \underset{\mathrm{id}}{\overset{\alpha}{\rightrightarrows}} A \otimes (A \otimes A).$$

In general, this diagram does *not* commute (see the following exercise). The reason why the diagram above is not a *formal* diagram is that it crucially relies on particular properties of **C** and A. In general, the associator is in the form $(A \otimes B) \otimes C \to A \otimes (B \otimes C)$, and there is no identity map parallel to it, and so a diagram like the one above cannot be formed "in general," i.e. just using the structure of a monoidal category with no further assumptions.

Exercise 6.1.17. Show that the diagram above does not commute in general. (*Hint*: Take a skeleton of **Set** with the cartesian product and A to be countably infinite. Show that if the diagram commuted, then there would be a unique endomorphism $A \to A$, which is a false statement.)

Let's now look at a further refinement of monoidal categories. Just as a monoidal category generalizes a monoid, a *symmetric* monoidal category generalizes a commutative monoid. Here is the definition.

Definition 6.1.18. A **symmetric monoidal category** is a monoidal category $(\mathbf{C}, \otimes, I)$, together with a natural isomorphism with components

$$A \otimes B \xrightarrow[\cong]{\beta_{A,B}} B \otimes A,$$

called the **braiding**, such that the following diagrams commute:

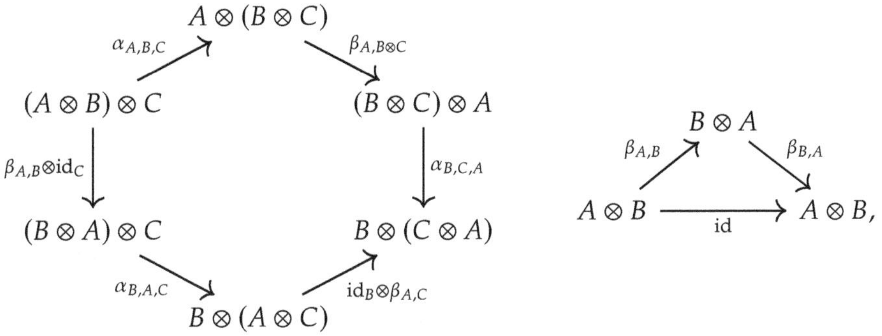

which are called, respectively, the **hexagon** and **involutivity** conditions. Note that the latter holds for all A and B if and only if $\beta_{A,B} = \beta_{B,A}^{-1}$.

There is also a notion of *braided monoidal category*, with a braiding which does not satisfy the involutivity condition and which satisfies a further, dual, hexagon condition. We will not treat that notion here.[2]

Example 6.1.19. $(\mathbf{Set}, \times, 1)$ is symmetric. The braiding is given by the map

$$\begin{array}{rcl} A \times B & \xrightarrow[\cong]{\beta} & B \times A \\ (a,b) & \longmapsto & (b,a). \end{array}$$

The hexagon and involutivity condition now look as follows for each $a, \in A$, $b \in B$, and $c \in C$:

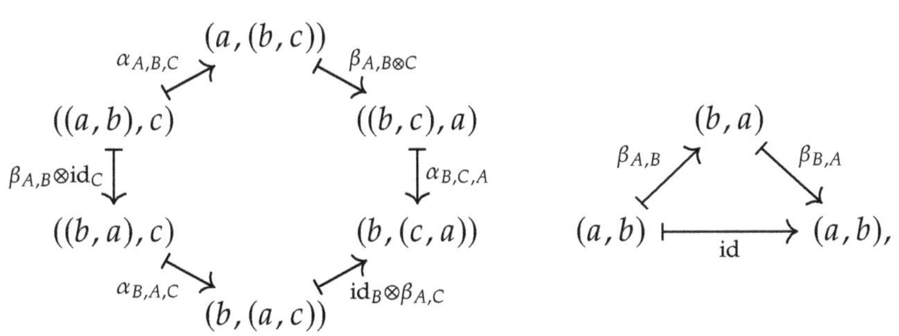

[2]See the nLab: https://ncatlab.org/nlab/show/braided+monoidal+category.

and they commute since in each diagram, each path arrives at the same element.

Exercise 6.1.20. Show that every cartesian monoidal category is symmetric. (*Hint*: Generalize the previous example.)

Just as for monoidal categories, there is a coherence result for symmetric monoidal categories, stating roughly that we can also "forget" about *braidings*, or treat them as if they were identities.

Theorem 6.1.21 (Coherence for symmetric monoidal categories). *In a symmetric monoidal category, every formal diagram freely built out of tensor products of objects and the unit (on objects) and out of associators, unitors, braidings, and their products (on morphisms) commutes.*

Therefore, just as for associators and unitors, in diagrams we label braidings (and their compositions and products with associators and unitors) simply by the symbol $\cong$, whenever this does not lead to ambiguity.

As we did for associators, here is an example of diagram that *does not* commute in general:

$$A \otimes A \underset{\mathrm{id}}{\overset{\beta}{\rightrightarrows}} A \otimes A.$$

Indeed, in $(\mathbf{Set}, \times, 1)$, one arrow would map (a, a') to (a', a), and the other arrow is the identity. The reason why this diagram is not *formal*, in this case, is that the identity arrow only exists between $A \otimes A$ and $A \otimes A$, and not, for example, between $A \otimes B$ and $B \otimes A$ with generic A and B, and so, the diagram above cannot be formed for $A \neq B$.

6.1.2 Parallel composition and string diagrams

Here is another interpretation for monoidal categories, often used in applications, for example, in physics, computer science, and probability.

Idea. *A monoidal category is a category in which morphisms, which one can think of as "processes" of some kind, can be composed not only* sequentially *(in the usual way) but also* in parallel, *meaning that they happen "at the same time," or at least "independently."*

Example 6.1.22 (sets and relations). Consider two functions $f : X \to Y$ and $g : A \to B$. Recall that the function $f \times g : X \times A \to Y \times B$ acts as follows:

$$\begin{aligned} X \times A &\longrightarrow Y \times B \\ (x, a) &\longmapsto (f(x), g(a)). \end{aligned}$$

There are many more functions $X \times A \to Y \times B$ than those in the form above: In general, for a function $h : X \times A \to Y \times B$, the Y-component of the result depends on *both* X and A, and the same holds for the B-component. The functions in the form $f \times g$, instead, are "separable," or "independent," or "non-signaling": The first output only depends on the first input, and the second output only depends on the second input. In other words, the function $f \times g$ is the process which describes f and g happening *in parallel.*

Whenever we interpret a monoidal category in this way, it is often helpful to use, in addition to the usual diagrams, another type of diagrams, called **string diagrams**, or, sometimes, **wiring diagrams**.[3] String diagrams are drawn as follows:

- Each object of our category, as well as its identity, is represented by a *wire*, as follows:

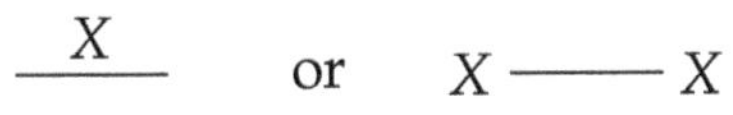

- Each morphism $f : X \to Y$ is represented by a *block* with the wire X as input (on the left) and the wire Y as output (on the right):

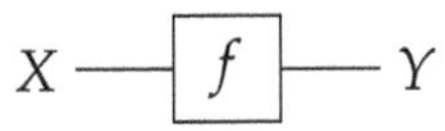

[3]Different authors use different terminology, and "string diagrams" and "wiring diagrams" are not always synonyms, see for example Refs. [CK17; FS19]. Here, we always use the term "string diagram."

Therefore, our string diagrams are always read from left to right. (Other texts have other conventions; for example, in physics, it is common to orient the diagrams from bottom to top.)

- The composition of morphisms $f : X \to Y$ and $g : Y \to Z$ is denoted as follows:

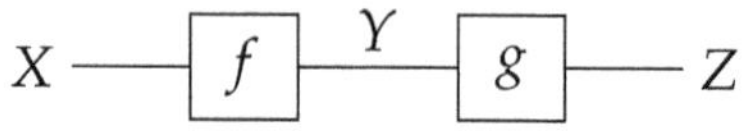

- The tensor product of two morphisms, $f \otimes g : X \otimes A \to Y \otimes B$, is denoted as follows:

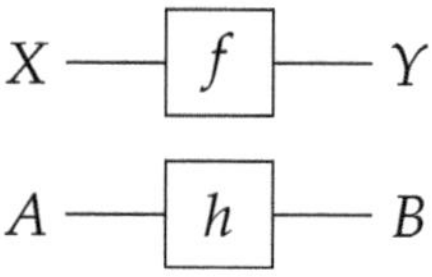

- A generic morphism in the form $X \otimes A \to Y \otimes B$ is denoted as follows:

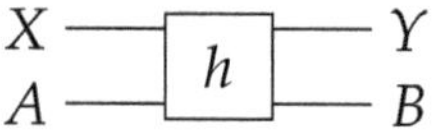

- The monoidal unit can be denoted by "no wire": For example, the morphisms $f : I \to A$, $g : A \to I$ and $h : I \to I$ are denoted as follows:

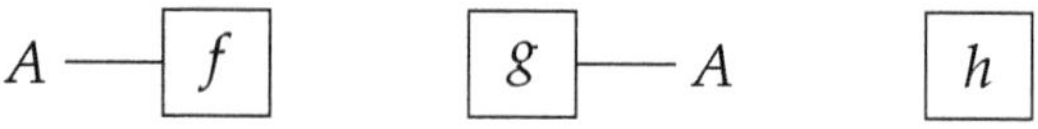

As one may have noticed, if one takes the tensor product of three or more morphisms, our string diagrams cannot distinguish between $(f \otimes g) \otimes h$ and $f \otimes (g \otimes h)$. Both are represented as follows:

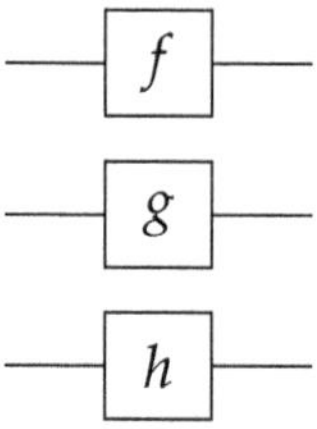

Similarly, our string diagrams cannot distinguish between $I \otimes f, f,$ and $f \otimes I$ since they are all represented as follows:

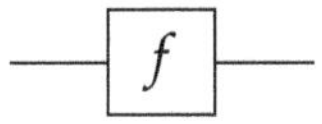

In other words, our string diagrams are *strictly associative and unital*, and so they are specifically suitable only for *strict* monoidal categories. Thanks to the coherence theorem, however, we can also use string diagrams to represent generic monoidal categories if we are careful. Let's illustrate this idea by means of an example. The string diagram

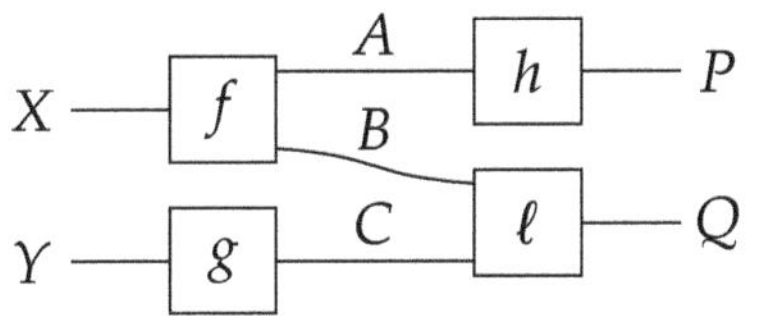

(6.1.1)

represents the sequential composition of the following tensor products:

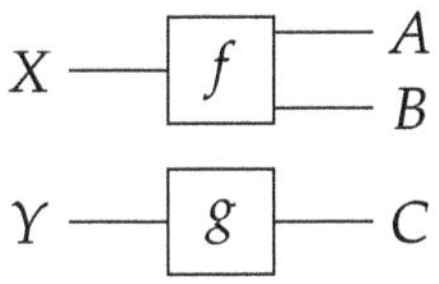

and

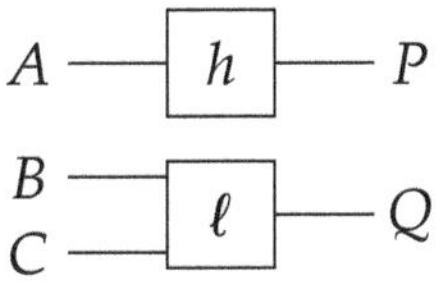

i.e.

$$X \otimes Y \xrightarrow{f \otimes g} (A \otimes B) \otimes C \quad \text{and} \quad A \otimes (B \otimes C) \xrightarrow{h \otimes \ell} P \otimes Q.$$

Now, technically, we cannot compose these two morphisms since the codomain of the first one is not *equal* to the domain of the second one. However, there is a unique canonical way of relating the two, namely, via the associator

$$X \otimes Y \xrightarrow{f \otimes g} (A \otimes B) \otimes C \xrightarrow{\alpha_{A,B,C}} A \otimes (B \otimes C) \xrightarrow{h \otimes \ell} P \otimes Q.$$

This composition is the one represented by the string diagram (6.1.1). This idea will be made precise in the form of a *strictification theorem of monoidal categories* (Theorem 6.3.25). It says that every monoidal category is *monoidally* equivalent to a strict one, and it implies that we can interpret the wires and blocks in a string diagram as representing particular *equivalence classes* of objects, called *cliques*, and morphisms between these (see Section 6.3.4).

Using Proposition 1.3.50, we can see that an equivalent condition to the joint functoriality of the tensor product is that for all morphisms $f : X \to Y$ and $g : A \to B$,

$$(\mathrm{id}_Y \otimes g) \circ (f \otimes \mathrm{id}_A) \;=\; (f \otimes \mathrm{id}_B) \circ (\mathrm{id}_X \otimes g);$$

that is, in terms of string diagrams,

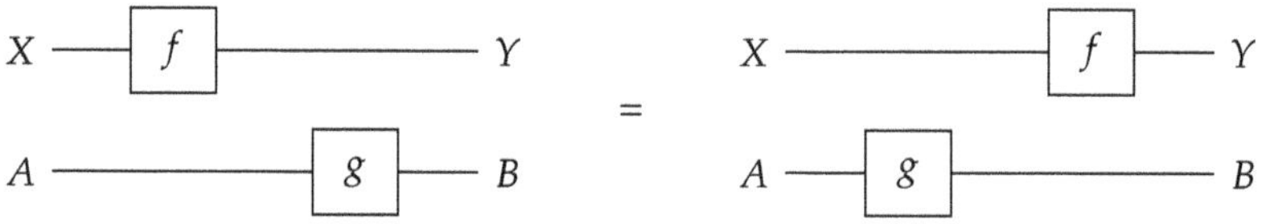

(6.1.2)

which we can interpret as the fact that f and g can happen in either order since they run on "different threads," and the output of one does not enter the input of the other one. (If you are familiar with special relativity, think of events with space-like separation.)

Similarly, consider the following diagram:

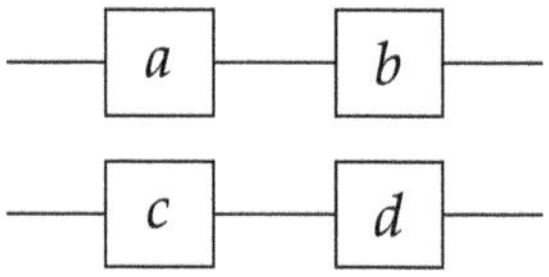

(6.1.3)

It can be seen either as the sequential composition of

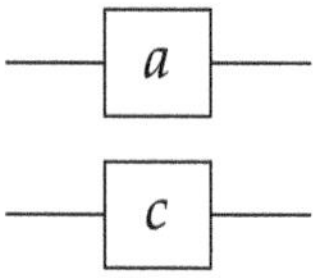

and

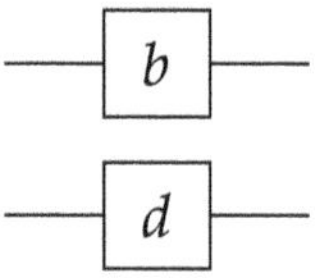

i.e. as $(b \otimes d) \circ (a \otimes c)$, or as the parallel composition (tensor product) of

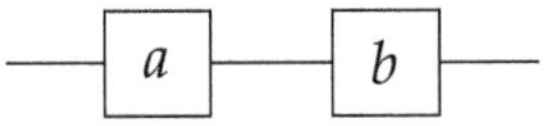

and

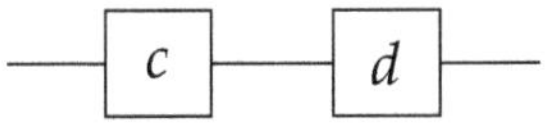

i.e. as $(b \circ a) \otimes (d \circ c)$. In a monoidal category, the diagram (6.1.3) is not ambiguous, thanks to the following **interchange law**[4]:

$$(b \otimes d) \circ (a \otimes c) \ = \ (b \circ a) \otimes (d \circ c),$$

which follows from the joint functoriality of the tensor product $\otimes$ (how?).

While associators and unitors are not usually written in string diagrams, the braiding of a symmetric monoidal category is, and it is represented as follows:

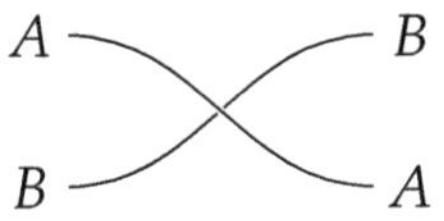

The involutivity condition states that we can "unknot" a double braiding to the identity

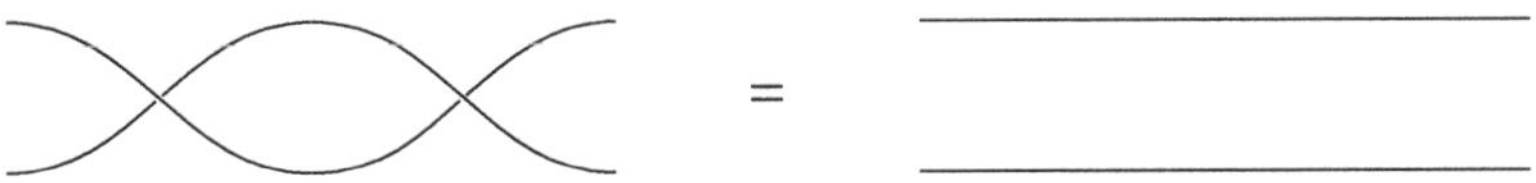

which geometrically would be impossible if we imagine these wires embedded in a three-dimensional space, but possible in four dimensions. (There is a specific link between monoidal categories and higher-dimensional geometry, but we will not explore that here.)

These are the basic building blocks of string diagrams — throughout this chapter, we occasionally make use of them, and on other occasions, we use traditional diagrams.

---

[4]Sometimes, the term "interchange law" denotes the related condition (6.1.2).

How to draw string diagrams in LaTeX: All the diagrams in this document have been drawn using the applet *TikZit*, which is based on PGF/TikZ.[5]

Additional interpretation: Before we conclude this section, we should also mention that there exists another important interpretation of monoidal categories: A monoidal category can be seen as a (weak version of a) 2-category, such as **Cat**, but with only a single object. (Compare with the fact that a monoid is equivalently a category with a single object.) This is however beyond the scope of this book — we refer the interested reader to more advanced material on higher category theory, such as the nLab.[6]

6.1.3 More examples of monoidal categories

Here are some more examples of monoidal categories. If it is unclear why they satisfy the axioms, we invite the reader to check them explicitly.

Example 6.1.23 (graph theory). In case you haven't done Exercise 3.2.5, here is the categorical product of multigraphs. Given multigraphs (for brevity, let's call them graphs here) G and H, the graph $G \times H$ has the following:

- As vertex set, we take the cartesian product of the sets of vertices, i.e. pairs (x, y), where x is a vertex of G and y is a vertex of H.
- As edge set, we take also the cartesian product: An edge $(x, y) \to (x', y')$ is a pair consisting of an edge $x \to x'$ of G and an edge $y \to y'$ of H.

This construction satisfies the universal property of products in the category **MGraph** (why?).

[5] https://tikzit.github.io/.

[6] https://ncatlab.org/nlab/show/k-tuply+monoidal+n-category.

This product is sometimes called the **cross product** of graphs and indeed denoted by $G \times H$ (for alternative terminology, see Remark 6.1.25). This name is suggestive of its structure: Consider the *interval graph* 2, i.e. the graph of two vertices, connected both ways, which we can represent simply as an undirected edge:

2 =  or

If we take the cross product 2×2, we indeed get a *cross*:

2×2 = 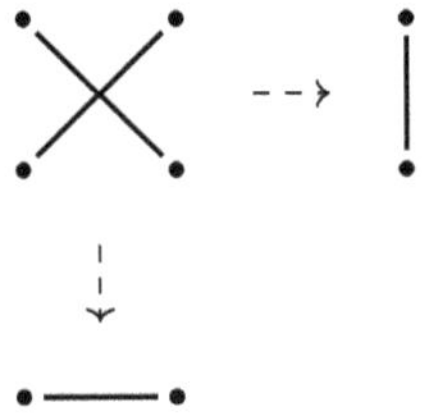

More generally, one can see the cross product of more complex graphs G and H as forming similar but more complex "crossings" (in particular, vertices in the form (x, y) and (x, y'), with the same x, are connected only when x has a loop).

The terminal graph 1 is the graph with one vertex and a single loop at it (why?). As one can see, $X \times 1 \cong 1 \times X \cong X$. This makes $(\mathbf{MGraph}, \times, 1)$ a cartesian monoidal category.

Example 6.1.24 (graph theory). Here is another possible tensor product on **MGraph**. (As we said before, graphs are such versatile objects that there are many possible structures one can build with them, depending on the specific application.) Given graphs G and H, the **box product** $G \,\square\, H$ is constructed as follows (see again Remark 6.1.25 for alternative terminology):

- As vertices, we take again the cartesian product, i.e. pairs (x, y) where x is a vertex of G and y is a vertex of H.
- There is an edge $(x, y) \to (x', y)$ (with the same y) for each edge $x \to x'$ of **G**.

- There is an edge $(x, y) \to (x, y')$ (with the same x) for each edge $y \to y'$ of **H**.

Just as for the cross product, this name is suggestive of its structure. This time, the box product of intervals $2 \mathbin{\square} 2$ looks indeed like a *box*:

$$2 \mathbin{\square} 2 \quad = $$

More generally, one can see that the box product of more complex graphs G and H as forming similar but more complex "boxes" (in particular, vertices in the form (x, y) and (x', y'), where both components are different, are never connected). Why does this *not* satisfy the universal property of products?

Consider now the graph G_0 with a single vertex and no edges. We have $X \mathbin{\square} G_0 \cong G_0 \mathbin{\square} X \cong X$. This, together with associativity and the triangle and pentagon conditions (why do they hold?) makes $(\mathbf{MGraph}, \square, G_0)$ a monoidal category, which is also symmetric (why?).

Remark 6.1.25. In the graph theory literature, the cross product of graphs is sometimes called the "tensor product," and the box product is sometimes called the "cartesian product." We will not use this terminology here since from a categorical perspective, it can be confusing: The *categorical* product, i.e. a category theorist's "cartesian product," is the cross product, while the box product is "just" another tensor product. The situation is similar for the internal homs, see Remark 6.5.7.[7]

Here is another prototypical example of a non-cartesian monoidal category, vector spaces with their *tensor* product:

[7]This conflicting terminology indicates that the category theory and graph theory communities do not interact very much (at least at the time of writing this book). In the author's opinion, this is very unfortunate, as the two communities could learn a lot from each other. (It is probably clear from reading this book that the author is a category theorist who really likes graph theory.)

Example 6.1.26 (linear algebra, commutative algebra). The category **Vect** of real vector spaces forms a monoidal category with the usual tensor product of vector spaces and with monoidal unit given by $\mathbb{R}$.

Recall the universal property of the tensor product from Section 2.3.2. It says that linear maps $A \otimes B \to C$ correspond bijectively to the maps $A \times B \to C$ from the cartesian product which are *bilinear*, i.e. separately linear in both arguments. The left unitor $\lambda : \mathbb{R} \otimes V \to V$ corresponds to the bilinear map $\lambda' : \mathbb{R} \times V \to V$ given by *scalar multiplication*, $(r, v) \mapsto rv$ (why is this linear in both arguments?). To show that this map is an isomorphism $\mathbb{R} \otimes V \cong V$, we can use the Yoneda lemma. Indeed, we can equivalently show that the natural transformation of components

$$\mathrm{Hom}_{\mathbf{Vect}}(V, A) \xrightarrow{-\circ\lambda} \mathrm{Hom}_{\mathbf{Vect}}(\mathbb{R} \otimes V, A)$$

is a natural bijection. Equivalently, the following map needs to be a bijection:

$$\mathrm{Hom}_{\mathbf{Vect}}(V, A) \xrightarrow{-\circ\lambda'} \mathrm{Bi}(\mathbb{R} \times V; A);$$

that is, we have to show that the bilinear maps out of $\mathbb{R} \times V$ are in bijective correspondence with the linear maps out of V via the assignment above. So, let $g : V \to A$ be linear. Precomposing with λ', i.e. adding a scalar multiplication, we obtain the bilinear map $(r, v) \mapsto g(rv) = rg(v)$. (Why is this bilinear?) The *number one* gives an inverse to this map: Let $f : \mathbb{R} \times V \to A$ be bilinear. By fixing its first argument as $1 \in \mathbb{R}$, the map $v \mapsto f(1, v)$ is linear, and this assignment inverts the one above. So, we have a natural bijection which, by the Yoneda lemma, makes the left unitor λ an isomorphism. The right unitor is similar, except that we write the scalar multiplication as a map $V \times \mathbb{R} \to V$.

The associator $(A \otimes B) \otimes C \to A \otimes (B \otimes C)$ can also be given in terms of Yoneda. It amounts to the fact that for a function of three variables $f : A \times B \times C \to D$, it is equivalent to say that it is separately bilinear in A and B and linear in C, or separately linear in A and bilinear in

B and C. Indeed, both situations simply mean that f is separately linear in all three variables. (Can you work this out explicitly?)

Also, the coherence conditions can be obtained by means of the universal property: For example, the pentagon condition says that there is only one way to be separately linear in four arguments. (You can try to work this out explicitly, but it might take quite some time and space. The triangle condition is easier; what does it say?)

This monoidal category is not cartesian since in general, $V \otimes W \not\cong V \times W$.

Example 6.1.27 (commutative algebra). More generally, the following "algebraic" tensor products give rise to non-cartesian monoidal categories:

- The tensor product of vector spaces (over the same field F), with F as monoidal unit.
- The tensor product of abelian groups, with $\mathbb{Z}$ as monoidal unit.
- The tensor product of modules over the same commutative ring R, with R as monoidal unit.

The proofs are all analogous to the case of **Vect**, which we saw in the previous example. Note that in all these examples, the algebraic structures are all "over *something*": modules over a ring, vector spaces over a field, and so on, and this "something" is the monoidal unit. As we saw in the previous example, the unitor $\lambda : I \otimes A \to A$ is a form of action or *multiplication by a scalar*, and algebraically, that's why all these structures are "over the monoidal unit."

More details on these examples and on why they are all analogous will be provided in Section 6.4.4.

Example 6.1.28. Let's now look at two tensor products in **Cat**. We already know that **Cat** has products (Definition 1.3.49), and the terminal category **1** has a single object with its identity. This makes **Cat** cartesian monoidal. Similar to graphs, there is a different product of categories which is occasionally useful, sometimes called the **funny**

tensor product of categories $\mathbf{C} \,\square\, \mathbf{D}$, analogous to the box product of graphs:

- The objects are pairs (C, D), where C is an object of $\mathbf{C}$ and $\mathbf{D}$ is an object of $\mathbf{D}$.
- There is a morphism $(C, D) \to (C, D')$ for each morphism $C \to C'$ of $\mathbf{C}$.
- There is a morphism $(C, D) \to (C', D)$ for each morphism $D \to D'$ of $\mathbf{D}$.
- A generic morphism $(C, D) \to (C', D')$ is a composition of finitely many morphisms in the two forms above, modulo the equations (commutative diagrams) that hold in $\mathbf{C}$ and $\mathbf{D}$ separately.

The monoidal unit is again the terminal category (why?).

Example 6.1.29 (probability). As we said in Example 5.1.10, a Kleisli morphism of the probability monad P on **Set** can be seen as a stochastic map (or equivalently a stochastic matrix, Example 5.5.9). We can construct a tensor product on the category which encodes the idea of "independent probabilistic transitions":

- On objects, $X \otimes Y$ is the cartesian product $X \times Y$.
- On morphisms, given stochastic matrices $f : X \to Y$ and $g : A \to B$ with entries $f(y|x)$ and $g(b|a)$, we can form the stochastic matrix $f \otimes g : X \otimes A \to Y \otimes B$ with entries

$$f \otimes g\,(y, b|a, x) \;:=\; f(y|x)\,g(b|a).$$

This encodes the idea that Y depends stochastically on X, but only on X, not on A, and similarly, B depends (stochastically) only on A. (Remember the notation in terms of string diagrams.)

In particular, if X and A are the monoidal unit 1 (the singleton set), stochastic matrices (i.e. vectors) $1 \to X$ and $1 \to A$ are (in bijection with) the elements of $\mathcal{P}X$ and $\mathcal{P}A$, i.e. finitely supported probability distributions on X and A (why?). In that case, the tensor product returns the product of probability distributions,

$$p \otimes q\,(y,b) \;=\; p(y)\,p(b),$$

which according to the laws of probability denotes independent events.

Similar considerations can be made for the Giry monad on the category of measurable spaces — in that case, we get the product of Markov kernels and of probability measures.

Exercise 6.1.30. Show that a category with coproducts and an initial object is canonically a symmetric monoidal category (sometimes called *cocartesian*). For graphs, as well as for categories, this encodes the idea of being "separately related."

Exercise 6.1.31. Show that the opposite category of a monoidal category is canonically monoidal.

6.1.4 Points or states

Thanks to the monoidal unit, in a monoidal category, there are particular morphisms that play the role of "points," or "elements," even when we are not in the category of sets.

Definition 6.1.32. Let X be an object in a monoidal category $(\mathbf{C}, \otimes, I)$. A **point**, or **state**, of X is a morphism $p : I \to X$.

Note that different authors differ on what they mean by "points" and "states."[8]

Example 6.1.33 (sets and relations). In $(\mathbf{Set}, \times, 1)$, points are exactly elements.

[8]For example, some authors require additional conditions on morphisms, or on the category, such as that I is terminal.

What is a "point of a graph"? We get a distinct notion, depending on the choice of monoidal structure, as follows.

Example 6.1.34 (graph theory). In $(\mathbf{MGraph}, \times, 1)$, a point is a vertex with a distinguished loop.

Example 6.1.35 (graph theory). In $(\mathbf{MGraph}, \square, G_0)$, a point is a vertex.

Which notion is the "correct" one? It depends on the context. Sometimes, a point should be just a vertex. Sometimes, we are in a situation where we want to keep track of the fact that "we are allowed to stay where we are." In that case, we need a distinguished loop at our vertex, encoding the idea of "staying." See also Section 6.5.1. (For reflexive graphs, there is no difference.)

Example 6.1.36 (linear algebra, commutative algebra). In $(\mathbf{Vect}, \otimes, \mathbb{R})$, a point of V is a linear map $\mathbb{R} \to V$. These are in bijection with the elements of V (seen as a set): Every $v \in V$ defines a linear map $\mathbb{R} \to V$ canonically by linear extension, $r \mapsto rv$. Conversely, any linear map $f : \mathbb{R} \to V$ can be evaluated at $1 \in \mathbb{R}$ to give a point $f(1) \in V$, and these assignments are mutually inverse. (Compare this with how we defined the unitor in Example 6.1.26.)

Note that it is crucial for the above to work that we take the monoidal structure $(\mathbf{Vect}, \otimes, \mathbb{R})$. If we instead took the cartesian product (or direct sum) $(\mathbf{Vect}, \times, 0)$, the only linear map $0 \to V$ is the zero map. (In particular, the points in this monoidal sense are not *always* the elements of the underlying set.)

Example 6.1.37 (commutative algebra). Similarly to the examples above, in the category $(\mathbf{AbGrp}, \otimes, \mathbb{Z})$, a point is a morphism $\mathbb{Z} \to G$, which uniquely identifies an element of G as a set. (Can you think of similar examples in other categories?)

Example 6.1.38 (probability). In the Kleisli category of the probability monad $\mathcal{P}$ on **Set**, as we saw above, Kleisli morphisms $I \to X$

(i.e. functions $I \to \mathcal{P}X$) are *probability distributions* on X. We can see these as *random points*, or *random states*, of X, following the philosophy of the previous chapter, where monads can formalize the idea of "generalized elements."

For the Giry monad, the situation is analogous.

Points in a monoidal category are a functorial construction: Given a point $p : I \to X$ and a morphism $f : X \to Y$, the composite $f \circ p : I \to Y$ is a point of Y, which we can denote suggestively as $f(p)$, as in the case of sets. The representable functor $\mathrm{Hom}_{\mathbf{C}}(I, -) : \mathbf{C} \to \mathbf{Set}$ assigns to each object X the set of all its points and to each morphism the induced function between points.

The dual to a point, a morphism $X \to I$, is sometimes called an *effect*. A possible interpretation of this name is that an effect makes something happen depending on the context (input) X, but with "no output." In a cartesian monoidal category, there is exactly one effect for each object (why?).

6.2 Monoids and Comonoids

One of the most fruitful ideas in the theory of monoidal categories is that we can generalize monoids to structures that live *internally* to a monoidal category. Their dual, comonoids, are an equally fruitful concept.

6.2.1 Internal monoids

Let's rewrite the definition of a monoid as follows. A monoid is a set M together with a distinguished element $e : 1 \to M$, the unit, and a map $m : M \times M \to M$, the multiplication, such that the following diagrams in **Set** commute, encoding associativity and unitality:

$$
\begin{array}{ccc}
M \times M \times M & \xrightarrow{m \times \mathrm{id}} & M \times M \\
\downarrow{\scriptstyle \mathrm{id} \times m} & & \downarrow{\scriptstyle m} \\
M \times M & \xrightarrow{m} & M
\end{array}
\qquad
\begin{array}{ccc}
1 \times M & \xrightarrow{e \times \mathrm{id}} & M \times M \\
& \searrow{\scriptstyle \cong} & \downarrow{\scriptstyle m} \\
& & M
\end{array}
\qquad
\begin{array}{ccc}
M \times 1 & \xrightarrow{\mathrm{id} \times e} & M \times M \\
& \searrow{\scriptstyle \cong} & \downarrow{\scriptstyle m} \\
& & M.
\end{array}
$$

Let's now generalize this to arbitrary monoidal categories.

Definition 6.2.1. Let $(\mathbf{C}, \otimes, I)$ be a monoidal category. A **monoid object**, or **internal monoid**, in $\mathbf{C}$ is an object M together with

- a distinguished morphism $e : I \to M$, called the **unit** map;
- a distinguished morphism $m : M \otimes M \to M$, called the **multiplication** map;

such that the following diagrams commute:

$$
\begin{array}{ccc}
& M \otimes M & \\
\overset{m \otimes \mathrm{id}}{\nearrow} & & \searrow^{m} \\
(M \otimes M) \otimes M & & M \\
\cong \downarrow & & \nearrow^{m} \\
M \otimes (M \otimes M) & & \\
\underset{\mathrm{id} \otimes m}{\searrow} & M \otimes M &
\end{array}
\qquad
\begin{array}{ccc}
I \otimes M & \xrightarrow{e \otimes \mathrm{id}} & M \otimes M \\
& \cong \searrow & \downarrow^{m} \\
& & M \\
& & \\
M \otimes I & \xrightarrow{\mathrm{id} \otimes e} & M \otimes M \\
& \cong \searrow & \downarrow^{m} \\
& & M.
\end{array}
$$

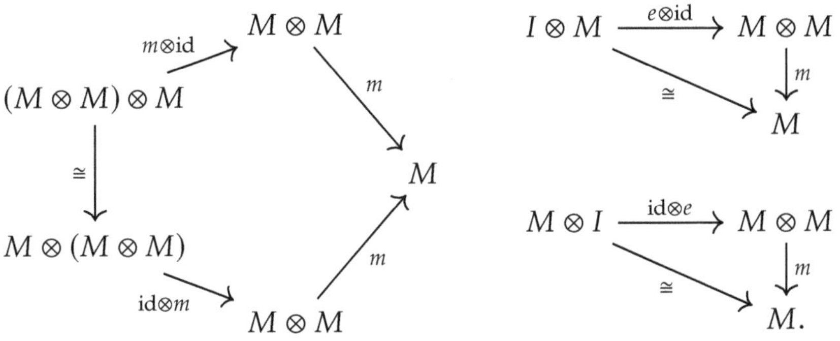

called, respectively, **associativity** and **left and right unitality**.

If $\mathbf{C}$ is *symmetric* monoidal, an internal monoid (M, m, e) is called **commutative** if the following diagram commutes:

$$
\begin{array}{ccc}
M \otimes M & & \\
\beta_{M,M} \downarrow \cong & \searrow^{m} & \\
& & M \\
M \otimes M & \nearrow_{m} &
\end{array}
$$

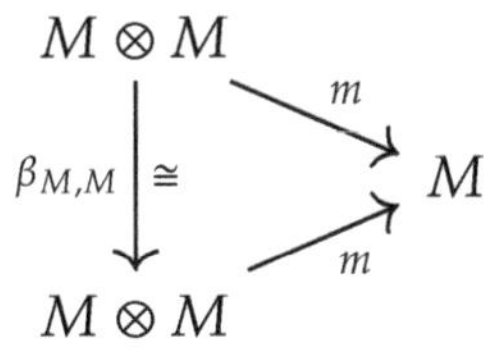

i.e. if the multiplication is invariant under braiding.

Sometimes, we call internal monoids simply "monoids" when it's clear from the context what we mean.

In terms of string diagrams, we can write an internal monoid as follows. First of all, we denote the unit and multiplication simply by bullets:

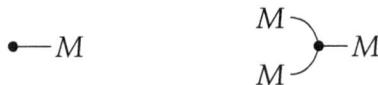

The unitality and associativity conditions can be expressed as

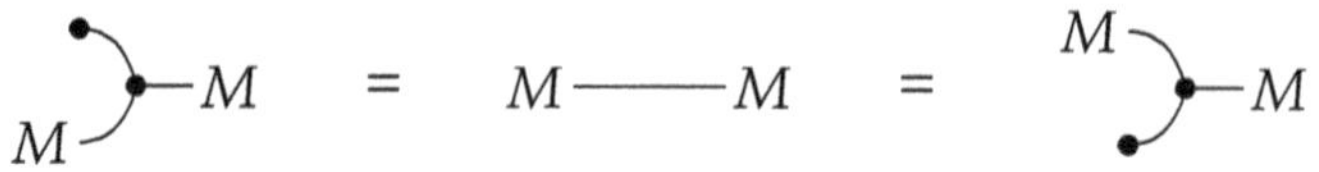

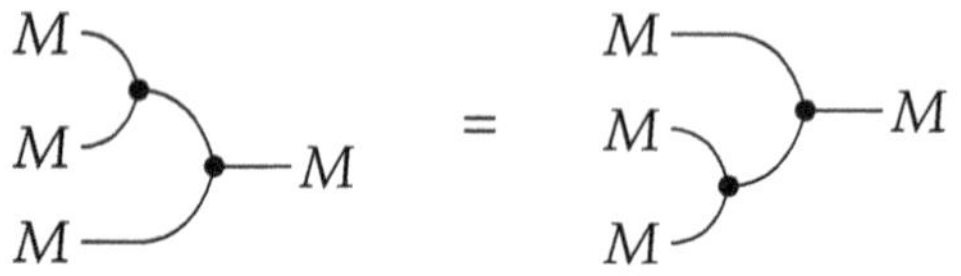

and for commutative monoids, we also have the following:

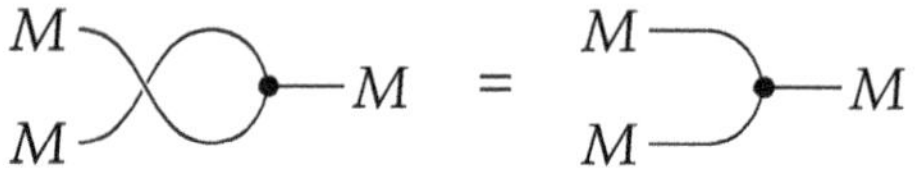

Example 6.2.2. In $(\mathbf{Set}, \times, 1)$, internal monoids are the usual monoids (see above). Similarly, internal commutative monoids are exactly commutative monoids since the commutativity condition states exactly that $a \cdot b = b \cdot a$.

Proposition 6.2.3. *In the category* $(\mathbf{Ab}, \otimes, \mathbb{Z})$ *of abelian groups with their tensor product, an internal monoid is exactly a* ring. *An internal commutative monoid is exactly a commutative ring.*

Proof. Let $(R, +, 0)$ be an abelian group.

First of all, we saw in Section 6.1.4 that a "point" $\mathbb{Z} \to R$ is simply an element of R as a set. Let's call this element 1.

Now, by the universal property of the *tensor* product of abelian groups, a group homomorphism $R \otimes R \to R$ is in bijection with a *biadditive* map $R \times R \to R$ (recall the analogous property for vector spaces in Section 2.3.2). In other words, the map $m : R \otimes R \to R$ corresponds bijectively to a map $R \times R \to R$ which is a group homomorphism in each argument separately. Let's denote this new map by a dot, $(r, s) \mapsto r \cdot s$. The fact that it is a group homomorphism in both arguments means that

$$(r + r') \cdot s = (r \cdot s) + (r' \cdot s), \quad 0 \cdot s = 0,$$

$$r \cdot (s + s') = (r \cdot s) + (r \cdot s'), \quad r \cdot 0 = 0.$$

In other words, this new operation must *distribute over the sum*.

The unitality requirement states that, in some sense, *the multiplication by a scalar is the correct one*: Recall from the analogous case of vector spaces (Example 6.1.26) that the left unitor $\lambda : \mathbb{Z} \otimes R \to R$ is in bijection with the biadditive map $\lambda' : \mathbb{Z} \times R \to R$ given by scalar multiplication. Now, the unitality diagram commutes,

$$\begin{array}{ccc} \mathbb{Z} \otimes R & \xrightarrow{e \otimes \mathrm{id}} & R \otimes R \\ & {\scriptstyle\lambda}\searrow^{\cong} & \downarrow{\scriptstyle m} \\ & & R, \end{array} \quad \text{or, equivalently,} \quad \begin{array}{ccc} \mathbb{Z} \times R & \xrightarrow{e \times \mathrm{id}} & R \times R \\ & {\scriptstyle\lambda'}\searrow^{\cong} & \downarrow{\scriptstyle \cdot} \\ & & R, \end{array}$$

and thus, if we multiply $r \in R$ by any $n \in \mathbb{Z}$ using our new map "$\cdot$," the result is the same as if we used the "old" scalar multiplication λ', i.e. $n \cdot r = nr$. In particular, if we multiply by $1 \in \mathbb{Z}$, it is the same as doing nothing (and this is a form of unitality). The right unitality requirement works analogously. The associativity condition for m corresponds now to associativity for the map "$\cdot$." □

Example 6.2.4 (linear algebra, commutative algebra). In $(\mathbf{Vect}, \otimes, \mathbb{R})$, a monoid object is an (associative, unital) **algebra**, and a commutative monoid object is a commutative algebra. An example of an algebra is the algebra of $n \times n$ matrices with fixed n. The reasoning is analogous to the above case of rings.

Example 6.2.5. Let $\mathbf{C}$ be any category. A monad is exactly a monoid object in the monoidal category of endofunctors $[\mathbf{C}, \mathbf{C}]$. Indeed, the unit and multiplication maps, which in this category are natural transformations, are exactly the unit and multiplication of the monad, and the associativity and unitality conditions are exactly the analogous ones for monads, as in Definition 5.0.1.

Example 6.2.6. What is a *monoid object in the category of monoids* (with the cartesian product)? Given a monoid $(M, \cdot, e)$, we need an additional monoid structure $(M, \bullet, i)$ such that the maps $i : 1 \to M$ and $\bullet : M \times M \to M$ are homomorphisms of monoids for the "old" multiplication. This, first of all, implies that $i = e$ since the only monoid homomorphism $1 \to M$ needs to map the unique element of 1 to the monoid unit. So, both monoid structures have the same unit e. The fact that $\bullet$ is a monoid homomorphism (in both variables jointly, not separately — *cartesian* product!) tells us that

$$(a \cdot b) \bullet (c \cdot d) = (a \bullet c) \cdot (b \bullet d).$$

Now, if we set $b = c = e$, we get

$$a \bullet d = (a \cdot e) \bullet (e \cdot d) = (a \bullet e) \cdot (e \bullet d) = a \cdot d,$$

which shows that the multiplication structures are the same, and there is no "other" multiplication. If now we set $a = d = e$, we get that

$$b \cdot c = (e \cdot b) \cdot (c \cdot e) = (e \cdot c) \cdot (b \cdot e) = c \cdot b,$$

so the multiplication must be commutative. In other words, *monoids in* **Mon** *are exactly commutative monoids*. Similarly, a monoid object in the category of groups is an abelian group. These statements, and other related ones, are sometimes called the **Eckmann–Hilton argument.**[9]

Example 6.2.7. A strict monoidal category is exactly an monoid object in **Cat**. (One may ask, what are non-strict monoidal categories an instance of? Those structures are called *pseudomonoids*, but we will not treat them here, besides obviously the case of actual monoidal categories.[10])

The fact that one can define monoids inside monoidal categories, which are a sort of "larger monoids," is an instance of what is sometimes called the *microcosm–macrocosm principle*: Very often in category theory, there are structures naturally living in similar but larger or higher structures. See for example the nLab page on this.[11] Another example is the fact that *commutative* monoids live in *symmetric* monoidal categories.

---

[9]https://ncatlab.org/nlab/show/Eckmann-Hilton+argument.

[10]See https://ncatlab.org/nlab/show/pseudomonoid.

[11]https://ncatlab.org/nlab/show/microcosm+principle.

Let's continue with more examples of internal monoids.

Example 6.2.8 (topology). A *topological monoid* is a monoid object in $(\mathbf{Top}, \times, 1)$, i.e. a monoid with a topology such that the multiplication map is (jointly) continuous. Topological groups are a special case (where the inversion map $g \mapsto g^{-1}$ has to be continuous too).

Example 6.2.9 (differential geometry). A *Lie monoid* is a monoid object in $(\mathbf{Mfd}, \times, 1)$, i.e. a monoid which is a manifold, such that the multiplication map is smooth. Lie groups are a special case (where the inversion map $g \mapsto g^{-1}$ has to be smooth too).

Exercise 6.2.10. Show that a monoid (G, m, e) in **Set** (and **Top** and **Mfd**) is a group if and only if the following associativity diagram is a pullback:

$$\begin{array}{ccc} G \times G \times G & \xrightarrow{\mathrm{id}_G \times m} & G \times G \\ \downarrow{\scriptstyle m \times \mathrm{id}_G} & & \downarrow{\scriptstyle m} \\ G \times G & \xrightarrow{m} & G. \end{array}$$

Exercise 6.2.11. Show that the unit I of a monoidal category $(\mathbf{C}, \otimes, I)$ has a canonical internal monoid structure, given by the coherence maps.

We saw in Example 5.1.7 that any monoid in **Set** gives rise to a monad, the *action* or *writer* monad, simply through the monoid structure. This generalizes to internal monoids as follows.

Definition 6.2.12. Let (M, m, e) be a monoid object in a monoidal category $(\mathbf{C}, \otimes, I)$. The (left-) **action monad** denoted by M is defined as follows:

- The functor is the left multiplication $A \mapsto M \otimes A$.
- The unit is given by the components

$$A \xrightarrow{\cong} 1 \otimes A \xrightarrow{e \otimes A} M \otimes A.$$

- The multiplication is given by the components

$$M \otimes (M \otimes A) \xrightarrow{\cong} (M \otimes M) \otimes A \xrightarrow{m \otimes A} M \otimes A.$$

The associativity and unitality conditions for the monad follow from the ones of the monoid.

One can equivalently define a right-action monad, and if M is commutative, we get isomorphic functors (and monads, once we define morphisms of monads).

So, *monads are a special case of monoids* (in the category of endofunctors), but also *monoids are a special case of monads* (via action monads). This sort of circular (but precisely defined) situation is very common in category theory; one can very often interpret a concept in terms of another concept and the other way around. This allows us to have many different points of view on the same structure, which makes the theory more interconnected.

Let's now look at the internal equivalent of monoid homomorphisms.

Definition 6.2.13. Let (M, m, e) and (N, m, e) be monoid objects in a monoidal category $(\mathbf{C}, \otimes, I)$. A **morphism of monoids** $M \to N$ is a morphism $f : M \to N$ of $\mathbf{C}$ such that the following diagrams commute:

$$\begin{array}{ccc} & & M \\ & \overset{e}{\nearrow} & \downarrow{\scriptstyle f} \\ I & \underset{e}{\searrow} & N \end{array} \qquad \begin{array}{ccc} M \otimes M & \xrightarrow{m} & M \\ \downarrow{\scriptstyle f \otimes f} & & \downarrow{\scriptstyle f} \\ N \otimes N & \xrightarrow{m} & N, \end{array}$$

called **unit** and **multiplication** conditions, respectively. Note that on the top of the diagrams, we have the units and multiplication of M and on the bottom those of N.

Thus, internal monoids in $\mathbf{C}$ form a category, which we call $\mathbf{Mon}(\mathbf{C})$.

Example 6.2.14 (several fields). Morphisms of internal monoids recover the following classical notions:

- In $(\mathbf{Set}, \times, 1)$, they are monoid homomorphisms.
- In $(\mathbf{AbGrp}, \otimes, \mathbb{Z})$, they are ring homomorphisms.

- In $(\mathbf{Vect}, \otimes, \mathbb{R})$, they are algebra homomorphisms.
- In $(\mathbf{Top}, \times, 1)$, they are continuous monoid homomorphisms.
- In $(\mathbf{Mfd}, \times, 1)$, they are smooth monoid homomorphisms.

Exercise 6.2.15. Give a possible definition of morphism between strict monoidal categories. (Functors between monoidal categories, in general, will be the topic of Section 6.3.)

Exercise 6.2.16. Give a possible definition of morphism of monads. The usual definition is more general since it allows us to change the base category; however, we will not treat it here.[12]

6.2.2 Internal modules

Just as monoids act on sets, forming for example G-sets, internal monoids can act too, in a suitably internal way.

Definition 6.2.17. Let (M, m, e) be an internal monoid in a monoidal category $(\mathbf{C}, \otimes, I)$. A (left) **module object**, or (left) **internal module**, over M is an object A of $\mathbf{C}$ together with a morphism $a : M \otimes A \to A$, called the **action map**, that makes the following diagrams commute:

$$
\begin{array}{ccc}
I \otimes A & \xrightarrow{e \otimes \mathrm{id}_A} & M \otimes A \\
& \cong \searrow & \downarrow a \\
& & A
\end{array}
$$

$$
\begin{array}{ccc}
(M \otimes M) \otimes A & \xrightarrow{m \otimes \mathrm{id}_A} & M \otimes A \\
\cong \downarrow & & \searrow a \\
M \otimes (M \otimes A) & & A \\
& \xrightarrow{\mathrm{id}_M \otimes a} M \otimes A \xrightarrow{a} &
\end{array}
$$

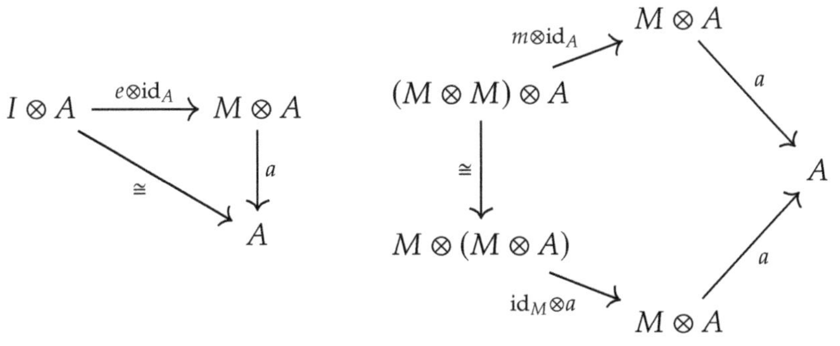

called **unit** and **multiplication** conditions, respectively.

---

[12]See the nLab article on monads, https://ncatlab.org/nlab/show/monad.

A right module object is an analogous structure, where the action map is in the form $A \otimes M \to A$ (and satisfies analogous conditions).

In string diagrams, we can denote the map a by a white dot (note that "left" is "top"):

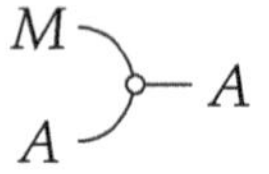

The unit and multiplication conditions now look as follows, respectively:

and

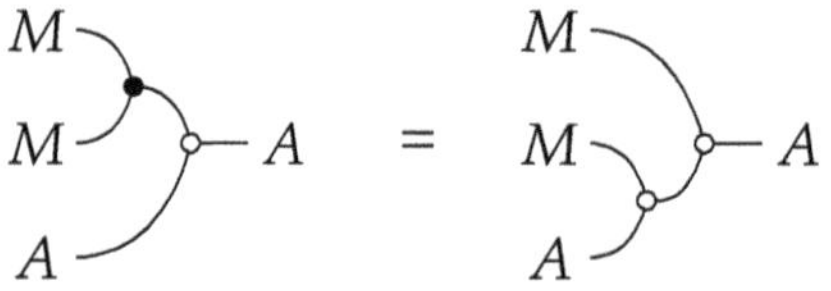

A possible interpretation of an M-module is either a "representation" or "action" of M, or as a situation where M indexes a type of dynamics.

Example 6.2.18 (group theory). In $(\mathbf{Set}, \times, 1)$, a module object A over a monoid (or group) M is exactly an action of M on A.

Proposition 6.2.19. *In* $(\mathbf{AbGrp}, \otimes, \mathbb{Z})$, *a left-module object A over a ring R is exactly a left-module in the ring-theoretic sense (hence the name). The analogous statement is true for right-module objects.*

Proof. First of all, a homomorphism $a : R \otimes A \to A$ is in bijection with a biadditive map $R \times A \to A$. Let's denote it again by "$\cdot$." Biadditivity means the following conditions:

$$(r + r') \cdot a = (r \cdot a) + (r' \cdot a), \quad 0 \cdot a = 0,$$
$$r \cdot (a + a') = (r \cdot a) + (r \cdot a'), \quad r \cdot 0 = 0.$$

Moreover, the unit and multiplication conditions of the module object are, just as for sets,

$$(r \cdot r') \cdot a = r \cdot (r' \cdot a), \quad 1 \cdot a = a,$$

where on the left, $r \cdot r'$ denotes the multiplication in R.

These are exactly the conditions that make A a left-R-module. Right-modules work analogously. □

Example 6.2.20. Internal modules in $(\mathbf{Vect}, \otimes, \mathbb{R})$ are sometimes called *modules over an algebra.* An example is given by how $n \times n$ matrices, which form an algebra, act on the vector space $\mathbb{R}^n$.

Example 6.2.21. A possible definition of *module over a monad,* which instantiates Definition 6.2.17, is the following. Given a monad (T, μ, η) on $\mathbf{C}$, a left-module over T is an endofunctor $A : \mathbf{C} \to \mathbf{C}$, together with a natural transformation $a : T \circ A \to A$, such that the following diagrams commute for each object X:

$$\begin{array}{ccc} AX & \xrightarrow{\eta_{AX}} & TAX \\ & \searrow^{\mathrm{id}} & \downarrow{\scriptstyle a_X} \\ & & AX \end{array} \qquad \begin{array}{ccc} TTAX & \xrightarrow{Ta_X} & TAX \\ \downarrow{\scriptstyle \mu_{AX}} & & \downarrow{\scriptstyle a_X} \\ TAX & \xrightarrow{a_X} & AX. \end{array}$$

We have that:

- if A is a constant functor, the definition recovers exactly the notion of algebra of a monad;
- the previous point is not the only example: for example, the functor T is itself such a module (why?).

The usual definition of module over a monad is more general, as it allows us to change the base category. We will not treat it here.[13]

Example 6.2.22 (topology, dynamical systems). In $(\mathbf{Top}, \times, 1)$, an internal module over a topological monoid M is a *continuous action* of the monoid M.

[13]See the nLab article, https://ncatlab.org/nlab/show/module+over+a+monad.

Example 6.2.23 (differential geometry, dynamical systems). In $(\mathbf{Mfd}, \times, 1)$, an internal module over a Lie monoid M is a *smooth action* of the monoid M.

Exercise 6.2.24. Show that in any monoidal category, the monoidal unit I (which, we have seen in Exercise 6.2.11, is a monoid object) makes every object A a left-module object through the unitor map.

Exercise 6.2.25. Let (M, m, e) be a monoid object in a monoidal category $(\mathbf{C}, \otimes, I)$. Show that an internal module over M is equivalently an algebra of the action monad induced by M.

The result above means, in particular, that the modules over a ring R (in the traditional sense) are equivalently the algebras of the monad on **AbGrp** defined by the internal monoid R.

Let's now turn to the morphisms of internal modules.

Definition 6.2.26. Let (A, a) and (B, b) be internal modules over an internal monoid (M, m, e) in $(\mathbf{C}, \otimes, I)$. A **morphism of internal modules** is a morphism $f : A \to B$ of $\mathbf{C}$ that makes the following diagram commute:

$$\begin{array}{ccc} M \otimes A & \xrightarrow{\mathrm{id}_M \otimes f} & M \otimes B \\ \downarrow{\scriptstyle a} & & \downarrow{\scriptstyle b} \\ A & \xrightarrow{\quad f \quad} & B, \end{array}$$

called the (internal) **equivariance** condition.

Thus, internal modules over M form a category, which we call M-**Mod**.

In terms of string diagrams,

Example 6.2.27 (several fields). The morphisms of internal modules recover the following classical notions:

- In $(\mathbf{Set}, \times, 1)$, given a monoid M, the morphism of M-modules are exactly equivariant maps.
- In $(\mathbf{AbGrp}, \otimes, \mathbb{Z})$, given a ring R, the morphisms of R-modules are exactly the module homomorphisms of ring theory.
- In $(\mathbf{Vect}, \otimes, \mathbb{R})$, they are algebra module homomorphisms.
- In $(\mathbf{Top}, \times, 1)$, they are continuous equivariant maps.
- In $(\mathbf{Mfd}, \times, 1)$, they are smooth equivariant maps.

Exercise 6.2.28. Instantiate the definition of an internal module in $(\mathbf{Cat}, \times, \mathbf{1})$. (The resulting structure is called a *strict actegory* — note the possibly confusing spelling, it's a portmanteau of "action" and "category."[14])

Exercise 6.2.29. If you've solved Exercise 6.2.16, show that the morphisms of algebras of a monad are a special case of morphisms of modules in this sense. What are other examples?

Exercise 6.2.30. Show that in a symmetric monoidal category, the left- and right-module objects over a commutative monoid object M are isomorphic in M-**Mod**. Interpret this in $(\mathbf{AbGrp}, \otimes, \mathbb{Z})$. (Why do we need the commutativity condition?)

Exercise 6.2.31 (important!). Show that an internal monoid is a module over itself. (*Hint*: Think of free algebras of a monad.)

Exercise 6.2.32 (several fields; important!). Let $f : (M, m, e) \to (P, m, e)$ be a morphism of internal monoids in a monoidal category $(\mathbf{C}, \otimes, I)$. Given a P-module (A, a), show that the map given by

$$M \otimes A \xrightarrow{f \otimes \mathrm{id}} P \otimes A \xrightarrow{a} A \tag{6.2.1}$$

equips A with an M-module structure as well. Show also that this way, a P-module morphism $A \to B$ is also an M-module morphism.

[14]See https://ncatlab.org/nlab/show/action+of+a+monoidal+category.

Conclude that this induces a functor $f^* : P\text{-}\mathbf{Mod} \to M\text{-}\mathbf{Mod}$ (note the direction). This functor is sometimes called the **restriction of scalars.**

Find concrete examples of this construction for $(\mathbf{C}, \otimes, I) = (\mathbf{Set}, \times, 1)$ and $(\mathbf{C}, \otimes, I) = (\mathbf{AbGrp}, \otimes, \mathbb{Z})$, and motivate the name "restriction of scalars" by looking at the ring homomorphism $\mathbb{Z} \to \mathbb{R}$. (*Hint*: Use string diagrams.)

Exercise 6.2.33 (several fields). In the context of the exercise above, show that by taking as A the monoid P itself, any morphism of monoids $f : M \to P$ makes P canonically an M-module and that the multiplication $P \otimes P \to P$ is a morphism of M-modules in the first argument, i.e.

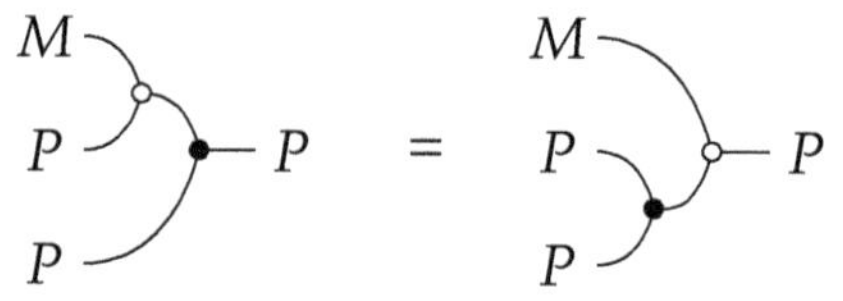

Show also that conversely, every M-module structure on P satisfying the equation above must be of the form,

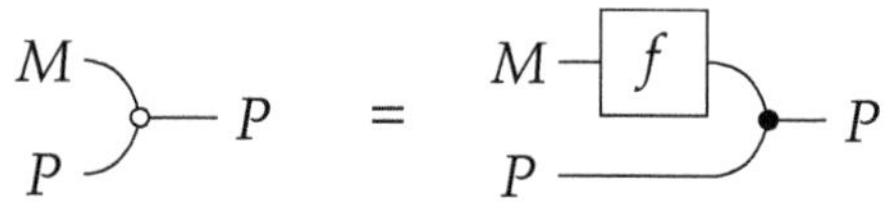

for a unique monoid morphism f. (*Hint*: To find f, precompose the module map with the unit of P. Using string diagrams will simplify the proofs significantly.)

6.2.3 Internal comonoids

Let's now look at the dual notion to internal monoids.

Definition 6.2.34. Let $(\mathbf{C}, \otimes, I)$ be a monoidal category. A **comonoid object** or **internal comonoid** in $\mathbf{C}$ is an object W together with

- a distinguished morphism $\partial : W \to I$, called the **counit** map;
- a distinguished morphism $c : W \to W \otimes W$, called the **comultiplication** map;

such that the following diagrams commute:

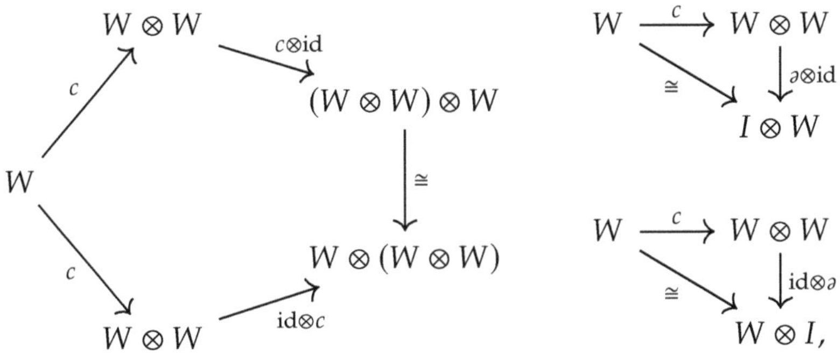

called, respectively, **coassociativity** and **left and right counitality**.

An internal comonoid (W, c, ∂) is called **cocommutative** (or sometimes simply **commutative**) if the following diagram commutes:

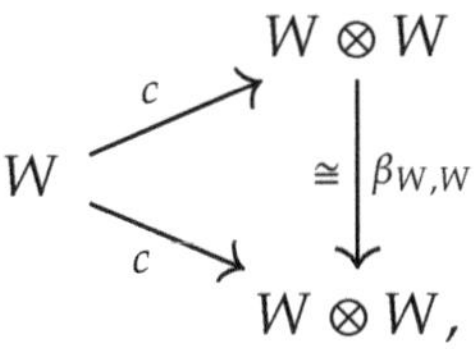

i.e. if the comultiplication is invariant under braiding.

In terms of string diagrams, we can write an internal comonoid as follows. The counit and comultiplication are maps what we can write in a dual way to the case of monoids:

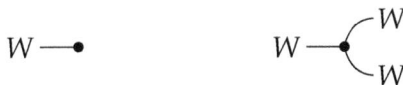

The counitality and coassociativity conditions are as follows:

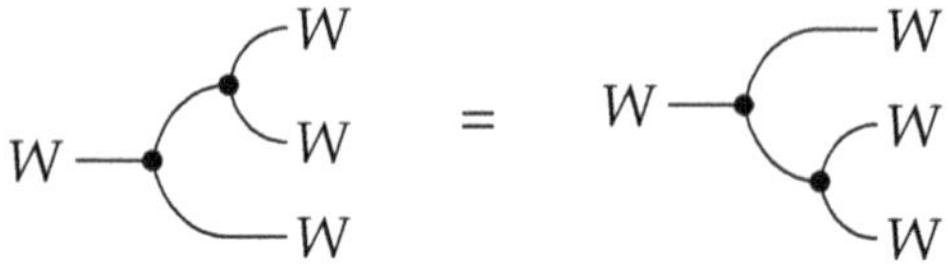

and for commutative comonoids, the following also holds:

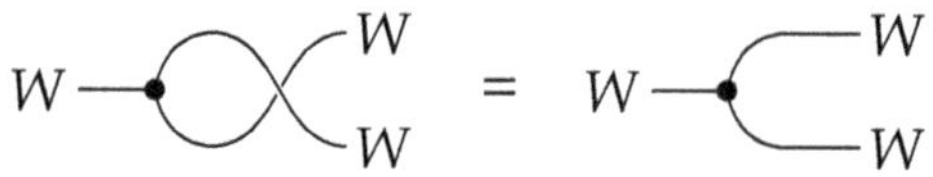

A commutative comonoid can be sometimes be interpreted as a situation where we can *copy and discard information*. In this situation, the counit is a "discard map" $X \to I$, which intuitively "forgets" the state where we are. The comultiplication is instead a "copy map" $X \to X \otimes X$, which intuitively "duplicates" our state or data. The counitality conditions tell us that copying and then discarding one of the copies is the same as doing nothing. The coassociativity condition tells us that copying one of the copies is the same as copying the other copy (so, in some sense, the copies are "perfect"). The commutativity condition tells us that the copies are interchangeable.

This is particularly evident in **Set**. Every set is canonically equipped with a comonoid structure as follows:

$$\begin{aligned} X &\xrightarrow{\text{copy}} X \times X \\ x &\longmapsto (x, x) \end{aligned} \qquad\qquad \begin{aligned} X &\xrightarrow{\ !\ } 1 \\ x &\longmapsto \bullet \end{aligned}$$

This is used implicitly whenever we use a variable twice, or whenever we don't use a variable. For example, given a function of two variables $f : X \times X \to Y$, the expression $f(x, x)$ can be formed via the following composition:

$$\begin{aligned} X &\xrightarrow{\text{copy}} X \times X \xrightarrow{\quad f \quad} Y \\ x &\longmapsto (x, x) \longmapsto f(x, x) \end{aligned}$$

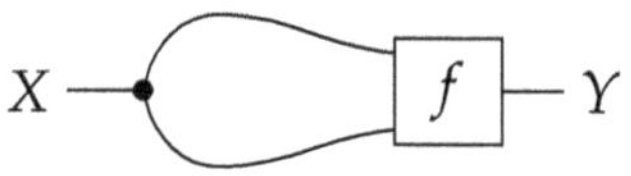

which uses the "copy" map to use the value x twice. Similarly, any function $g : X \to Z$ can also be seen as a function $g : X \times Y \to Z$ which "does not use the variable Y":

$$\begin{array}{ccccc} X \times Y & \xrightarrow{\mathrm{id}\times !} & X \times 1 \cong X & \xrightarrow{g} & Z \\ (x, y) & \longmapsto & (x, \bullet) \mapsto x & \longmapsto & g(x), \end{array}$$

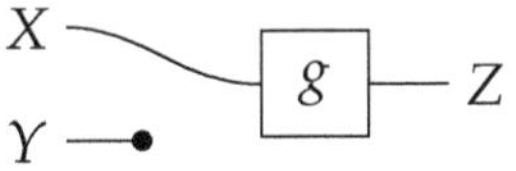

i.e. which uses the "discard" map to get rid of the value y. As a particular case, a single point $p : I \to Y$ can be seen as a (constant) function $X \to Y$.

Comonoids may be confusing structures at first because as we have seen above, in **Set** (and similar categories), every set is canonically a comonoid, and so it may seem that the structure is trivial. (There is more to this, as we will see in Proposition 6.2.45.) The importance of comonoids becomes more apparent when we turn to other monoidal categories, and the canonical example of that is vector spaces.

Example 6.2.35 (linear algebra, quantum physics). A comonoid object in $(\mathbf{Vect}, \otimes, \mathbb{R})$ is called a **coalgebra** (since it's dual to an associative unital algebra). Coalgebras are a building block for more complex structures such as *Hopf algebras* and *Frobenius algebras*, but we will not treat them here.[15] Note that this time, the comultiplication map $V \to V \otimes V$ cannot be defined using bilinear maps since the tensor product is the codomain, not the domain.

A standard way of equipping a vector space with a coalgebra structure is picking a basis. So, let V be a vector space, and let $E = \{e_i\}_{i \in I}$ be a basis of V. Every vector $v \in V$ can be expressed uniquely as a finite linear combination:

$$v = \sum_i f^i \, e_i.$$

[15] See https://ncatlab.org/nlab/show/Hopf+algebra and https://ncatlab.org/nlab/show/Frobenius+algebra.

Now, define the coalgebra structure maps as follows:

$$\begin{array}{ccc} V & \longrightarrow & V \otimes V \\ \sum_i f^i e_i & \longmapsto & \sum_i f^i e_i \otimes e_i \end{array} \qquad \begin{array}{ccc} V & \longrightarrow & \mathbb{R} \\ \sum_i f^i e_i & \longmapsto & \sum_i f^i. \end{array}$$

One can check that these are coassociative and counital.

Also, in the example above, the coalgebra structure has to do with "copying and discarding information," in this case, copying the *values of the coordinates*. However, differently from the case of **Set**, we had to *choose* a basis and pick *which* aspects or information we want to manipulate, copy, and discard. In quantum physics, a comonoid structure can be interpreted as a particular experimental context, or a particular setup for detecting something specific, such as the spin of an electron along one particular axis, which cannot be extended to a larger setup without losing consistency. (We saw in Example 1.4.17, and we will see again in Example 6.2.42, that this assignment is not compatible with changes of basis.)

Exercise 6.2.36 (linear algebra). Show that the coalgebra structure maps of the previous example are given by linearly extending the comonoid structure maps copy : $E \to E \times E$ and $! : E \to 1$ of E as a comonoid in **Set**, viewing E as a basis of V, $E \times E$ as a basis of $V \otimes V$, and 1 as a basis of $\mathbb{R}$.

The idea of "copying and discarding information" was central in the construction of the *reader comonad* (Example 5.3.1). We can construct a similar monad in any monoidal category whenever we have a comonoid object (E, c, ∂), which we can consider an object of "extra data" that can be copied and discarded.

Definition 6.2.37. Let (E, c, ∂) be an internal comonoid in a monoidal category $(\mathbf{C}, \otimes, I)$. The **reader comonad** induced by E has as functor, $- \otimes E$, and the counit and comultiplication are defined based on ∂ and c.

Note that in **Set**, we didn't have to require that E is a comonoid (Example 5.3.1) since every set is canonically a comonoid. In a general monoidal category, we have to require it specifically.

Let's now define the morphisms.

Definition 6.2.38. Let (X, c, ∂) and (Y, c, ∂) be comonoid objects in a monoidal category $(\mathbf{C}, \otimes, I)$. A **morphism of comonoids** $X \to Y$ is a morphism $f : X \to Y$ of $\mathbf{C}$ such that the following diagrams commute:

$$\begin{array}{ccc} X & \searrow^{\partial} & \\ {\scriptstyle f}\downarrow & & I \\ Y & \nearrow_{\partial} & \end{array} \qquad \begin{array}{ccc} X & \xrightarrow{c} & X \otimes X \\ \downarrow{\scriptstyle f} & & \downarrow{\scriptstyle f\otimes f} \\ Y & \xrightarrow{c} & Y \otimes Y, \end{array}$$

called, respectively, **counit** and **comultiplication** conditions.

Thus, comonoids form a category, which we denote by **Comon**(**C**). In terms of string diagrams, these read as follows:

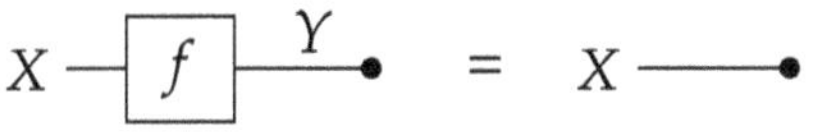

and

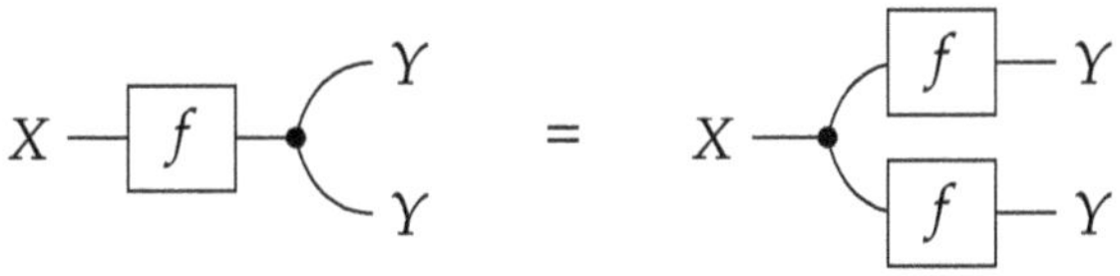

A possible interpretation, again in terms of experiments, is that f is like a process that we can perform with no effects:

- Performing the process and discarding the result is the same as discarding the system altogether (the *act* of performing it didn't change anything in the background).
- Repeating the same process twice, with the same input, gives the same result.

Once again, comonoids in **Set** behave rather trivially, as seen in the following example.

Example 6.2.39. Let X and Y be sets with their canonical comonoid structure. Then, every function $f : X \to Y$ is a morphism of comonoids: For all $x \in X$,

$$!(f(x)) = \bullet = !(x)$$

and

$$\text{copy}(f(x)) = (f(x), f(x)) = (f \times f)(x, x) = (f \times f)(\text{copy}(x)).$$

For the case of vector spaces, instead, not all maps commute with the coalgebra structure.

Exercise 6.2.40 (linear algebra; important!). Let V be a vector space, and let (C, c, ∂) be a coalgebra on V given by some basis E. Consider now $\mathbb{R}$ with its canonical coalgebra structure given by the one-element basis $\{1\}$. Show that a vector $v \in V$ is an element of E if and only if the corresponding point $\mathbb{R} \to V$ a is morphism of coalgebras.

Exercise 6.2.41 (linear algebra). Generalize the previous exercise by replacing $\mathbb{R}$ with a vector space A, with the coalgebra structure given by a basis.

Example 6.2.42 (linear algebra, quantum physics). Here is a way to make more precise the idea in Example 1.4.17 that "broadcasting" or "copying" is not compatible with changes of basis. Consider the vector space $\mathbb{R}^2$, and the following bases, orthonormal, and 45° ($\pi/4$ rad) apart:

$$E = \{e_1, e_2\} = \left\{ \begin{pmatrix} 1 \\ 0 \end{pmatrix}, \begin{pmatrix} 0 \\ 1 \end{pmatrix} \right\}; \quad F = \{f_1, f_2\} = \left\{ \begin{pmatrix} 1/\sqrt{2} \\ 1/\sqrt{2} \end{pmatrix}, \begin{pmatrix} -1/\sqrt{2} \\ 1/\sqrt{2} \end{pmatrix} \right\}.$$

The identity map $\mathbb{R}^2 \to \mathbb{R}^2$ is a morphism of **Vect**, but it is not a morphism of *coalgebras* (between the coalgebras induced by the two bases). (Show this using the previous exercise.)

In other words, the *system* is the same, but the *experiment* has changed, and the two experiments are not compatible. In **Set**, this does not happen: Every set has only one aspect that can be observed, there are no incompatible experiments.

In quantum physics, a similar situation happens (with Hilbert spaces over $\mathbb{C}$) for the spin of an electron along different axes. (This also happens for position and momentum — Heisenberg's indeterminacy principle — but that situation is not technically about *bases* of a vector space.)

Example 6.2.43 (probability theory). The Kleisli category of the probability monad $\mathcal{P}$, just as any other Kleisli category, inherits the morphisms from **Set** via the monoidal unit and adds new ones (in this case, "random" ones). If we write the Kleisli morphisms $X \to Y$ as stochastic matrices, with entries $f(y|x)$, the copy and discard maps inherited from **Set** are the following:

$$\mathrm{copy}(x', x''|x) = \begin{cases} 1 & x' = x'' = x; \\ 0 & \text{otherwise}, \end{cases} \qquad !(\bullet|x) = 1.$$

Let's now see which stochastic maps are comonoid morphisms for this comonoid structure. Given $f : X \to Y$ with entries $f(y|x)$, the counitality condition says that for all $x \in X$,

$$\sum_{y \in Y} f(y|x) = 1.$$

This is true for all stochastic matrices f: The sum over each column is one (one can interpret this as the fact that probabilities are normalized). The comultiplication condition says that for all $x \in X$ and $y', y'' \in Y$,

$$\sum_{y \in Y} \mathrm{copy}(y', y''|y)\, f(y|x) = \sum_{x', x''} f(y'|x')\, f(y''|x'')\, \mathrm{copy}(x', x''|x).$$

Equivalently, for all $x, y \in Y$,

$$f(y|x) = f(x|y)^2,$$

which means that $f(y|x)$ has to be either 0 or 1, nothing in between. In other words, the comonoid morphisms are precisely those Kleisli morphisms which are inherited from **Set**, the *deterministic* ones. All the "random" ones are not comonoid morphisms.

In classical probability, akin to quantum physics, there are processes which, when performed twice, may give different results (think of flipping a coin). In some sense, the *randomness* of a process can be seen as a failure of being a comonoid morphism. There is however an important difference between classical probability and quantum physics: In classical probability, while there is randomness, there are no incompatible experiments on the same system.

Exercise 6.2.44 (probability theory, linear algebra). Show that in the Kleisli category of the probability monad $\mathcal{P}$, with the comonoid structures given as in the previous example, every isomorphism is also an isomorphism of comonoids. (*Hint*: Use the fact that the entries of a stochastic matrix are non-negative.)

While here we have focused on discrete probability distributions, similar statements are true about more general monads.

The following proposition can be interpreted as the fact that cartesian monoidal categories, such as **Set**, are precisely those monoidal categories where everything is canonically copyable and discardable.

Proposition 6.2.45. *A symmetric monoidal category* $(\mathbf{C}, \otimes, I)$ *is* cartesian *monoidal if and only if:*

- *every object has a unique comonoid structure;*
- *every morphism is a morphism of comonoids.*

Moreover, in that case, every comonoid is commutative.

Proof. Consider a cartesian monoidal category $(\mathbf{C}, \times, 1)$. By the universal property of the terminal object, there is only one possible counit map $X \to 1$. If we want counitality to hold, the only possible comultiplication map is the unique map appearing in the following diagram: (Why?)

$$
\begin{array}{ccc}
& X & \\
{\scriptstyle id}\swarrow & \downarrow & \searrow{\scriptstyle id} \\
X \xleftarrow{\ p_1\ } & X \otimes X & \xrightarrow{\ p_2\ } X.
\end{array}
$$

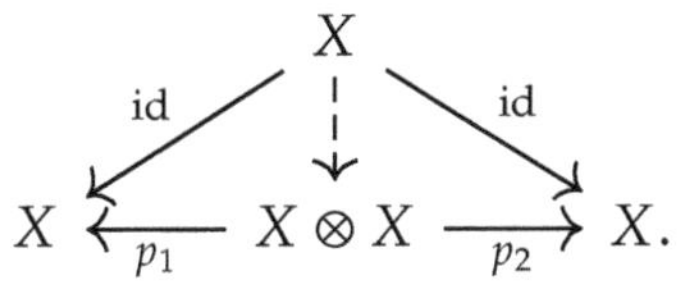

Coassociativity and commutativity hold by the universal property of products. (Why is every morphism of $\mathbf{C}$ a morphism of comonoids?)

Conversely, suppose that every object has a unique comonoid structure and that every morphism is a comonoid morphism. Denote by $p_1 : X \otimes Y \to X$ the map

$$
X \otimes Y \xrightarrow{\ id \otimes \partial\ } X \otimes I \xrightarrow{\ \cong\ } X,
$$

and define $p_2 : X \otimes Y \to Y$ similarly. In order to show that p_1 and p_2 are product projections, we have to show that in the diagram

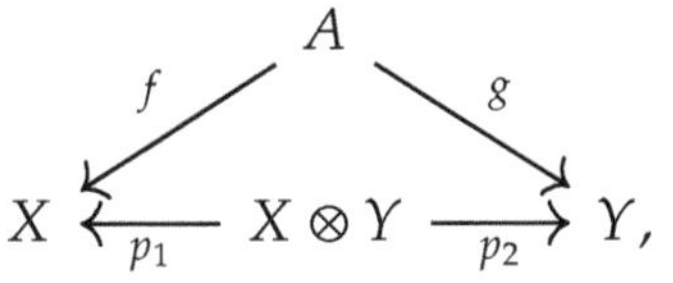

$$
\begin{array}{ccc}
& A & \\
{\scriptstyle f}\swarrow & & \searrow{\scriptstyle g} \\
X \xleftarrow{\ p_1\ } & X \otimes Y & \xrightarrow{\ p_2\ } Y,
\end{array}
\tag{6.2.2}
$$

there exists a unique map $A \to X \otimes Y$ that makes the diagram commute. For existence, consider now the map

$$
A \xrightarrow{\ c\ } A \otimes A \xrightarrow{\ f \otimes g\ } X \otimes Y.
$$

Since f and g, by hypothesis, commute with the counits, this map makes the diagram (6.2.2) commute. For uniqueness, see the next exercise.

Let's now show that I is terminal. First of all, note that the unitors give a comonoid structure on I since the comonoid axioms hold by coherence. By uniqueness (assumed by hypothesis), this is the only comonoid structure on I. In particular, the counit $I \to I$ is the identity. Consider now maps $r, s : X \to I$. Since they are morphisms of comonoids, we have

$$r = \mathrm{id}_I \circ r = \partial_I \circ r = \partial_X = \partial_I \circ s = \mathrm{id}_I \circ s = s.$$

Therefore, I is terminal, and hence $(\mathbf{C}, \otimes, I)$ is cartesian monoidal. □

Exercise 6.2.46. Let $h : A \to X \otimes Y$ be a map that makes (6.2.2) commute, and suppose that h is a morphism of comonoids, where the comultiplication of $X \otimes Y$ is given as follows:

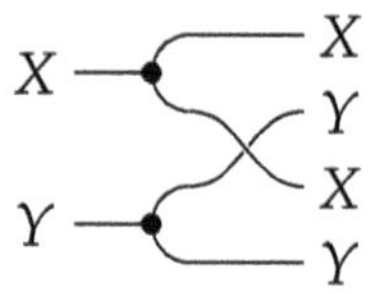

Show that $h = (f \otimes g) \circ c$. (Why does this complete the proof above?)

6.2.4 Internal comodules

Definition 6.2.47. Let (W, c, ∂) be an internal comonoid in a monoidal category $(\mathbf{C}, \otimes, I)$. A (left) **comodule object**, or (left) **internal comodule**, over W is an object A of $\mathbf{C}$ together with a morphism $a : A \to W \otimes A$, called the **coaction map**, making the following diagrams commute:

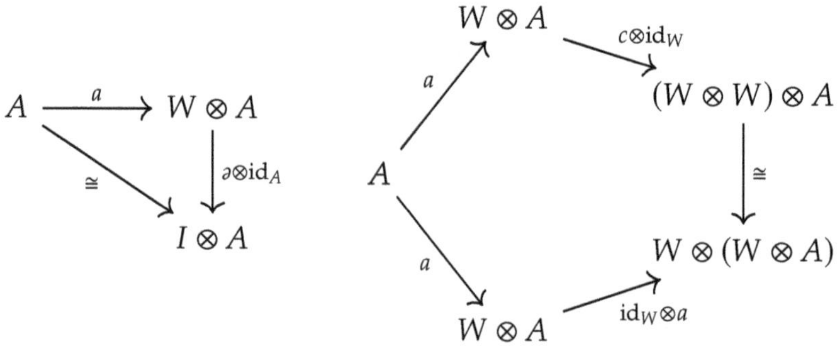

called **counit** and **comultiplication** conditions, respectively.

A right comodule object is an analogous structure, where the coaction map is in the form $A \to A \otimes W$ (and satisfying analogous conditions).

In terms of string diagrams, we can again use a white dot, but in the other direction,

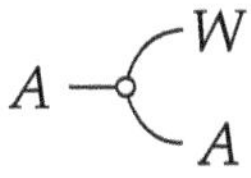

and the conditions are

and

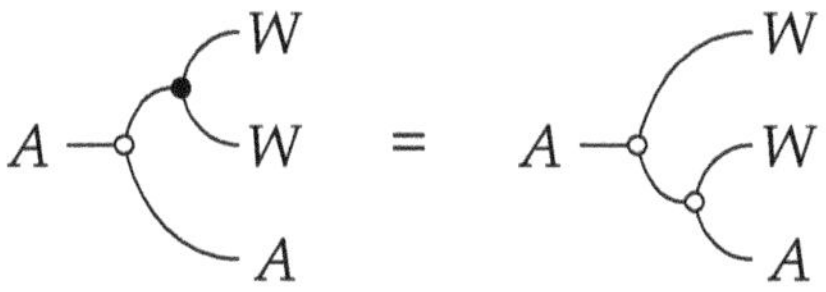

Let's see what these are in **Set**. Let X be a set, which is canonically a comonoid. An internal comodule amounts first of all to a map $a : A \to X \times A$, and by the universal property of the product, this amounts to a pair of maps $f : A \to X$ and $g : A \to A$. The counitality constraint forces us to have $g = \mathrm{id}_A$. With this choice, the map $(f, \mathrm{id}_A) : A \to X \times A$ satisfies associativity automatically (why?).

The argument given above also works for the following, more general statement.

Proposition 6.2.48. *In a cartesian monoidal category, an internal comodule A over an internal comonoid (i.e. an object X) is specified by a unique morphism $A \to X$.*

Exercise 5.4.15 asked to prove something analogous for the reader comonad. It is now subsumed by the argument above, together with the following exercise.

Exercise 6.2.49. Show that internal comodules over an internal comonoid are equivalently coalgebras of the associated reader comonad.

Exercise 6.2.50 (probability, linear algebra). Show that in the Kleisli category of the probability monad $\mathcal{P}$ on **Set**, similar to the case of sets and functions, any comodule $a : A \to A \otimes X$ over the comonoid X (with the copy and discard maps induced from **Set**) is uniquely specified by a *deterministic* morphism $A \to X$. (*Hint*: Use string diagrams.)

The situations above can be understood in the abstract, dualizing Exercises 6.2.32 and 6.2.33, as follows.

Corollary 6.2.51. *Let* (A, c, ∂) *and* (X, c, ∂) *be internal comonoids in a monoidal category* $(\mathbf{C}, \otimes, I)$*. Then, for every* X*-comodule structure on* A *(denoted by a white dot), the following conditions are equivalent:*

(a)

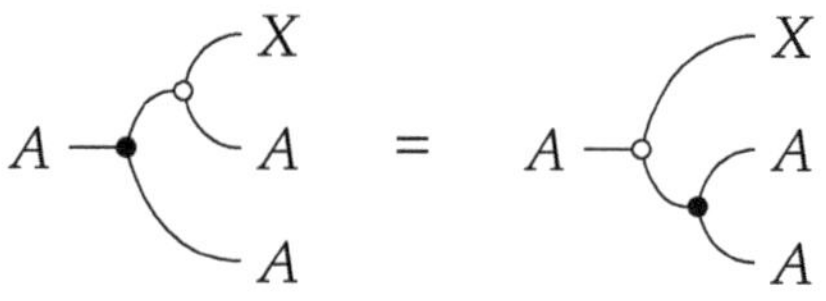

(b) *There exists a comonoid homomorphism* $f : A \to X$ *such that*

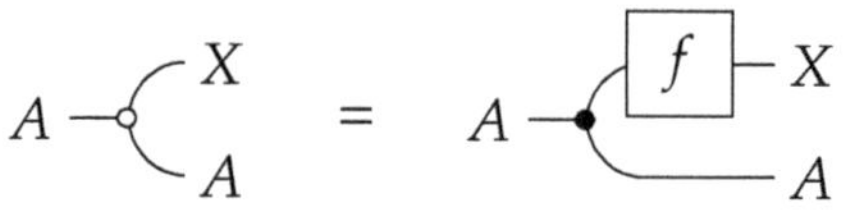

In cartesian monoidal categories and in the Kleisli category of $\mathcal{P}$, all internal coalgebras are in the form above.

We can interpret this map f in terms of algebraic geometry by viewing the map $f : A \to X$ as a *bundle*, or as setting the space A to being "*over* X" in some way.

Example 6.2.52 (algebraic geometry). The opposite category to commutative rings, $\mathbf{CRing}^{\mathrm{op}}$, is called the category of *affine schemes*.

To avoid confusion, for a ring R, denote by R^{op} the same ring but taken as an object of the opposite category. We can think of R^{op} as a *topological space* of a certain kind,[16] and of a ring homomorphism $f : R \to S$ as a *continuous function* $f^{\mathrm{op}} : S^{\mathrm{op}} \to R^{\mathrm{op}}$, which we can see as a sort of "bundle" over R^{op} (in a very loose way).

We saw in the corollary above that, similarly to how a morphism of internal modules (i.e. rings) $f : R \to S$ makes S into an R-module canonically, a morphism of internal comonoids (i.e. spaces) $S^{\mathrm{op}} \to R^{\mathrm{op}}$ makes S^{op} a comodule over R^{op}. Therefore, in some cases, one can view the expressions "module *over* a certain ring" and "bundle *over* a certain space" as dual to one another, with the word "over" playing a dual role.

Exercise 6.2.53. Let V be a vector space equipped with a specified basis. What are the comodules over the corresponding coalgebra?

Definition 6.2.54. Let (A, a) and (B, b) be internal comodules over an internal comonoid (W, c, ∂) in $(\mathbf{C}, \otimes, I)$. A **morphism of internal comodules** is a morphism $f : A \to B$ of $\mathbf{C}$ that makes the following diagram commute:

$$\begin{array}{ccc} A & \xrightarrow{\ f\ } & B \\ \downarrow{\scriptstyle a} & & \downarrow{\scriptstyle b} \\ W \otimes A & \xrightarrow{\mathrm{id}_W \otimes f} & W \otimes B, \end{array}$$

called (internal) **coequivariance** condition.

In terms of string diagrams,

[16]The *spectrum* of R is the set of its prime ideals with the Zariski topology. See the Stacks project, https://stacks.math.columbia.edu/tag/00DY and https://stacks.math.columbia.edu/tag/01HR.

This way, internal comodules form a category, which we denote by W-**Comod**.

Exercise 6.2.55. Suppose that, in a monoidal category $(\mathbf{C}, \otimes, I)$, every morphism is canonically equipped with a comonoid structure (such as in **Set**, or in the Kleisli category of $\mathcal{P}$, or in the category of vector spaces each equipped with a distinguished basis). Fix a comonoid W. Using Corollary 6.2.51, compare the category W-**Comod** and the coslice category $W/\mathrm{id}_{\mathbf{Comon}(\mathbf{C})}$ of morphisms from W (recall Exercise 3.2.49).

6.3 Monoidal Functors

Here we study some notions of functors between monoidal categories. The simplest example of a monoidal functor is a *strict* monoidal functor, defined as follows.

Definition 6.3.1. Let $(\mathbf{C}, \otimes, I)$ and $(\mathbf{D}, \otimes, I)$ be monoidal categories. A **strict-monoidal functor** is a functor $F : \mathbf{C} \to \mathbf{D}$ such that

$$FI = I, \qquad F(A \otimes B) = FA \otimes FB, \qquad F(f \otimes g) = Ff \otimes Fg,$$

for all objects A, B and morphisms and for all f, g of $\mathbf{C}$, and such that

$$F(\lambda) = \lambda, \qquad F(\rho) = \rho, \qquad F(\alpha) = \alpha.$$

If we view strict monoidal categories as internal monoids in **Cat**, this is the corresponding notion of monoid morphism (but it is also defined for non-strict monoidal categories).

Just as strict monoidal categories are too restrictive to capture, for example, the cartesian product of sets, strict monoidal functors are too restrictive to capture, for example, the fact that a continuous functor preserves categorical products. Let's now give more general notions of monoidal functors. As usual, we start with precise definitions, and then we present an intuitive understanding and some examples.

6.3.1 Lax, colax, and strong-monoidal functors

Definition 6.3.2. Let $(\mathbf{C}, \otimes, I)$ and $(\mathbf{D}, \otimes, I)$ be monoidal categories. A **lax-monoidal functor**[17] is a functor $F : \mathbf{C} \to \mathbf{D}$ that consists of

- a distinguished morphism $e : I \to FI$ of $\mathbf{D}$, from the unit of $\mathbf{D}$ to the image of the unit of $\mathbf{C}$, called the **unit** map;
- a natural transformation between the functors $\mathbf{C} \times \mathbf{C} \to \mathbf{D}$ with components
$$\nabla : FA \otimes FB \to F(A \otimes B),$$
from the tensor product in $\mathbf{D}$ of images to the image of the tensor product in $\mathbf{C}$, called the **multiplication** map,

such that the following diagrams commute:

$$\begin{array}{ccc}
(FA \otimes FB) \otimes FC & \xrightarrow{\cong} & FA \otimes (FB \otimes FC) \\
\big\downarrow{\scriptstyle \nabla_{A,B} \otimes \mathrm{id}} & & \big\downarrow{\scriptstyle \mathrm{id} \otimes \nabla_{B,C}} \\
F(A \otimes B) \otimes FC & & FA \otimes F(B \otimes C) \\
\big\downarrow{\scriptstyle \nabla_{A \otimes B, C}} & & \big\downarrow{\scriptstyle \nabla_{A, B \otimes C}} \\
F((A \otimes B) \otimes C) & \xrightarrow{\cong} & F(A \otimes (B \otimes C))
\end{array}$$

$$\begin{array}{ccc}
I \otimes FA & \xrightarrow{e \otimes \mathrm{id}} & FI \otimes FA \\
\big\downarrow{\scriptstyle \cong} & & \big\downarrow{\scriptstyle \nabla_{I,A}} \\
FA & \xleftarrow[\cong]{} & F(I \otimes A)
\end{array}
\qquad
\begin{array}{ccc}
FA \otimes I & \xrightarrow{\mathrm{id} \otimes e} & FA \otimes FI \\
\big\downarrow{\scriptstyle \cong} & & \big\downarrow{\scriptstyle \nabla_{A,I}} \\
FA & \xleftarrow[\cong]{} & F(A \otimes I),
\end{array}$$

called, respectively, **associativity** and **left and right unitality**.

Definition 6.3.3. A **strong-monoidal functor** is a lax-monoidal functor (F, ∇, e) in which the unit e and multiplication maps ∇ are isomorphisms:

$$I \xrightarrow[\cong]{e} FI, \qquad FA \otimes FB \xrightarrow[\cong]{\nabla} F(A \otimes B).$$

---

[17]We use the hyphen since there is also a notion of *lax functor* (https://ncatlab.org/nlab/show/lax+functor). Our structure is *an ordinary functor which is laxly monoidal.*

Definition 6.3.4. Let $(\mathbf{C}, \otimes, I)$ and $(\mathbf{D}, \otimes, I)$ be monoidal categories. A **colax-monoidal functor**[18] is a functor $F : \mathbf{C} \to \mathbf{D}$ that consists of

- a distinguished morphism $\partial : FI \to I$ of $\mathbf{D}$, called the **counit** map;
- a natural transformation between functors $\mathbf{C} \times \mathbf{C} \to \mathbf{D}$ with components
$$\Delta : F(A \otimes B) \to FA \otimes FB,$$
called the **comultiplication** map

such that the following diagrams commute:

$$\begin{array}{ccc} F((A \otimes B) \otimes C) & \xrightarrow{\cong} & F(A \otimes (B \otimes C)) \\ \downarrow{\scriptstyle \Delta_{A \otimes B, C}} & & \downarrow{\scriptstyle \Delta_{A, B \otimes C}} \\ F(A \otimes B) \otimes FC & & FA \otimes F(B \otimes C) \\ \downarrow{\scriptstyle \Delta_{A,B} \otimes \mathrm{id}} & & \downarrow{\scriptstyle \mathrm{id} \otimes \Delta_{B,C}} \\ (FA \otimes FB) \otimes FC & \xrightarrow{\cong} & FA \otimes (FB \otimes FC) \end{array}$$

$$\begin{array}{ccc} F(I \otimes A) & \xrightarrow{\Delta_{I,A}} & FI \otimes FA \\ \downarrow{\scriptstyle \cong} & & \downarrow{\scriptstyle \partial \otimes \mathrm{id}} \\ FA & \xleftarrow[\cong]{} & I \otimes FA \end{array}$$

$$\begin{array}{ccc} F(A \otimes I) & \xrightarrow{\Delta_{A,I}} & FA \otimes FI \\ \downarrow{\scriptstyle \cong} & & \downarrow{\scriptstyle \mathrm{id} \otimes \partial} \\ FA & \xleftarrow[\cong]{} & FA \otimes I, \end{array}$$

called, respectively, **coassociativity** and **left and right counitality**.

Exercise 6.3.5. Show that a strong-monoidal functor is equivalently a colax-monoidal functor in which the counit and comultiplication maps are isomorphisms.

Beware that often people simply write "monoidal functor," and depending on the author, this might mean either "strict" or "strong" or "lax" (usually not "colax").

Between symmetric monoidal categories, we often require an additional condition, as given in the following.

[18] Alternative terminology: *oplax*-monoidal functor, or *co*monoidal functor.

Definition 6.3.6. We say that a lax-monoidal functor (F, ∇, e) between symmetric monoidal categories is **symmetric** if the multiplication satisfies the **braiding** condition:

$$\begin{array}{ccc} FA \otimes FB & \xrightarrow{\cong} & FB \otimes FA \\ \downarrow{\scriptstyle \nabla} & & \downarrow{\scriptstyle \nabla} \\ F(A \otimes B) & \xrightarrow{\cong} & F(B \otimes A). \end{array}$$

We say that a colax-monoidal functor (F, Δ, ∂) between symmetric monoidal categories is **symmetric** if the comultiplication satisfies the **braiding** condition:

$$\begin{array}{ccc} F(A \otimes B) & \xrightarrow{\cong} & F(B \otimes A) \\ \downarrow{\scriptstyle \Delta} & & \downarrow{\scriptstyle \Delta} \\ FA \otimes FB & \xrightarrow{\cong} & FB \otimes FA. \end{array}$$

One can represent these ideas by means of a generalization of string diagrams. In this book, we will not use this formalism, but it's worth mentioning since it can simplify some derivations. One can represent the application of a functor by a shading that covers everything in its image. For example, given a functor $F : \mathbf{C} \to \mathbf{D}$, we denote an object D or a morphism $d : D \to D'$ of $\mathbf{D}$ as usual:

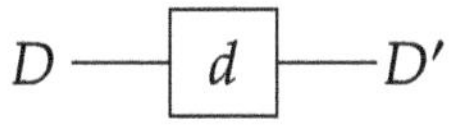

However, we draw an object in the form FC or a morphism in the form $Fc : FC \to FC'$ coming from $\mathbf{C}$ as follows, representing F with a shading:

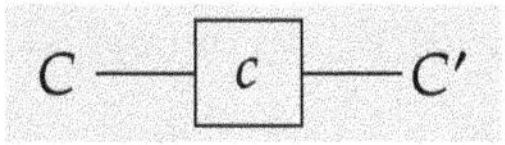

This way, for example, we can represent the map $\nabla : FA \otimes FB \to F(A \otimes B)$ as follows:

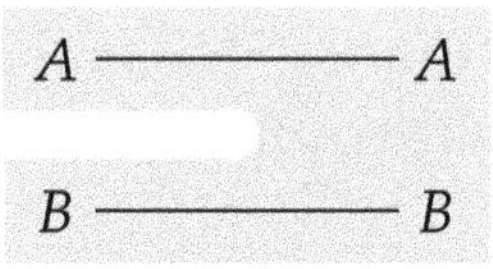

and the map $e : I \to FI$ as

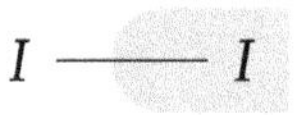 or, omitting I as usual, simply as

(What do the unitality and associativity conditions look like in terms of these?)

6.3.2 Examples, interpretation, and more on complexity

Let's interpret the definitions of monoidal functors. One possible way, following the ideas of Section 3.3.1, is that lax- and colax-monoidal functors encode a more general notion of complex behavior. We saw in Section 3.3.1 that by the universal properties of products and coproducts (when they exist), any functor $F : \mathbf{C} \to \mathbf{D}$ has "comparison maps" in $\mathbf{D}$ as follows:

$$F(X \times Y) \longrightarrow FX \times FY \qquad FX \sqcup FY \longrightarrow F(X \sqcup Y). \tag{6.3.1}$$

The interpretation that we gave was that whenever we form "more complex" objects ($X \times Y$ and $X \sqcup Y$) out of "simpler" ones (X and Y), a functor F may fail to see the whole as the composition of its parts and may only be able to compare them in one direction. The idea of monoidal functors is similar, but for a more general way of building more complex objects out of simpler ones, through a monoidal product $X \otimes Y$ which is not necessarily $X \times Y$ or $X \sqcup Y$:

- A lax-monoidal functor has a comparison map $FX \otimes FY \to F(X \otimes Y)$ similar to the one for coproducts, but without having to assume that the monoidal structure is given by coproducts.
- A colax-monoidal functor has a comparison map $F(X \otimes Y) \to FX \otimes FY$ similar to the one for categorical products, but without having to assume that the monoidal structure is cartesian.
- For a strong-monoidal functor, the comparison map is an isomorphism, and so it generalizes functors that preserve finite products or coproducts.

Moreover, these comparison maps need to satisfy the unitality and associativity conditions. We didn't have to require that for products and coproducts because once again, in that case, it follows from their universal properties.

Exercise 6.3.7. Suppose that **C** and **D** have all finite products and coproducts, as well as initial and terminal objects (so that they are cartesian and *co*-cartesian monoidal categories). Prove that the maps in (6.3.1), obtained via the universal property of products and coproducts, make F a symmetric colax- and a symmetric lax-monoidal functor, respectively, for $\otimes = \times$ and $\otimes = \sqcup$.

Let's now see some examples.

Example 6.3.8 (sets and relations). We saw in Example 3.3.8 that if P denotes the power set functor, the canonical comparison map $P(X \times Y) \to PX \times PY$ is not an isomorphism. As we have noted, this is related to the fact that, for example, many different regions of the plane $\mathbb{R}^2$ have the same pair of projections onto the axes. Now, while this is true, *there is a canonical choice* of the region of $\mathbb{R}^2$, given two projections. Given $A \subseteq \mathbb{R}$ and $B \subseteq \mathbb{R}$, we can form the *rectangle* $A \times B := (A \times \mathbb{R}) \cap (\mathbb{R} \times \mathbb{R}) \subseteq \mathbb{R}^2$:

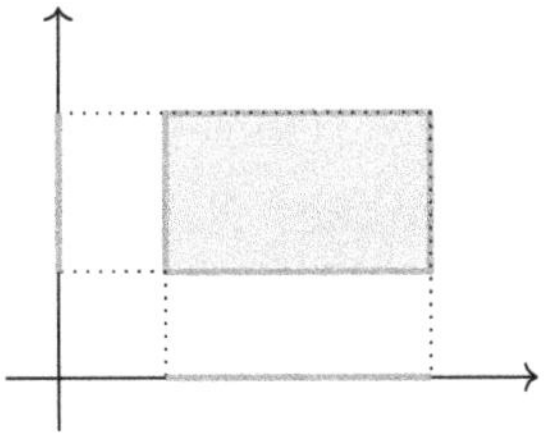

(The same idea works even if A and B are not intervals.) More generally, given $A \in PX$ and $B \in PY$, we can form the set $A \times B \in P(X \times Y)$ defined in the same way. This gives a morphism $\nabla : PX \times PY \to P(X \times Y)$, and it equips P with the structure of a lax-monoidal functor $(\mathbf{Set}, \times, 1) \to (\mathbf{Set}, \times, 1)$. (Try to prove this. The unit map is the singleton inclusion, and unitality and associativity follow from the analogous properties of the cartesian product of *subsets*.)

Of course, ∇ is not an inverse to $\Delta : P(X \times Y) \to PX \times PY$ (since Δ is not an isomorphism). We simply have that P is both lax-monoidal and colax-monoidal with two different maps. (There is a notion of *bilax-monoidal functor*, but we will not treat it here.[19])

The map $\nabla : PX \times PY \to P(X \times Y)$ does not come from the universal property of limits (it goes in the wrong direction). The idea of lax-monoidal functors is an abstraction of all the maps similar to ∇.

Let's also remark that ∇ can be interpreted as a notion of "default case," or "no knowledge situation." If we only know the projections of a figure with no way of saying how X and Y interact, it is a canonical, consistent choice, compatible with taking further products (by associativity). Of course, in principle, a functor may admit *several* such consistent choices, i.e. several lax-monoidal structures, but very often there is only one.

Example 6.3.9 (probability). Similar to the case above, we saw in Example 3.3.10 that for the probability functor $\mathcal{P}$, the canonical map $\mathcal{P}(X \times Y) \to \mathcal{P}X \times \mathcal{P}Y$ is not an isomorphism, and the interpretation was that a joint probability measure contains, in general, more information than the pair of its marginals. Just as above, given a pair of marginals (p, q) on X and Y, we can form the *product probability*

$$(p \otimes q)(x, y) \;:=\; p(x)\, q(y),$$

the joint distribution that makes X and Y independent. (This is how we denoted it in Example 6.1.29, and it's not a coincidence, see Example 6.4.41.) This gives a map $\mathcal{P}X \times \mathcal{P}Y \to \mathcal{P}(X \times Y)$ that makes P a lax-monoidal functor (again, prove this).

Once again, this map does not come from any universal property. It is just a consistent choice of forming a joint distribution from its marginals (and in this case, it reflects having "minimal knowledge" since it is the choice of maximal entropy).

[19]See [AM10, Section 3.1].

Exercise 6.3.10. Show that the lax-monoidal structures given in the previous two examples are symmetric.

Exercise 6.3.11 (algebra, computer science; important!). Define a lax-monoidal structure on the list functor L by generalizing the following idea:

$$([a_1, a_2], [b_1, b_2]) \longmapsto \big[(a_1, b_1), (a_1, b_2), (a_2, b_1), (a_2, b_2)\big].$$

Show that the resulting lax-monoidal structure is *not* symmetric.

We conclude this section with a few structural remarks in the form of exercises.

Exercise 6.3.12 (important!). Prove that:

- the identity functor is canonically strong-monoidal;
- the composition of lax-monoidal functors (F, ∇_F, e_F) and (G, ∇_G, e_G) is canonically lax-monoidal, with structure maps

$$GFA \otimes GFB \xrightarrow{\nabla_G} G(FA \otimes FB) \xrightarrow{G\nabla_F} GF(A \otimes B)$$

  and

$$I \xrightarrow{e_G} GI \xrightarrow{Ge_F} GFI.$$

- the composition of colax-monoidal functors is canonically colax-monoidal, with structure maps dual to the ones above.

Conclude that the strong-monoidal, lax-monoidal, and colax-monoidal functors form categories whose objects are monoidal categories.

Exercise 6.3.13 (important!). Prove that:

- the category $\mathbf{1}$ has a unique monoidal structure;
- internal monoids in $\mathbf{C}$ are exactly lax-monoidal functors $\mathbf{1} \to \mathbf{C}$;
- internal comonoids in $\mathbf{C}$ are exactly colax-monoidal functors $\mathbf{1} \to \mathbf{C}$.

Conclude that the image of an internal monoid under a lax-monoidal functor is canonically an internal monoid and the image of an internal comonoid under a colax-monoidal functor is canonically an internal comonoid.

Exercise 6.3.14 (important!). Let $(F, \nabla, e) : (\mathbf{C}, \otimes, I) \to (\mathbf{D}, \otimes, I)$ be lax-monoidal:

- Show that the map $\nabla : FI \otimes FI \to F(I \otimes I) \cong FI$ makes FI, an internal monoid in $\mathbf{D}$, commutative if F is symmetric. (*Hint*: Recall Exercise 6.2.11.)
- Show that for every object A of $\mathbf{C}$, the map $\nabla : FI \otimes FA \to F(I \otimes A) \cong FA$ makes FA an internal FI-module.

Exercise 6.3.15 (probability, group theory). The set $\mathbb{R}$ of real numbers is a monoid (even a group) under addition. Since the probability functor $\mathcal{P}$ is lax-monoidal (Example 6.3.9), the set $\mathcal{P}\mathbb{R}$ of (discrete) probability distributions on $\mathbb{R}$ is a monoid too. Then:

- show that this monoid is given by taking *convolutions* of probability measures;
- show that this monoid is not a group (even if $\mathbb{R}$ is a group);
- if you can, interpret the failure of being a group in terms of probability (what happens to the variance when we convolve two distributions)?

Define, more generally, the convolution of (discrete) probability distributions on any group.

Exercise 6.3.16 (sets and relations, group theory). Let G be a group. Define the *convolution of sets* to be the monoid structure on PG induced by the power set functor P, and write down the construction explicitly.

6.3.3 Monoidal transformations and monoidal equivalences

Just as categories have arrows (natural transformations) between arrows (functors), so do monoidal categories. (As we said, we have a *2-category*, but that's beyond the scope of this book.[20])

Definition 6.3.17. Let (F, ∇, e) and (G, ∇, e) be lax-monoidal functors $(\mathbf{C}, \otimes, I) \to (\mathbf{D}, \otimes, I)$. A **lax-monoidal natural transformation** is a natural transformation $\gamma : F \Rightarrow G$ that makes the following diagrams commute (for all $A, B \in \mathbf{C}$):

$$\begin{array}{ccc} & \overset{e}{\nearrow} & FI \\ I & & \downarrow{\scriptstyle \gamma_I} \\ & \underset{e}{\searrow} & GI \end{array} \qquad \begin{array}{ccc} FA \otimes FB & \xrightarrow{\nabla} & F(A \otimes B) \\ \downarrow{\scriptstyle \gamma_A \otimes \gamma_B} & & \downarrow{\scriptstyle \gamma_{A \otimes B}} \\ GA \otimes GB & \xrightarrow{\nabla} & G(A \otimes B), \end{array}$$

called, respectively, the **unit** and **multiplication** conditions.

Note that on the top of the diagrams, there are the unit and multiplication maps of F and on the bottom the ones of G. Also, on the left of the diagrams, there are the unit object and the multiplication functors of $\mathbf{D}$ and on the right the (images of) the ones of $\mathbf{C}$.

Definition 6.3.18. Let (F, Δ, ∂) and (G, Δ, ∂) be colax-monoidal functors $(\mathbf{C}, \otimes, I) \to (\mathbf{D}, \otimes, I)$. A **colax-monoidal natural transformation** is a natural transformation $\gamma : F \Rightarrow G$ that makes the following diagrams commute (for all $A, B \in \mathbf{C}$):

$$\begin{array}{ccc} FI & \overset{\partial}{\searrow} & \\ {\scriptstyle \gamma_I}\downarrow & & I \\ GI & \underset{\partial}{\nearrow} & \end{array} \qquad \begin{array}{ccc} F(A \otimes B) & \xrightarrow{\Delta_F} & FA \otimes FB \\ \downarrow{\scriptstyle \gamma_{A \otimes B}} & & \downarrow{\scriptstyle \gamma_A \otimes \gamma_B} \\ G(A \otimes B) & \xrightarrow{\Delta_G} & GA \otimes GB, \end{array}$$

called, respectively, the **counit** and **comultiplication** conditions.

[20]Once again, see https://ncatlab.org/nlab/show/2-category as well as [JY21; Lac09] for more, but I recommend to finish this book first.

Exercise 6.3.19. How would you represent these conditions in terms of the *shadings* of Section 6.3.1?

Exercise 6.3.20. Show that the natural transformations (units of the monads) $\sigma : X \to PX$ and $\delta : X \to \mathcal{P}X$ are lax-monoidal (using the maps ∇ defined in Section 6.3.2).

Similarly to how two groups can be isomorphic as sets but have nonisomorphic group structure, two categories can be equivalent but still fail to have *equivalent monoidal structures*. Let's define this notion of equivalence, which requires all three levels (monoidal categories, functors, and natural transformations).

Definition 6.3.21. Let $(\mathbf{C}, \otimes, I)$ and $(\mathbf{D}, \otimes, I)$ be monoidal categories. A **monoidal equivalence** between $\mathbf{C}$ and $\mathbf{D}$ consists of a pair of lax-monoidal functors $F : \mathbf{C} \to \mathbf{D}$ and $G : \mathbf{D} \to \mathbf{C}$ and lax-monoidal natural isomorphisms $\eta : G \circ F \Rightarrow \mathrm{id}_{\mathbf{C}}$ and $\varepsilon : F \circ G \Rightarrow \mathrm{id}_{\mathbf{D}}$.

We can equivalently restate the definition with "colax" everywhere instead of "lax" (why?). Less trivially, we can also equivalently replace "lax" in the definition above with "strong." Here's why: Recall that between ordinary categories, a functor $F : \mathbf{C} \to \mathbf{D}$ induces an equivalence of categories if and only if it is fully faithful and essentially surjective (Theorem 1.5.16). The following is a monoidal version of that statement.

Theorem 6.3.22. *A lax- (or colax-) monoidal functor* $(F, \nabla, e) : (\mathbf{C}, \otimes, I_{\mathbf{C}}) \to (\mathbf{D}, \otimes, I_{\mathbf{D}})$ *induces a monoidal equivalence of categories if and only if it is fully faithful, essentially surjective, and strong-monoidal.*

For reasons of space, we leave some details in the proof for the reader to check (see the following exercises).

Proof. First of all, let (F, G) be a monoidal equivalence. Then, F is in particular fully faithful and essentially surjective. To prove it is strong-monoidal, let's use the fact that η and ε are lax-monoidal natural transformations, and together with Exercise 6.3.12, we get that the following diagrams must commute, for all X and Y of $\mathbf{C}$ and A and B of $\mathbf{D}$:

$$
\begin{array}{ccc}
GFX \otimes GFY & & \\
{\scriptstyle \nabla_G}\downarrow & \searrow^{\eta\otimes\eta}_{\cong} & \\
G(FX \otimes FY) & & X \otimes Y \\
{\scriptstyle G\nabla_F}\downarrow & \nearrow^{\cong}_{\eta} & \\
GF(X \otimes Y) & &
\end{array}
\qquad\qquad
\begin{array}{ccc}
FGA \otimes FGB & & \\
{\scriptstyle \nabla_F}\downarrow & \searrow^{\varepsilon\otimes\varepsilon}_{\cong} & \\
F(GA \otimes GB) & & A \otimes B \\
{\scriptstyle F\nabla_G}\downarrow & \nearrow^{\cong}_{\varepsilon} & \\
FG(A \otimes B) & &
\end{array}
$$

Now, since the maps η and ε are isomorphisms, the diagram on the right implies that all the maps ∇_F are split mono, and the diagram on the left implies that all the maps $G\nabla_F$ are split epi. However, since G is an equivalence, so also ∇_F are split epi, and hence they are isomorphisms. For the unit maps e_F, we can work analogously. Therefore, F (and also G) is strong-monoidal.

Conversely, suppose that F is fully faithful, essentially surjective, and strong-monoidal. Then, it has a pseudoinverse G that makes it an equivalence of categories. To prove that this equivalence is monoidal, it suffices to make G strong-monoidal too and show that the natural isomorphisms $\eta : G \circ F \Rightarrow \mathrm{id}_{\mathbf{C}}$ and $\varepsilon : F \circ G \Rightarrow \mathrm{id}_{\mathbf{D}}$ are (lax-) monoidal. We construct a unit map $e_G : I_{\mathbf{C}} \to GI_{\mathbf{D}}$ as follows: Since F is fully faithful, this morphism (of $\mathbf{C}$) is uniquely specified by a morphism $FI_{\mathbf{C}} \to FGI_{\mathbf{D}}$ of $\mathbf{D}$, and we take

$$FI_{\mathbf{C}} \xrightarrow[\cong]{e_F^{-1}} I_D \xrightarrow[\cong]{\varepsilon^{-1}} FGI_{\mathbf{D}}.$$

Similarly, as the multiplication map $GX \otimes GY \to G(X \otimes Y)$ we take the one uniquely specified by the following composition:

$$F(GX \otimes GY) \xrightarrow[\cong]{\nabla_F^{-1}} FGX \otimes FGY \xrightarrow[\cong]{\varepsilon\otimes\varepsilon} X \otimes Y \xrightarrow[\cong]{\varepsilon^{-1}} FG(X \otimes Y).$$

This makes G a strong-monoidal functor, and it makes η and ε monoidal (see the following exercises). □

In order to solve the following exercises, it helps to keep in mind the proof of Theorem 1.5.16.

Exercise 6.3.23. Show the associativity and unitality conditions for G.

Exercise 6.3.24. Show that η and ε are monoidal natural transformations.

6.3.4 Strictification

We now have all the ingredients to discuss strictification: Not only is every monoidal category equivalent to a strict one, it is equivalent in a *monoidal* way to a strict one.

Theorem 6.3.25 (Strictification of monoidal categories). *Every monoidal category is monoidally equivalent to a strict monoidal category.*

We prove this theorem using the idea of a *clique*, which makes precise the idea of "class of objects which are all uniquely isomorphic in a specified way."[21]

Definition 6.3.26. A groupoid **G** is called **contractible** if it is nonempty and connected, and every diagram commutes.

Equivalently, a contractible groupoid is a category where any two objects are connected by a unique isomorphism, and there are no other morphisms.

Exercise 6.3.27. Show that:

- in a contractible groupoid, the only endomorphisms $A \to A$ are identities;
- a category is a contractible groupoid if and only if it is equivalent to **1**.

[21] This is similar to cliques in graph theory, except that it needs to be *commutative* in a certain way.

Definition 6.3.28. A **clique** in a category $\mathbf{C}$ is a functor $F : \mathbf{G} \to \mathbf{C}$ from a contractible groupoid $\mathbf{G}$.

We say that an object X of $\mathbf{C}$ is **in the clique** $F : \mathbf{G} \to \mathbf{C}$ if $X = FA$ for some A of G.

We can view a clique as a perfectly commutative diagram in $\mathbf{C}$ (without size restrictions) where all the arrows are isomorphisms.

Warning. Keep in mind that, just as for usual diagrams, saying that the diagram commutes does not mean that its image in $\mathbf{C}$ is again a contractible groupoid. Let's see a trivial example. If the functor F "reuses an object," i.e. if for $A \neq A'$ in $\mathbf{G}$, we have that $F(A)$ and $F(A')$ are both *equal* to an object X of $\mathbf{C}$, then X will appear *twice* in the diagram, not once. In particular, as A and A' are connected by a unique isomorphism $u : A \to A'$ of $\mathbf{G}$, this gives an isomorphism $X \to X$ as in the diagram on the left,

$$X \xrightarrow{Fu} X \qquad\qquad X \circlearrowright Fu$$

not as on the right. Indeed, in our convention, the diagram on the right commutes if and only if Fu is the identity, which in general we cannot assume.

> **Exercise 6.3.29.** Let $D : \mathbf{J} \to \mathbf{C}$ be a diagram in $\mathbf{C}$, and suppose that it has a limit. Prove that all the limit objects of D, together with the isomorphisms between them commuting with the respective limit cones, form a clique.

Definition 6.3.30. Let $F : \mathbf{A} \to \mathbf{C}$ and $G : \mathbf{B} \to \mathbf{C}$ be cliques in $\mathbf{C}$. A **morphism of cliques** is a morphism $f_{A,B} : FA \to GB$ of $\mathbf{C}$, for each object A of $\mathbf{A}$ and B of $\mathbf{B}$, such that for each unique isomorphism $u : A \to A'$ of $\mathbf{A}$ and $v : B \to B'$ of $\mathbf{B}$, the following diagram commutes:

$$\begin{array}{ccc} FA & \xrightarrow{f_{A,B}} & GB \\ {\scriptstyle\cong}\big\downarrow{\scriptstyle Fu} & & {\scriptstyle\cong}\big\downarrow{\scriptstyle Gv} \\ FA' & \xrightarrow{f_{A',B'}} & GB'. \end{array}$$

In other words, a morphism of cliques joins two cliques into a joint, perfectly commutative diagram, but where now not all arrows are necessarily isomorphisms:

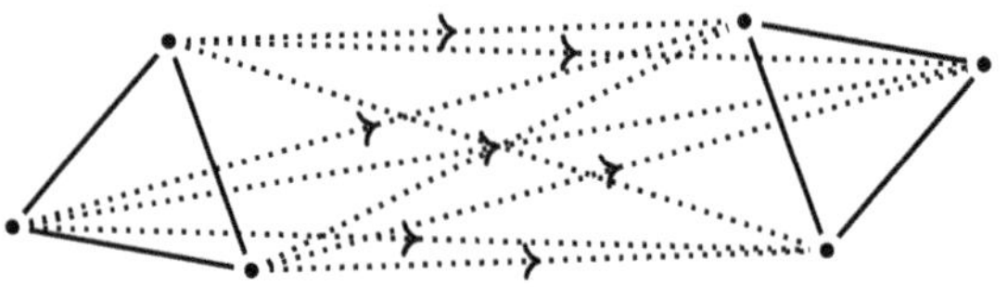

Thus, cliques form a category, which we denote by **Clique**(**C**). Note that by Lemma 1.5.17, any morphism of cliques is uniquely determined by a single morphism from an object in the first clique to an object in the second clique (why?).

Exercise 6.3.31. Let $D, E : \mathbf{J} \to \mathbf{C}$ be (generic) diagrams in $\mathbf{C}$, and suppose that both limits exist. Show that a natural transformation $D \Rightarrow E$ induces a morphism of cliques between the clique of limits of D (as in Exercise 6.3.29) and the one of E.

We are now ready to prove the strictification theorem.

Proof of Theorem 6.3.25. Let $(\mathbf{C}, \otimes, I)$ be a monoidal category. Given a finite list $G = [A_1, \ldots, A_n]$ of objects of $\mathbf{C}$, possibly repeated, consider the set of all formal expressions that we can build out of the A_i, using all of them exactly once and in their order, together with tensors and units. For example, for $[A, B, C]$, a possible such expression is

$$((I \otimes A) \otimes (B \otimes C)) \otimes I.$$

Note that, as we did in Chapters 4 and 5, we are not taking the *results* of these expressions, but the actual expressions. Turn this set into a contractible groupoid by adding a unique arrow between any of its elements, and call it **G**. Define now the functor $F : \mathbf{G} \to \mathbf{C}$ that, on objects, evaluates each of these expressions to their actual result. On morphisms, by the coherence theorem for monoidal categories, there is a unique isomorphism between any two of these results that can be built out of associators and unitors. This makes F a functor and a clique. For the empty list, note that we can still form expressions such as I, $I \otimes I$, and so on.

Define now the category **MClique**(**C**) ("monoidal cliques") as the full subcategory of **Clique**(**C**) of all the cliques in the form constructed above.

Let's now construct an equivalence $E : \mathbf{C} \to \mathbf{MClique}(\mathbf{C})$. On objects, we assign to each object A of **C** the clique $E(A)$ on the length-one list $[A]$ (which contains A as well as $(A \otimes I) \otimes I$, etc.). By Lemma 1.5.17, this assignment is fully faithful, as a morphism between the clique defined by A and the one defined by B is uniquely specified by the morphism $A \to B$ of **C**. To see that this functor is essentially surjective, consider the monoidal clique on the list $G = [A_1, \dots, A_n]$. Denote by R the (actual result of the) tensor product $((A_1 \otimes A_2) \otimes \cdots \otimes A_n)$ in **C**, with all the brackets associated to the left. Then, the clique of G is isomorphic to $E(R)$. Therefore, E is an equivalence.

Let's now make **MClique**(**C**) a strict monoidal category. Given $G = [A_1, \dots, A_n]$ and $H = [B_1, \dots, B_m]$, define the tensor product of the respective cliques to be the clique generated by the concatenation of the lists, $[A_1, \dots, A_n, B_1, \dots, B_m]$, and as the monoidal unit, take the clique defined by the empty list. This category is strict monoidal: The tensor product of the cliques defined by $[A_1, \dots, A_n]$ and by the empty list is the clique defined by $[A_1, \dots, A_n]$ (not just isomorphic, they are the same clique). The same thing can be said about tensoring with the unit on the left and with associating.

Let's now show that the functor $E : \mathbf{C} \to \mathbf{MClique}(\mathbf{C})$ is strong-monoidal. First of all, for the monoidal unit, $E(I)$ is exactly the clique defined by $[I]$. This is isomorphic to the monoidal unit of **MClique**(**C**) since I is in the clique defined by the empty list. On tensor products, let $C = A \otimes B$. The clique $E(A) \otimes E(B)$ contains the element $A \otimes B = C$, and so the identity $A \otimes B \to C$ induces (on all the other objects) an isomorphism of cliques, $\Delta : E(C) \to E(A) \otimes E(B)$. The associativity and unitality conditions now follow from the fact that all the arrows in those diagrams can be seen as arrows of the same clique (why?), and so the diagrams commute. This makes **C** monoidally equivalent to **MClique**(**C**). □

6.4 Monads on Monoidal Categories

Monads and monoidal structures on categories can interact, giving a very rich and useful theory. In this section, we present its basic ideas. For more in-depth material, see for example [Bra14; Sea13] and the series of papers where most of these concepts were developed, Refs. [Koc70; Koc71; Koc72; Koc75].[22]

6.4.1 Monoidal monads

Definition 6.4.1. Let $(\mathbf{C}, \otimes, I)$ be a monoidal category. A **monoidal monad** on **C** consists of a monad (T, μ, η), where the functor T has the structure of a lax-monoidal functor (T, ∇, e) and where μ and ∇ are monoidal natural transformations.

Explicitly, and using Exercise 6.3.12, this means that a monoidal monad is a structure (T, μ, η, ∇) where (T, μ, η) is a monad, T is a lax-monoidal functor where the unit map given by the unit of the monad $\eta : I \to TI$ (why?), and the multiplication map ∇ makes the following two diagrams commute:

$$\begin{array}{ccc} & & TA \otimes TB \\ & \overset{\eta_A \otimes \eta_B}{\nearrow} & \downarrow{\scriptstyle \nabla_{A,B}} \\ A \otimes B & \underset{\eta_{A \otimes B}}{\searrow} & T(A \otimes B) \end{array} \qquad \begin{array}{ccc} TTA \otimes TTB & \xrightarrow{\mu_A \otimes \mu_B} & TA \otimes TB \\ \downarrow{\scriptstyle \nabla_{TA,TB}} & & \\ T(TA \otimes TB) & & \downarrow{\scriptstyle \nabla_{A,B}} \\ \downarrow{\scriptstyle T\nabla_{A,B}} & & \\ TT(A \otimes B) & \xrightarrow{\mu_{A \otimes B}} & T(A \otimes B), \end{array}$$

[22]Note that in those papers, functors are applied and composed on the *right*.

which we call the **unit** and **multiplication** conditions, respectively.

Exercise 6.4.2. In Section 6.3.1, we saw that one can represent monoidal functors in terms of "shaded" string diagrams. Express the conditions above in terms of shading. (*Hint*: You need *double shading* and some way to represent the unit and multiplication of the monad.)

Exercise 6.4.3. What should a monoidal *comonad* be? (*Hint*: We are not simply reversing all the arrows. We want the underlying functor to be lax-monoidal, not colax.)

Here are some examples in the form of exercises.

Exercise 6.4.4 (sets and relations). Show that the power set monad is monoidal (using the lax-monoidal structure of Example 6.3.8).

Exercise 6.4.5 (probability). Show that the probability monad on **Set** is monoidal (using the lax-monoidal structure of Example 6.3.9).

A canonical example of a non-monoidal monad, whose functor is still lax-monoidal, is the list monad.

Exercise 6.4.6 (algebra, computer science; important!). Show that the list monad is *not* a monoidal monad (using the lax-monoidal structure of Exercise 6.3.11). (*Hint*: Test the multiplication condition on the following tuple of nested lists:

$$([[a_{11}, a_{12}], [a_{21}, a_{22}]], [[b_{11}, b_{12}], [b_{21}, b_{22}]]),$$

and remember that the ordering matters.)

The result of the previous exercise might be surprising: What does the multiplication have to do with *ordering*? Recall from Exercise 6.3.11 that the "ordering" also prevented the list functor from being symmetric. This is not a coincidence, and in the following section, we explore the connection between the two.

6.4.2 Strong monads

A *strength* on a monad is a way of being "monoidal in one variable." This will be made more concrete later in this section. Let's now look at the precise definition.

Definition 6.4.7. Let $(\mathbf{C}, \otimes, I)$ be a monoidal category, and let $F : \mathbf{C} \to \mathbf{C}$ be an endofunctor. A (left) **strength** on F is a natural transformation with components

$$A \otimes FB \xrightarrow{\ell_{A,B}} F(A \otimes B)$$

that makes the following diagrams commute:

$$\begin{array}{ccc} I \otimes FA & \xrightarrow{\ell_{I,A}} & F(I \otimes A) \\ & {\scriptstyle\cong}\searrow \quad \swarrow{\scriptstyle\cong} & \\ & FA & \end{array} \qquad \begin{array}{ccccc} (A \otimes B) \otimes FC & & \xrightarrow{\ell_{A\otimes B,C}} & & F((A \otimes B) \otimes C) \\ \downarrow{\scriptstyle\cong} & & & & \downarrow{\scriptstyle\cong} \\ A \otimes (B \otimes FC) & \xrightarrow{\mathrm{id}_A \otimes \ell_{B,C}} & A \otimes F(B \otimes C) & \xrightarrow{\ell_{A,B\otimes C}} & F(A \otimes (B \otimes C)), \end{array}$$

called **unitality** and **associativity** conditions, respectively. We call (F, ℓ) a **strong functor**.

A right strength[23] is a natural transformation with components $r_{A,B} : FA \otimes B \to F(A \otimes B)$ that satisfies analogous conditions.

Definition 6.4.8. Let $(\mathbf{C}, \otimes, I)$ be a monoidal category, and let (T, μ, η) be a monad on $\mathbf{C}$. A (left) **strength** on (T, μ, η) is a strength ℓ on the endofunctor T compatible with the monad structure by making in addition the following diagrams commute:

$$\begin{array}{ccc} & A \otimes B & \\ {\scriptstyle\mathrm{id}_A \otimes \eta_B}\swarrow & & \searrow{\scriptstyle\eta_{A\otimes B}} \\ A \otimes TB & \xrightarrow{\ell_{A,B}} & T(A \otimes B) \end{array} \qquad \begin{array}{ccccc} A \otimes TTB & \xrightarrow{\ell_{A,TB}} & T(A \otimes TB) & \xrightarrow{T\ell_{A,B}} & TT(A \otimes B) \\ \downarrow{\scriptstyle\mathrm{id}_A \otimes \mu_B} & & & & \downarrow{\scriptstyle\mu_{A\otimes B}} \\ A \otimes TB & & \xrightarrow{\ell_{A,B}} & & T(A \otimes B), \end{array}$$

[23]Sometimes, a right strength is called a "costrength," but we will not use that term here.

called **unit** and **multiplication** conditions, respectively. We call (T, μ, η, ℓ) a **strong monad**.

A right strength on (T, μ, η) is a right strength on the endofunctor T that satisfies analogous conditions.

Exercise 6.4.9. Express these diagrams in terms of the shaded string diagrams of Section 6.3.1. (*Hint*: You would require a way to depict the map $\ell : A \otimes TB \to T(A \otimes B)$.)

Exercise 6.4.10. Define the strength of a comonad. (*Hint*: Just as for the monoidal structure, we are not simply reversing all the arrows, we want a map in the direction $A \otimes CB \to C(A \otimes B)$.)

Exercise 6.4.11. Write down explicitly the axioms for a right strength on a monad.

Show that in a symmetric monoidal category, from a left strength $\ell : A \otimes TB \to T(A \otimes B)$, we can canonically obtain a right strength $r : TA \otimes B \to T(A \otimes B)$ using the braiding, as follows:

$$TA \otimes B \xrightarrow[\cong]{\beta} B \otimes TA \xrightarrow{\ell} T(B \otimes A) \xrightarrow[\cong]{T\beta} T(A \otimes B).$$

Here is a possible intuition for a strength. We mentioned in Chapter 5 that a monad T is often a way of forming either "generalized elements" or "formal expressions" and that the unit $\eta : A \to TA$ is often the "inclusion of ordinary elements into generalized elements," or a way of forming "one-term, trivial expressions" out of single elements. In this vein, a strength $\ell : A \otimes TB \to T(A \otimes B)$ can be seen as a way of forming generalized elements or formal expressions in $A \otimes B$, but *which are trivial in* A. We start from an ordinary element of A and a generalized element (or formal expression) of B, and combine them in a way that keeps the A-part of the expression "ungeneralized."

Here are some examples to illustrate this idea.

Example 6.4.12 (algebra, computer science). The list monad on $(\mathbf{Set}, \times, 1)$ has the following strength:

$$\begin{aligned} A \times LB &\xrightarrow{\quad \ell \quad} L(A \times B) \\ (a, [b_1, \dots, b_n]) &\longmapsto \big[(a, b_1), \dots, (a, b_n)\big]. \end{aligned}$$

(Why is this a strength?)

We see that in some sense, the list on the right-hand side, in the variable A, is "just a."

Example 6.4.13 (sets and relations). The power set monad on **Set** has the following strength:

$$\begin{aligned} A \times PB &\xrightarrow{\quad \ell \quad} L(A \times P) \\ (a, S) &\longmapsto a \times S := \{(a, s) : s \in S\}. \end{aligned}$$

Once again, this is "just one element" in the variable A.

One might wonder, since we have seen that the unit of the monad encodes this idea of "trivial formal expressions," whether we can define a strength in terms of the unit of the monad, applied to one of the two variables. This is indeed the case, and that's exactly what happened in the example above (can you see why?). Here is a similar construction for the probability monad.

Example 6.4.14 (probability). The probability monad on **Set** has the following strength:

$$\begin{aligned} A \times \mathcal{P}B &\xrightarrow{\ \ell\ } \mathcal{P}(A \times B) \\ (a, p) &\longmapsto \delta_a \otimes p, \end{aligned}$$

where

$$(\delta_a \otimes p)(a', b) = \begin{cases} p(b), & a = a'; \\ 0, & a \neq a'. \end{cases}$$

As we can see, this is "deterministic in the variable A." According to the law of a joint random variable, this distribution gives zero entropy (or zero variance if $A = \mathbb{R}$) to the marginal A.

One can do this in general for every monoidal monad.

Exercise 6.4.15. Let (T, μ, η, ∇) be a monoidal monad on a category **C**. Show that the morphism ℓ_∇ given by

$$A \otimes TB \xrightarrow{\eta\otimes\mathrm{id}} TA \otimes TB \xrightarrow{\nabla} T(A \otimes B)$$

is a strength on the monad T. (*Hint*: You may need quite some space to draw the necessary diagrams and decompose them. Also, for some, it may be helpful to use shaded string diagrams, as in Section 6.3.1, suitably generalized.)

As we have seen, the monads P and $\mathcal{P}$ (but not L) are monoidal, and the strengths that we have constructed can be seen as arising from the statement above. In this sense, a strong monad is like a monad that's "monoidal in only one variable" or "in both variables separately" if there is a right strength too (such as if we are in a symmetric monoidal category). The list monad is such an example: It is "separately monoidal in both variables" (i.e. strong), but not "jointly monoidal" (i.e. not monoidal in the precise sense). Compare this with the discussion on joint and separate functoriality (Section 1.3.8) and on linear versus bilinear maps (Section 2.3.2 and Example 6.1.26).

It turns out that in fact *every monad on* **Set** *is strong*. (We will not prove this since it would take us to *enriched category theory*, which is beyond the scope of this book. See the nLab,[24] and the references therein.) Because of this, in many situations (such as in Haskell), when people use the term "monad" on a category similar to **Set**, they often implicitly mean "strong monad."

Here is an exercise giving a class of examples outside of **Set**.

[24]https://ncatlab.org/nlab/show/strong+monad.

Exercise 6.4.16. Let (M, m, e) be an internal monoid in a symmetric monoidal category $(\mathbf{C}, \otimes, I)$, and recall that the functor $M \otimes -$ has a canonical monad structure. Show that the morphism

$$A \otimes (M \otimes B) \xrightarrow{\cong} (A \otimes M) \otimes B \xrightarrow[\cong]{\beta \otimes B} (M \otimes A) \otimes B \xrightarrow{\cong} M \otimes (A \otimes B)$$

gives the components of a strength for the monad $M \otimes -$. (*Hint*: This is really just a statement about coherence. If you want to check manually, you can use string diagrams — no need for shadings here.)

Give concrete examples of this map for $(\mathbf{Set}, \times, 1)$ and $(\mathbf{AbGrp}, \otimes, \mathbb{Z})$.

A monad comes equipped with a notion of *morphism of algebras* (Definition 5.2.11), which in some sense respects the formal operations encoded by the monad (for example, algebras of the list monad are monoids, and morphisms of algebras are monoid homomorphisms). With strong monads, we can also talk about morphisms which are "morphisms of algebras separately in one or both variables."

Definition 6.4.17. Let (T, μ, η) be a monad on $(\mathbf{C}, \otimes, I)$ with left strength ℓ and right strength r. Let (A, a), (B, b), and (C, c) be T-algebras. Given a morphism $f : A \otimes B \to C$ of $\mathbf{C}$, consider the following diagrams:

$$\begin{array}{ccccc} TA \otimes B & \xrightarrow{r_{A,B}} & T(A \otimes B) & \xrightarrow{Tf} & TC \\ \downarrow{\scriptstyle a \otimes \mathrm{id}} & & & & \downarrow{\scriptstyle c} \\ A \otimes B & & \xrightarrow{\quad f \quad} & & C \end{array} \qquad \begin{array}{ccccc} A \otimes TB & \xrightarrow{\ell_{A,B}} & T(A \otimes B) & \xrightarrow{Tf} & TC \\ \downarrow{\scriptstyle \mathrm{id} \otimes b} & & & & \downarrow{\scriptstyle c} \\ A \otimes B & & \xrightarrow{\quad f \quad} & & C. \end{array} \qquad (6.4.1)$$

We say that:

- f is a T-**morphism in** A if the diagram on the left commutes;
- f is a T-**morphism in** B if the diagram on the right commutes;
- f is a T-**morphism in** A **and** B **separately**, or a T-**bimorphism**, if both commute.

Note that for the left diagram to make sense, we don't actually need the right strength to exist, and we also don't need A to be a T-algebra. A similar statement is true for the right diagram. (Also note that the notion of T-morphism in the *left* argument uses the *right* strength, and vice versa.)

Example 6.4.18 (linear algebra). Let T be the monad of real vector spaces on **Set** (recall Exercise 5.2.16), and take $\mathbb{R}$ as algebra. Then, T-bimorphisms are exactly bilinear maps. For example:

- the map $(a, b) \mapsto ab$ is a T-bimorphism, and it is bilinear;
- the map $(a, b) \mapsto ab^2$ is a T-morphism in a, i.e. it is linear in a, but not in b;
- the map $(a, b) \mapsto a + b$ is not separately linear in any sense. (It is *jointly* linear, but that's a different notion — see Exercise 6.4.24.)

(Check these statements. Can you prove the general case?)

Example 6.4.19. Algebras of the list monad L are monoids. Take now the monoid $(\mathbb{N}, +)$, and let X be any set. X is not an L-algebra, but the free algebra (LX, μ) is. Now, the map

$$\begin{aligned} \mathbb{N} \times X &\longrightarrow LX \\ (n, x) &\longmapsto \underbrace{[x, \dots, x]}_{n \text{ times}} \end{aligned}$$

is an L-morphism in the first argument. (The second argument is not even an algebra.)

Exercise 6.4.20. Show that the right strength $r : TA \otimes B \to T(A \otimes B)$ is a T-morphism in the first variable (between free algebras) and that the left strength $\ell : A \otimes TB \to T(A \otimes B)$ is a T-morphism in the second variable.

Example 6.4.21. For any monoidal monad, with strength given as in Exercise 6.4.15, the multiplication map $\nabla : TA \otimes TB \to T(A \otimes B)$

is a T-bimorphism of free algebras. Indeed, the left diagram of Equation (6.4.2) can be decomposed as follows:

$$
\begin{array}{ccccccc}
TA \otimes TTB & \xrightarrow{\eta\otimes\mathrm{id}} & TTA \otimes TTB & \xrightarrow{\nabla} & T(TA \otimes TB) & \xrightarrow{T\nabla} & TT(A \otimes B) \\
{\scriptstyle TA\otimes\mathrm{id}}\downarrow & \swarrow{\scriptstyle \mu\otimes\mu} & & & & & \downarrow{\scriptstyle \mu} \\
TA \otimes TB & & & \xrightarrow{\qquad\qquad\nabla\qquad\qquad} & & & T(A \otimes B),
\end{array}
$$

where the left triangle commutes by the unitality of the monad, and the remaining region is the multiplication condition that makes T a monoidal monad. The right diagram works analogously.

The fact above can be interpreted as "forming the product of formal expressions is in some sense bilinear" (for example, think of forming a product of probabilities).

For this to hold, though, it is crucial that the monad is monoidal.

Exercise 6.4.22. Show that for the list monad L, the map ∇ of Exercise 6.3.11 is *not* an L-bimorphism.

Often, the map ∇ is moreover a *universal* bimorphism, see Exercise 6.4.39.

Here is another exercise, generalizing what we saw in Example 6.1.26 for vector spaces.

Exercise 6.4.23. Show that given T-algebras A and B, the T-bimorphisms $TI \otimes A \to B$ are in bijection with the T-morphisms $A \to B$. (*Hint*: Recall Exercise 6.3.14.)

Before we end this section, Example 6.4.18 may have left you wondering about the *joint* linearity of the map $(a, b) \mapsto a + b$: Can we write it in terms of monads? The answer is in the following exercise.

Exercise 6.4.24. Let (T, μ, η, Δ) be a *comonoidal* monad on $(\mathbf{C}, \otimes, I)$, i.e. a monad where the functor and the natural transformations are colax-monoidal. Define the *direct product of algebras* via the map

$$T(A \otimes B) \xrightarrow{\Delta} TA \otimes TB \xrightarrow{a \otimes b} A \otimes B.$$

Show that jointly linear maps between the vector spaces $A \times B \to C$ are morphisms of algebras for this structure. What else does this construction generalize?

(We will not look at direct products or comonoidal monads any further.)

6.4.3 Commutative monads

Let's now look even more deeply at what sets L apart from monads such as P and $\mathcal{P}$. Just as monoids can be commutative, so can monads, according to the following definition.

Definition 6.4.25. A strong monad (T, μ, η, ℓ) on a symmetric monoidal category is called **commutative** if the following diagram commutes:

$$\begin{array}{ccccc}
 & & TA \otimes TB & & \\
 & \swarrow^{\ell_{TA,B}} & & \searrow^{r_{A,TB}} & \\
T(TA \otimes B) & & & & T(A \otimes TB) \\
\downarrow{\scriptstyle Tr_{A,B}} & & & & \downarrow{\scriptstyle T\ell_{A,B}} \\
TT(A \otimes B) & & & & TT(A \otimes B) \\
 & \searrow_{\mu_{A,B}} & & \swarrow_{\mu_{A,B}} & \\
 & & T(A \otimes B) & &
\end{array} \tag{6.4.2}$$

where the right strength is the one induced via ℓ and the braiding.

(One also refers to ℓ as a *commutative strength*.)

This condition has the interpretation that *the formal operations encoded by the monad commute with one another*. Here are some examples (and exercises).

Example 6.4.26 (algebra, computer science). We have that:

- the list monad (whose algebras are monoids) is *not* commutative.
- the monad of Example 5.2.1, whose algebras are *commutative* monoids, is commutative.

Let's illustrate this via an example. Consider the tuple of lists

$$([a_1, a_2], [b_1, b_2]),$$

which is an element of $LA \times LB$. The left path of diagram (6.4.2) gives

$$\begin{aligned} ([a_1, a_2], [b_1, b_2]) &\longmapsto \big[([a_1, a_2], b_1), ([a_1, a_2], b_2)\big] \\ &\longmapsto \begin{bmatrix} [(a_1, b_1), (a_2, b_1)], \\ [(a_1, b_2), (a_2, b_2)] \end{bmatrix} \\ &\longmapsto [(a_1, b_1), (a_2, b_1), (a_1, b_2), (a_2, b_2)], \end{aligned}$$

where for convenience, we wrote a doubly nested list as a matrix. The right path of diagram (6.4.2) gives

$$\begin{aligned} ([a_1, a_2], [b_1, b_2]) &\longmapsto \big[([a_1, a_2], b_1), ([a_1, a_2], b_2)\big] \\ &\longmapsto \begin{bmatrix} [(a_1, b_1), (a_1, b_2)], \\ [(a_1, b_1), (a_2, b_2)] \end{bmatrix} \\ &\longmapsto [(a_1, b_1), (a_1, b_2), (a_2, b_1), (a_2, b_2)]. \end{aligned}$$

We see that these lists differ by their *ordering* since in the middle lines, the two matrices are transpose to each other. If the ordering does not matter, the two results are equal. (Compare with Exercise 6.3.11.)

In other words, the commutativity condition is about *whether reading a doubly formal expression row-first or column-first, we get the same result or not*. (Can you turn this into a proof of the fact that the monad of commutative monoids is commutative?)

Free monoids are not the only structures that can be commutative.

Exercise 6.4.27 (algebra; important!). Let M be an internal monoid in a symmetric monoidal category. Recall that the action monad $M \otimes -$ has a canonical strength given by Exercise 6.4.16.

Show that the monad $M \otimes -$ is commutative if and only if M is commutative as an internal comonoid.

Instantiating the statement above, we get in particular that:

- a monoid is commutative if and only if its action monad is commutative;
- a ring is (multiplicatively) commutative if and only if the corresponding monad on **AbGrp** is commutative;
- an algebra is (multiplicatively) commutative if and only if the corresponding monad on **Vect** is commutative (for example, think of the algebra of diagonal matrices).

Exercise 6.4.28 (sets and relations, probability). Show that:

- the power set monad is commutative;
- the probability monad on **Set** is commutative.

The analogous statement for the more general Giry monad on **Meas** is related to Fubini's theorem.

The main theorem in this section is the following. Recall that in Proposition 1.3.50, we gave a condition, the *interchange law*, which makes a separately functorial assignment jointly functorial. This theorem is analogous and tells that the commutativity condition (6.4.2) makes a "separately monoidal" monad (i.e. strong with both strengths) "jointly monoidal" (i.e. monoidal).

Theorem 6.4.29. *Let* $(\mathbf{C}, \otimes, I)$ *be a symmetric monoidal category, and let* (T, μ, η) *be a monad on* $\mathbf{C}$:

(a) *Given a commutative strength* ℓ *on* (T, μ, η)*, the map* $\nabla_\ell : TA \otimes TB \to T(A \otimes B)$ *obtained by composing the arrows on either side of diagram* (6.4.2) *makes* (T, μ, η) *a symmetric monoidal monad.*

(b) *The assignment* $\ell \mapsto \nabla_\ell$ *given above and the assignment* $\nabla \mapsto \ell_\nabla$ *of Exercise* 6.4.15 *are mutually inverse.*

In other words, symmetric monoidal monads and commutative monads are really the same structure. This can be seen as the reason why, for example, the list monad is not monoidal.

For reasons of space, we cannot include the proof. We invite the interested reader to try to prove this at least partially, with the caveat about the sizes of the resulting diagrams. (Using shaded string diagrams can help, but it takes up space too.) The full proof can be found in [GLN08, Appendix A], where the map ∇ is called *mediator*.

Exercise 6.4.30 (important!). Let T be a commutative monad, and let (A, a), (B, b) and (C, c) be T-algebras. Using Theorem 6.4.29, show that a morphism $f : A \otimes B \to C$ is a T-bimorphism if and only if the following diagram commutes:

$$\begin{array}{ccccc} TA \otimes TB & \xrightarrow{\nabla_{A,B}} & T(A \otimes B) & \xrightarrow{Tf} & TC \\ \downarrow{\scriptstyle a\otimes b} & & & & \downarrow{\scriptstyle c} \\ A \otimes B & & \xrightarrow{\quad f \quad} & & C. \end{array} \tag{6.4.3}$$

6.4.4 The tensor product of algebras and of Kleisli morphisms

Let's now construct tensor products in the Eilenberg–Moore and Kleisli categories of a commutative (or, equivalently, symmetric monoidal) monad. These products generalize the tensor product of vector spaces, abelian groups, modules, and more.

Recall from Definition 6.4.17 the notion of T-bimorphism, and denote by $\mathrm{Bi}(A, B; C)$ the set of T-bimorphisms $A \otimes B \to C$. Composing a bimorphism with a T-morphism we get again a bimorphism (why?). This gives a functor $\mathrm{Bi}(A, B; -) : \mathbf{C}^T \to \mathbf{Set}$, acting on morphisms via postcomposition, analogous to what we did for vector spaces in Section 2.3.2. With the case of vector spaces in mind, we can then define the tensor product of algebras A and B, if it exists and up to isomorphism, as the object representing the functor $\mathrm{Bi}(A, B; -)$:

Definition 6.4.31. Let (T, μ, η, ∇) be a commutative monad on a symmetric monoidal category $(\mathbf{C}, \otimes, I)$, and let (A, a) and (B, b) be T-algebras. The **tensor product of algebras** $A \otimes_T B$, if it exists, is a T-algebra giving a natural bijection

$$\mathrm{Hom}_{\mathbf{C}^T}(A \otimes_T B, C) \;\cong\; \mathrm{Bi}(A, B; C).$$

More explicitly, $A \otimes_T B$ (if it exists) comes equipped with a *universal bimorphism* $p : A \otimes B \to A \otimes_T B$ such that for every bimorphism $f : A \otimes B \to C$, there exists a unique T-morphism $\tilde{f} : A \otimes_T B \to C$ that makes the following diagram of $\mathbf{C}$ commute:

$$\begin{array}{ccc} A \otimes B & & \\ {\scriptstyle p}\big\downarrow & \searrow^{f} & \\ A \otimes_T B & \underset{\tilde{f}}{\dashrightarrow} & C. \end{array}$$

In many cases, such as in cocomplete categories, this object exists.

Theorem 6.4.32. *A representing object $A \otimes_T B$ of $\mathrm{Bi}(A, B; -)$ is given, if it exists, by the coequalizer in $\mathbf{C}^T$ of the following pair:*

$$\begin{array}{c} T(TA \otimes TB) \xrightarrow{T\nabla_{A,B}} TT(A \otimes B) \xrightarrow{\mu_{A\otimes B}} T(A \otimes B), \\ T(TA \otimes TB) \xrightarrow[T(a\otimes b)]{\qquad\qquad\qquad\qquad} T(A \otimes B). \end{array} \tag{6.4.4}$$

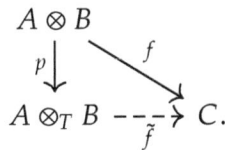

The statement above can be proven by means of the following lemma, whose proof is left as a guided exercise.

Lemma 6.4.33. *Let (A, a), (B, b), and (C, c) be T-algebras. A morphism $f : A \otimes B \to C$ of $\mathbf{C}$ is a bimorphism if and only if the composites in the following diagram are equal:*

$$T(TA \otimes TB) \underset{T(a\otimes b)}{\overset{T\nabla_{A,B}}{\rightrightarrows}} TT(A \otimes B) \xrightarrow{\mu_{A,B}} T(A \otimes B) \xrightarrow{Tf} TC \xrightarrow{c} C. \tag{6.4.5}$$

Exercise 6.4.34. Prove the statement above. (*Hint*: In one direction, decompose diagram (6.4.5). In the other direction, precompose (6.4.5) with η, and show that it gives a decomposition of diagram (6.4.3).)

Proof of Theorem 6.4.32. Recall that the Eilenberg–Moore adjunction gives a natural bijection as follows:

$$\begin{array}{ccc} \mathrm{Hom}_{\mathbf{C}^T}(T(A \otimes B), C) & \xrightarrow{\cong} & \mathrm{Hom}_{\mathbf{C}}(A \otimes B, C) \\ f^\flat & \longmapsto & f^\flat \circ \eta_{A\otimes B} \\ \shortparallel & & \shortparallel \\ c \circ Tf^\sharp & \longleftarrow & f^\sharp. \end{array}$$

Now, denote by P the diagram giving the parallel pair (6.4.4) of Theorem 6.4.32, and denote by $q : T(A\otimes B) \to A\otimes_T B$ the coequalizing map. By the universal property of the coequalizer $A \otimes_T B$, we have a natural bijection

$$\begin{array}{ccc} \mathrm{Hom}_{\mathbf{C}^T}(A \otimes_T B, C) & \xrightarrow{\cong} & \mathrm{Cone}(P, C) \\ g & \longmapsto & g \circ q, \end{array}$$

where an element of $\mathrm{Cone}(P, C)$ is an arrow $h : T(A \otimes B) \to C$ such that $h \circ T(a \otimes b) = h \circ \mu_{A,B} \circ T\nabla_{A,B}$. As cones are just particular arrows $T(A \otimes B) \to C$, we have the following inclusion:

$$\mathrm{Cone}(P, C) \hookrightarrow \mathrm{Hom}_{\mathbf{C}^T}(T(A \otimes B), C) \xrightarrow{\cong} \mathrm{Hom}_{\mathbf{C}}(A \otimes B, C).$$

Lemma 6.4.33 can now be restated as follows: $f^\flat$ is in $\mathrm{Cone}(P, C)$ if and only if $f^\sharp$ is a bimorphism. In other words, the bijection is restricted to form the following commutative diagram:

$$\begin{array}{ccc} \mathrm{Cone}(P,C) & \xrightarrow{\cong} & \mathrm{Bi}(A,B;C) \\ \downarrow & & \downarrow \\ \mathrm{Hom}_{\mathbf{C}^T}(T(A\otimes B),C) & \xrightarrow{\cong} & \mathrm{Hom}_{\mathbf{C}}(A\otimes B, C). \end{array}$$

Composing with the universal property of the coequalizer, we get the natural bijection

$$\begin{array}{ccccc} \mathrm{Hom}_{\mathbf{C}^T}(A\otimes_T B, C) & \xrightarrow{\cong} & \mathrm{Cone}(P,C) & \xrightarrow{\cong} & \mathrm{Bi}(A,B;C) \\ & & \downarrow & & \downarrow \\ & & \mathrm{Hom}_{\mathbf{C}^T}(T(A\otimes B),C) & \xrightarrow{\cong} & \mathrm{Hom}_{\mathbf{C}}(A\otimes B, C) \\ g & \longmapsto & g\circ q & \longmapsto & g\circ q\circ \eta_{A\otimes B} \end{array}$$

between the T-morphisms out of $A \otimes_T B$ and the T-bimorphisms out of $A\otimes B$ by precomposing with the map $q\circ\eta_{A\otimes B} : A\otimes B \to A\otimes_T B$. □

This makes $\mathbf{C}^T$ a symmetric monoidal category, provided all the necessary coequalizers exist and that they are preserved by the functor $A\otimes_T -$ (see the following). The latter property often holds, for example in a *closed* monoidal category, see Section 6.5.

Let's see how. The monoidal unit for $\otimes_T$ is the free algebra TI, where I is the monoidal unit of $\mathbf{C}$. Compare this with the fact that the monoidal unit in $(\mathbf{Vect}, \otimes, \mathbb{R})$ is $\mathbb{R}$, the free vector space on a singleton, the unit in $(\mathbf{AbGrp}, \otimes, \mathbb{Z})$ is $\mathbb{Z}$, and the unit for the tensor product of R-modules over a ring R is R. (Why is it always a monoid object?)

Exercise 6.4.35. Using Exercise 6.4.23, show that for every T-algebra, $TI \otimes_T A \cong A$.

Just as in Example 6.1.26, this gives the unitors. Associators, again similarly, will be given in terms of "trimorphisms" (that's where the coequalizer preservation condition comes in), and the coherence conditions work in the same way as for vector spaces. (How do you define a "trimorphism"?) The full details of this construction can be found, for example [Bra14, Section 6.4], in which T-algebras are called *modules*.

This construction generalizes the tensor products of not only vector spaces, modules, and abelian groups (which could all actually be seen as particular modules, why?) but also more general structures, such as the following example.

Exercise 6.4.36. The monad of commutative monoids (Example 5.2.1) is commutative (recall Example 6.4.26). Define the tensor product of commutative monoids.

Exercise 6.4.37. Recall that the algebras of the probability monad P on **Set** are *convex spaces* (Example 5.2.9). Define the tensor product of convex spaces, and compare it with the one of vector spaces.

Let's now turn to the Kleisli category. On objects, we can keep the same tensor product $\otimes$ of $(\mathbf{C}, \otimes, I)$. On morphisms, given Kleisli morphisms $f : X \to Y$ and $g : A \to B$ (i.e. morphisms $f^\sharp : X \to TY$ and $g^\sharp : A \to TB$ of $\mathbf{C}$), we take the morphism corresponding to the map $X \otimes A \to T(Y \otimes B)$ of $\mathbf{C}$ given by

$$X \otimes A \xrightarrow{f^\sharp \otimes g^\sharp} TY \otimes TB \xrightarrow{\nabla} T(Y \otimes B).$$

Note that on objects, we inherit the associators and unitors from $\mathbf{C}$ with their coherences. What is not *a priori* clear, instead, is that this assignment is (jointly) functorial on Kleisli morphisms.

Exercise 6.4.38. Show that the tensor product constructed above satisfies the interchange law (on Kleisli morphisms).

Recall from Proposition 5.5.8 that we can view the Kleisli category equivalently as the category of *free* T-algebras. If we do so, the two tensor products we defined turn out to agree, as shown in the following.

Exercise 6.4.39. We saw in Example 6.4.21 that the map $\nabla : TX \otimes TY \to T(X \otimes Y)$ is a T-bimorphism. Show that it is a *universal* one, in the sense that every bimorphism out of $TX \otimes TY$ factors uniquely through ∇. Conclude that $TX \otimes_T TY \cong T(X \otimes Y)$. Use this to show that, if we write the Kleisli category as a category of free algebras, the two tensor products coincide.

Example 6.4.40 (linear algebra). We saw in Exercise 5.2.27 that every vector space is a *free* algebra of the vector space monad. The condition $TX \otimes_T TY \cong T(X \otimes Y)$ of the exercise above becomes, for sets X and Y, $TX \otimes TY \cong T(X \times Y)$, i.e. the cartesian product of the bases gives a basis of the tensor product. (Recall the informal definition of tensor product, Definition 2.3.12.)

Example 6.4.41 (probability). For the Kleisli category of the probability monad on **Set** we defined the tensor product of Kleisli morphisms in Example 6.1.29 and a tensor product of probability measures (giving a lax-monoidal structure) in Example 6.3.9. As one can see directly (they both simply multiply the probabilities), the two constructions coincide, and in the free case, they coincide with the construction in Exercise 6.4.37.

We conclude with the following exercise, which once again shows how a strong monad can be seen as "separately monoidal, but not jointly" and also shows why we need commutativity to define the tensor product.

Exercise 6.4.42 (difficult!). Let (T, μ, η, ℓ) be a strong but not commutative monad on a symmetric monoidal category $(\mathbf{C}, \otimes, I)$. Show that if we try to define a tensor product on the Kleisli category analogously as above, but using one of the two (inequivalent) paths of diagram (6.4.2), the resulting "tensor product" is separately functorial, but not jointly: The interchange law fails. (The resulting structure, which is more general than a monoidal category, is called a *premonoidal category*.)

6.5 Closed Monoidal Categories

A closed monoidal category is, roughly, "when there are objects of that category which can play the role of hom-sets."

Definition 6.5.1. A symmetric monoidal category $(\mathbf{C}, \otimes, I)$ is called **closed** if for every object A, the functor $- \otimes A : \mathbf{C} \to \mathbf{C}$ has a right-adjoint.

Explicitly, $(\mathbf{C}, \otimes, I)$ is closed monoidal if there exists a functor

$$\begin{aligned} \mathbf{C}^{\mathrm{op}} \times \mathbf{C} &\xrightarrow{[-,-]} \mathbf{C} \\ (A, B) &\longmapsto [A, B] \end{aligned}$$

and a bijection with the components

$$\mathrm{Hom}_{\mathbf{C}}(A \otimes B, C) \xrightarrow{\cong} \mathrm{Hom}_{\mathbf{C}}(A, [B, C]) \qquad (6.5.1)$$

natural in A and C.[25]

We call the object $[B, C]$ the **internal hom**, or **hom-object**, between B and C; we call the functor $[-, -]$ the **internal hom functor**, and we call the adjunction (6.5.1) the **hom-tensor adjunction**.

We sometimes call a closed symmetric monoidal category simply a *closed monoidal category*.[26]

The bijective function in (6.5.1) is sometimes called the procedure of *currying* (named after Haskell Curry) and its inverse *uncurrying*.

A closed monoidal category where the monoidal structure is cartesian is called **cartesian closed monoidal category** (or sometimes simply **cartesian closed**). In that case, the internal hom $[A, B]$ is often denoted by B^A and called the **exponential object**.

---

[25]For naturality in B, see Proposition 6.5.16.

[26]Closed symmetric monoidal categories are not the most general example of categories with an internal hom: There also exist *non-symmetric* monoidal closed categories, which can be "left-closed" and "right-closed," as well as the notion of *closed category* without the monoidal part. We will not treat those cases here.

6.5.1 Examples of internal homs

The main interpretation of closed monoidal categories is that there is an "object of morphisms" between any two objects. As we have seen throughout this book, in any (locally small) category **C**, morphisms between any two objects form a *set*, $\mathrm{Hom}_{\mathbf{C}}(A, B)$, regardless of what the objects of the category are. In a closed monoidal category, the internal hom $[A, B]$ is itself an object of **C**.

Example 6.5.2 (sets and relations). The category **Set** is cartesian-closed. Indeed, we have the adjunction

$$\mathrm{Hom}_{\mathbf{Set}}(A \times B, C) \xrightarrow{\cong} \mathrm{Hom}_{\mathbf{Set}}(A, C^B),$$

where $C^B = \mathrm{Hom}_{\mathbf{Set}}(B, C)$ (mind the possibly confusing notation: in **Set**, the internal hom is the usual hom-set since sets *are* the objects of our category). Concretely, what this says is that a function of two variables $f : A \times B \to C$ can be equivalently seen as a function $f^\sharp$ that takes one variable, fixes it, and returns a function of the second variable:

$$\begin{aligned} A &\xrightarrow{f^\sharp} \mathrm{Hom}_{\mathbf{Set}}(B, C) \\ a &\longmapsto \big(b \mapsto f(a, b)\big). \end{aligned}$$

Example 6.5.3 (linear algebra). As you probably know, matrices of a fixed size form a vector space. More generally, linear maps $V \to W$ form themselves a vector space with the pointwise sum and multiplication by a scalar. For example, given $f, g : V \to W$, we have

$$(f + g)(v) := f(v) + g(v).$$

Denoting this vector space of linear maps by $[V, W]$, we have the following hom-tensor adjunction:

$$\mathrm{Hom}_{\mathbf{Vect}}(V \otimes W, U) \cong \mathrm{Hom}_{\mathbf{Vect}}(V, [W, U]).$$

Let's see how this works. A linear map $f : V \otimes W \to U$ corresponds to a *bi*linear map $f' : V \times W \to U$. Fix now a vector v of V. The map

$f_v^\sharp : W \to U$ defined by $w \mapsto f'(v, w)$ is again linear in w by the bilinearity of f (which implies linearity in w). Moreover, since f is also linear in v, the assignment $V \to [W, U]$ given by $v \mapsto f_v^\sharp$ is linear too, and all the linear maps $V \to [U, W]$ can be seen as arising in this way and uniquely so (why?). This makes $(\mathbf{Vect}, \otimes, \mathbb{R})$ monoidal closed.

Example 6.5.4 (commutative algebra). Just as **Vect**, the following categories of *algebras of a commutative monad* are monoidal closed:

- $(\mathbf{Vect}_F, \otimes, F)$ for a field F, constructed similarly as for the real case;
- $(\mathbf{AbGrp}, \otimes, \mathbb{Z})$, with the internal hom $[G, H]$ given by the group homomorphisms $G \to H$ equipped with the pointwise group structure;
- $(\mathbf{R\text{-}Mod}, \otimes_R, R)$ for a commutative ring R, with $[M, N]$ given by the R-linear maps $M \to N$ equipped with the pointwise module structure;
- $(\mathbf{CRing}, \otimes, \mathbb{Z})$, with $[R, S]$ given by the ring homomorphisms $R \to S$ equipped with the pointwise ring structure.

Just as the tensor products of those categories are all instances of the same construction (recall Section 6.4.4), the internal homs are all instances of the same construction (as we will see in Section 6.5.5).

Example 6.5.5 (graph theory). The category $(\mathbf{MGraph}, \times, 1)$ is cartesian-closed. The resulting exponential object (which is *not* the usual "graph exponential," see Remark 6.5.7) is the graph H^G where:

- vertices are *all maps between vertices Vert*$(G) \to$ *Vert*(H) (not graph homomorphisms — if you are unsatisfied with this fact, wait until Example 6.5.12)[27];

[27]Note that, moreover, for non-simple graphs, graph homomorphisms are not just a subset of all maps between vertices, but they have extra structure because the assignment from edges to edges in general involves a choice.

- if G and H are simple graphs, given $f, g : Vert(G) \to Vert(H)$ there is an edge $f \to g$ if and only if for all vertices x, x' of G, we have that $x \to x'$ in G implies $f(x) \to g(x')$ in H;
- for general G and H, given $f, g : Vert(G) \to Vert(H)$ there is an edge $f \to g$ for each assignment from edges $x \to x'$ of G to edges $f(x) \to g(x')$. That is,

$$\mathrm{Edge}(f,g) := \prod_{x,x' \in \mathrm{Vert}(G)} \mathrm{Edge}\big(f(x), g(x')\big)^{\mathrm{Edge}(x,x')},$$

where $Edge(x, x')$ denotes the set of edges $x \to x'$, and so on.

Let's now show that we have a bijection

$$\mathrm{Hom}_{\mathbf{MGraph}}(F \times G, H) \cong \mathrm{Hom}_{\mathbf{MGraph}}(F, H^G).$$

Given a graph homomorphism $f : F \times G \to H$, in particular we have an assignment between vertices, $f_0 : Vert(F) \times Vert(G) \to Vert(H)$. We can partially apply this, and for any vertex w of F, we get a function $f^\sharp_w : Vert(G) \to Vert(H)$. On edges, recall from Example 6.1.23 that an edge $(w, x) \to (w', x')$ in $F \times G$ is a pair of edges $w \to w'$ and $x \to x'$. The homomorphism f will map this edge of $F \times G$ to an edge $f(w, x) \to f(w', x')$ of H. Therefore, for each edge $w \to w'$ of F, the functions $f^\sharp_w$ and $f^\sharp_{w'}$ have a way of assigning to each edge $x \to x'$ an edge $f^\sharp_w(x) \to f^\sharp_{w'}(x')$ in H. That is, each edge $w \to w'$ is mapped canonically to an edge $f^\sharp_w \to f^\sharp_{w'}$ of H^G. This makes the assignment $w \mapsto f^\sharp_w$ a graph homomorphism $F \to H^G$, and every graph homomorphism $F \to H^G$ arises uniquely in this way (why?).

Example 6.5.6 (graph theory). The cat $(\mathbf{MGraph}, \square, G_0)$ is monoidal-closed too. Given graphs G and H, the internal hom $[G, H]$ is constructed as follows:

- The vertices, this time, are the graph homomorphisms $G \to H$.
- If H is simple, given graph homomorphisms $f, g : G \to H$, we have an edge $f \to g$ if and only if for each vertex x of G, there is an edge $f(x) \to g(x)$.

- For general H, given graph homomorphisms $f, g : G \to H$, there is an edge $f \to g$ for each way of assigning edges $f(x) \to g(x)$ for all vertices x of G. In other words,

$$\mathrm{Edge}(f, g) := \prod_{x \in Vert(G)} \mathrm{Edge}\big(f(x), g(x)\big).$$

Note that, differently from the previous example, $[G, H]$ does not actually use the edges of G (except in the choice of its vertices since they are graph homomorphisms). Compare this with the pointwise order of monotone maps (Example 1.3.2). Let's now show that we have a bijection

$$\mathrm{Hom}_{\mathbf{MGraph}}(F \,\square\, G, H) \;\cong\; \mathrm{Hom}_{\mathbf{MGraph}}(F, [G, H]).$$

Let $f : F \,\square\, G \to H$ be a graph homomorphism. Once again, on vertices, we have an assignment $f_0 : Vert(F) \times Vert(G) \to Vert(H)$, which we can partially evaluate at a vertex w of W, getting a function $f^\sharp_w : Vert(G) \to Vert(H)$. On edges, recall from Example 6.1.24 that the only edges in $F \,\square\, G$ are either in the form $(w, x) \to (w, x')$, for some edge $x \to x'$ of G, or in the form $(w, x) \to (w', x)$, for some edge $w \to w'$ of F. Using the first case, since f is a graph homomorphism, we have that for each vertex w of F, the functions $f^\sharp_w$ map the edges $x \to x'$ of G to the edges $f^\sharp_w(x) \to f^\sharp_w(x')$ of H. Thus, each $f^\sharp_w$ is a graph homomorphism, and the assignment $w \mapsto f^\sharp_w$ is a well-defined function $Vert(F) \to Vert([G, H])$. Using the second case, since f is a graph homomorphism, we have that for each vertex x of G, each edge $w \to w'$ of F gives a specified edge $f^\sharp_w(x) \to f^\sharp_{w'}(x)$. This makes the assignment $w \to f^\sharp_w$ a graph homomorphism, and once again, every such graph homomorphism arises uniquely in this way (why?).

Remark 6.5.7. In graph theory, sometimes the construction $[G, H]$ of Example 6.5.6 is called the "graph exponential." Similar to the case of "cartesian" products (Remark 6.1.25), this terminology is incompatible with the categorical framework: The exponential object in the sense of category theory, i.e. the right-adjoint to the *categorical* product, is rather the construction H^G of Example 6.5.5. To avoid ambiguity, we will not use the term "graph exponential."

Example 6.5.8. The category $(\mathbf{Cat}, \times, \mathbf{1})$ is cartesian-closed. The exponential object is given by the functor category $[\mathbf{C}, \mathbf{D}]$ — this is an abstract way of stating Exercise 1.4.27.

Exercise 6.5.9. Show that the *funny tensor product* of categories given in Example 6.1.28 makes **Cat** monoidal closed too, but in a different way.

Exercise 6.5.10. Consider the infinite interval $[0, \infty]$ as a poset category with arrows pointing *downward* in the order. Show that this is a closed monoidal category where:

- the tensor product is the sum of numbers, and 0 the monoidal unit;
- the internal hom is given by

$$[x, y] := \begin{cases} y - x, & x \leq y; \\ 0, & x > y, \end{cases}$$

where by convention, $\infty - \infty = 0$.

(A possible way to interpret this is that the number $[x, y]$ is the "cost of going up from x to y," and "falling down" is free.)

Exercise 6.5.11 (analysis, geometry). Show that the category $\mathbf{Lip}_1$ of metric spaces and 1-Lipschitz maps is closed monoidal, where:

- the tensor product $X \otimes Y$ of metric spaces is given by the cartesian product, with the metric

$$d\big((x, y), (x', y')\big) := d(x, x') + d(y, y');$$

- the internal hom $[X, Y]$ is given by the set of all 1-Lipschitz maps, equipped with the L^∞-metric

$$d(f,g) := \sup_{x \in X} d\big(f(x), g(x)\big).$$

One might now wonder how the internal hom $[A, B]$ relates to the hom-set $\mathrm{Hom}_{\mathbf{C}}(A, B)$. Recall from Section 6.1.4 that in a monoidal category, the monoidal unit I gives a canonical "forgetful" functor $\mathrm{Hom}_{\mathbf{C}}(I, -) : \mathbf{C} \to \mathbf{Set}$, which tests our objects for "points" or "states." If we apply this to internal homs, in some sense, we test our internal homs for "arrows": The hom-tensor adjunction gives a bijection

$$\mathrm{Hom}_{\mathbf{C}}(I, [A, B]) \cong \mathrm{Hom}_{\mathbf{C}}(I \otimes A, B) \cong \mathrm{Hom}_{\mathbf{C}}(A, B)$$

so that "testing an internal hom for elements" returns the usual hom-set, i.e. the set of arrows. For example, the vectors of the internal hom $[V, W]$ in **Vect** are exactly the linear maps $V \to W$ (why?). Since we can always recover the hom-set from the internal hom, in particular, *the internal hom always contains more (or equal) information than the hom-set*.

It should be kept in mind that the intuition of "testing for elements" does not always apply immediately, or at least it requires some care. For example, consider the following case of graphs.

Example 6.5.12. Let's test internal homs of graphs for "arrows":

- For the graph $[G, H]$ of Example 6.5.6, the vertices of $[G, H]$ are indeed the graph homomorphisms.
- Instead, for the graph H^G of Example 6.5.5, the vertices of H^G are the *maps between the vertices* of G and H, not graph homomorphisms.

The reason is the following: Recall from Section 6.1.4 that for $(\mathbf{MGraph}, \square, G_0)$, the points of a graph G, in the sense of morphisms from the monoidal unit $G_0 \to G$, are exactly the vertices. Instead, for $(\mathbf{MGraph}, \times, 1)$, the points are vertices *together with a loop*. In H^G, this gives us an assignment $f_0 : \mathit{Vert}(G) \to \mathit{Vert}(H)$ which has an edge to itself in H^G, i.e. a way of assigning to each edge $x \to x'$ of G an edge $f_0(x) \to f_0(x')$ of H. But this is exactly a graph homomorphism. So, even if the *vertices* of H^G are the maps between vertices, the *points* of H^G are the graph homomorphisms.

To conclude this section, we ask, *what is not monoidal closed*? Two famous examples of non-closed monoidal categories are the cartesian products in **Top** and **Meas**. The details of why this is the case are however too technical for this book; we refer the interested reader to the nLab[28] for the case of **Top** and [Aum61] for the case of **Meas**.

However, the two categories are canonically *monoidal* closed.

Exercise 6.5.13 (topology). Given topological spaces X and Y, define a topology on the set $X \times Y$ which makes a function $f : X \times Y \to Z$ continuous if and only if it is *separately* continuous in each argument, i.e. if $x \mapsto f(x, y)$ is continuous for every $y \in Y$ and $y \mapsto f(x, y)$ is continuous for every $x \in X$. Show that this is a monoidal structure on **Top** and that it is closed, where the internal hom $[X, Y]$ is the set of continuous functions $X \to Y$ equipped with the topology of pointwise convergence. Can you see a similarity with the situation of Example 6.5.3, with continuity in place of linearity?

Exercise 6.5.14 (measure theory). Construct a monoidal structure on **Meas** similar to the one of **Top** in the exercise above.

[28] See https://ncatlab.org/nlab/show/convenient+category+of+topological+spaces.

6.5.2 Functoriality of the internal hom

Just as the usual hom-sets are parts of a functor, the same is true for internal homs.

Proposition 6.5.15. *Let* $\mathbf{C}$ *be a closed monoidal category. For each object* B *of* $\mathbf{C}$*, the assignment* $C \mapsto [B, C]$ *is part of a functor* $[B, -] : \mathbf{C} \to \mathbf{C}$*.*

Proof. Let $c : C \to C'$ be a morphism of $\mathbf{C}$, and let A and B be objects. By Lemma 1.5.17, there exists a unique map that makes the following diagram commute:

$$\begin{array}{ccc} \mathrm{Hom}_{\mathbf{C}}(A \otimes B, C) & \xrightarrow{\cong} & \mathrm{Hom}_{\mathbf{C}}(A, [B, C]) \\ \downarrow{\scriptstyle c\circ -} & & \vdots \\ \mathrm{Hom}_{\mathbf{C}}(A \otimes B, C') & \xrightarrow{\cong} & \mathrm{Hom}_{\mathbf{C}}(A, [B, C']). \end{array} \tag{6.5.2}$$

All the arrows in the diagram above are natural in A, so we have a natural transformation $\mathrm{Hom}_{\mathbf{C}}(-, [B, C]) \Rightarrow \mathrm{Hom}_{\mathbf{C}}(-, [B, C'])$. By the Yoneda embedding theorem, this specifies a unique morphism $[B, C] \to [B, C']$. Denote this morphism by $[B, c]$. The assignment $g \mapsto [B, c]$ is functorial since the dashed arrow in the diagram (6.5.2) is the identity if c is the identity and since diagrams in the form of (6.5.2) can be stacked vertically. □

For a concrete interpretation of what the morphism $[B, c] : [B, C] \to [B, C']$ does, it is helpful to apply the canonical forgetful functor to **Set**:

$$\begin{array}{ccc} \mathrm{Hom}_{\mathbf{C}}(I, [B, C]) & \xrightarrow{\cong} & \mathrm{Hom}_{\mathbf{C}}(B, C) \\ \downarrow{\scriptstyle [B,c]\circ -} & & \downarrow{\scriptstyle c\circ -} \\ \mathrm{Hom}_{\mathbf{C}}(I, [B, C']) & \xrightarrow{\cong} & \mathrm{Hom}_{\mathbf{C}}(B, C'). \end{array}$$

We see that, in some sense, the morphism $[B, c]$ is the internal version of *postcomposition with* c.

Proposition 6.5.16. *Let* $\mathbf{C}$ *be a closed monoidal category. For each object* C *of* $\mathbf{C}$*, the assignment* $B \mapsto [B, C]$ *is part of a functor* $[-, C] : \mathbf{C}^{\mathrm{op}} \to \mathbf{C}$*,* *making the bijection* (6.5.1) *natural in* B *as well. Explicitly, given* $b : B \to B'$*, the following diagram commutes:*

$$\begin{array}{ccc} \mathrm{Hom}_{\mathbf{C}}(A \otimes B', C) & \xrightarrow{\cong} & \mathrm{Hom}_{\mathbf{C}}(A, [B', C]) \\ \downarrow{\scriptstyle -\circ(A\otimes b)} & & \downarrow{\scriptstyle [b,C]\circ -} \\ \mathrm{Hom}_{\mathbf{C}}(A \otimes B, C) & \xrightarrow{\cong} & \mathrm{Hom}_{\mathbf{C}}(A, [B, C]). \end{array} \tag{6.5.3}$$

Exercise 6.5.17. Prove the above proposition. (*Hint*: It is analogous, and somewhat dual, to Proposition 6.5.15.)

Given $b : B \to B'$, we can interpret $[b, C] : [B', C] \to [B, C]$ as an internal version of *precomposition with* b:

$$\begin{array}{ccc} \mathrm{Hom}_{\mathbf{C}}(I, [B', C]) & \xrightarrow{\cong} & \mathrm{Hom}_{\mathbf{C}}(B', C) \\ \downarrow{\scriptstyle [b,C]\circ -} & & \downarrow{\scriptstyle -\circ b} \\ \mathrm{Hom}_{\mathbf{C}}(I, [B, C]) & \xrightarrow{\cong} & \mathrm{Hom}_{\mathbf{C}}(B, C). \end{array}$$

Exercise 6.5.18. Prove that the internal hom is a functor of the two variables $[-, -] : \mathbf{C}^{\mathrm{op}} \times \mathbf{C} \to \mathbf{C}$ jointly. (*Hint*: Recall Proposition 1.3.51.)

6.5.3 Evaluation and coevaluation

It is helpful to take a look at the unit and counit of the hom-tensor adjunction (6.5.1). Instantiating the general case of Section 4.2, the counit is a morphism

$$[B, C] \otimes B \xrightarrow{\mathrm{ev}_{B,C}} C$$

natural in C, which we call the **evaluation**. For its interpretation, let's look at what this map does in **Set**:

(a) It takes an arrow $g : B \to C$.

(b) It takes an element b of B.

(c) It *evaluates* g at b, returning an element $g(b)$ of C.

In other categories, such as **Vect**, the interpretation is similar.

Let's now look at the unit, a map

$$A \xrightarrow{\mathrm{coev}_{A,B}} [B, A \otimes B]$$

natural in A, which we call the **coevaluation**. Again, let's look at what it does in **Set**:

(a) It takes an element $a \in A$.

(b) It returns a map $B \to A \times B$ which assigns to $b \in B$ the pair $(a, b) \in A \times B$. In other words, it maps the set B to the "slice" $\{a\} \times B \subseteq A \times B$.

Just as for the evaluation, in other categories the interpretation is similar.

The unit and counit of the adjunction allow us to express the bijection (6.5.1) more explicitly, as follows:

$$\begin{array}{ccc} \mathrm{Hom}_{\mathbf{C}}(A \otimes B, C) & \xrightarrow{\cong} & \mathrm{Hom}_{\mathbf{C}}(A, [B, C]) \\ f^{\flat} & \longmapsto & [B, f^{\flat}] \circ \mathrm{coev}_{A,B} \\ \rotatebox{90}{$\in$} & & \rotatebox{90}{$\in$} \\ \mathrm{ev}_{B,C} \circ (f^{\sharp} \otimes B) & \longleftarrow\!\shortmid & f^{\sharp}, \end{array} \tag{6.5.4}$$

and this is often convenient when writing proofs. Also quite useful are the triangle identities, which in this context look as follows.

Corollary 6.5.19. *In a closed monoidal category, the following diagrams commute for all objects A, B, and C:*

$$\begin{array}{ccc} A \otimes B & \xrightarrow{\mathrm{coev} \otimes B} & [B, A \otimes B] \otimes B \\ & \searrow^{\mathrm{id}} & \downarrow \mathrm{ev} \\ & & A \otimes B \end{array} \qquad \begin{array}{ccc} [B, C] & \xrightarrow{\mathrm{coev}} & [B, [B, C] \otimes B] \\ & \searrow^{\mathrm{id}} & \downarrow [B,\mathrm{ev}] \\ & & [B, C]. \end{array}$$

Yet another useful fact for proofs is instantiating Lemma 4.2.6 for the specific case of the hom-tensor adjunction, as follows.

Corollary 6.5.20. *In a closed monoidal category, given the morphisms $f^\flat : A \otimes B \to C$ and $g^\flat : A' \otimes B \to C'$, or, equivalently, $f^\sharp : A' \to [B, C']$ and $g^\sharp : A \to [B, C]$, and given $h : A \to A'$ of $\mathbf{C}$ and $k : C \to C'$, the diagram on the left commutes if and only if the diagram on the right commutes:*

$$\begin{array}{ccc} A \otimes B & \xrightarrow{f^\flat} & C \\ \downarrow{\scriptstyle h\otimes B} & & \downarrow{\scriptstyle k} \\ A' \otimes B & \xrightarrow{g^\flat} & C' \end{array} \qquad\qquad \begin{array}{ccc} A & \xrightarrow{f^\sharp} & [B, C] \\ \downarrow{\scriptstyle h} & & \downarrow{\scriptstyle [B,k]} \\ A' & \xrightarrow{g^\sharp} & [B, C']. \end{array}$$

Now, back to the general theory. We saw that $\mathrm{coev}_{A,B} : A \to [B, A \otimes B]$ and $\mathrm{ev}_{B,C} : [B, C] \otimes B \to C$ are natural in A and C, respectively. One may ask, what about in B? In B, the two maps satisfy a property called **extranaturality**, which in this setting one can think of as an "internal" version of naturality. Let's see the precise statement: in this work, we will not consider extranaturality any deeper.[29]

Proposition 6.5.21 (extranaturality of evaluation and coevaluation). *Let $\mathbf{C}$ be a closed monoidal category. For all objects A and C and for all morphisms $b : B \to B'$, the following diagrams commute:*

$$\begin{array}{ccc} [B', C] \otimes B & \xrightarrow{[B',C]\otimes b} & [B', C] \otimes B' \\ \downarrow{\scriptstyle [b,C]\otimes B} & & \downarrow{\scriptstyle \mathrm{ev}_{B',C}} \\ [B, C] \otimes B & \xrightarrow{\mathrm{ev}_{B,C}} & C \end{array} \qquad \begin{array}{ccc} A & \xrightarrow{\mathrm{coev}_{A,B}} & [B, A \otimes B] \\ \downarrow{\scriptstyle \mathrm{coev}_{A,B'}} & & \downarrow{\scriptstyle [B,A\otimes b]} \\ [B', A \otimes B'] & \xrightarrow{[b,A\otimes B']} & [B, A \otimes B']. \end{array} \qquad (6.5.5)$$

Proof. Let's prove the statement for the evaluation map. For the coevaluation map, the proof is (dually) analogous. Let's instantiate

[29] We refer the interested reader to see for example, [Lor21].

the naturality condition (6.5.3) using the Yoneda trick, taking the case of $A = [B', C]$:

$$\begin{array}{ccc} \mathrm{Hom}_{\mathsf{C}}([B',C]\otimes B', C) & \xleftarrow{\cong} & \mathrm{Hom}_{\mathsf{C}}([B',C],[B',C]) \\ \Big\downarrow {\scriptstyle -\circ([B',C]\otimes b)} & & \Big\downarrow {\scriptstyle [b,C]\circ -} \\ \mathrm{Hom}_{\mathsf{C}}([B',C]\otimes B, C) & \xleftarrow{\cong} & \mathrm{Hom}_{\mathsf{C}}([B',C],[B,C]). \end{array}$$

Starting with the identity at the top-right corner and using (6.5.4), we get that

$$\mathrm{ev}_{B',C} \circ ([B',C]\otimes b) \;=\; \mathrm{ev}_{B,C} \circ ([b,C]\otimes B).$$

This is exactly saying that the first diagram in Equation (6.5.5) commutes. □

Exercise 6.5.22. Prove explicitly that the second diagram in (6.5.5) commutes.

Exercise 6.5.23. Let A be an object in a closed monoidal category. Show that $[A, A]$ has a canonical internal monoid structure, compatible with the monoid structure of $\mathrm{Hom}_{\mathsf{C}}(A, A)$. (*Hint*: Try to express, internally, what it means to compose morphisms.)

6.5.4 Strong monads on closed monoidal categories

In a closed monoidal category, the left and right strengths of a monad acquire additional significance. This is achieved by forming, from the strength maps

$$A \otimes TB \xrightarrow{\ \ell\ } T(A \otimes B) \quad \text{and} \quad TA \otimes B \xrightarrow{\ r\ } T(A \otimes B),$$

these other maps:

$$[A,B] \otimes TA \xrightarrow{\ \ell'\ } TB \quad \text{and} \quad T[A,B] \otimes A \xrightarrow{\ r'\ } TB.$$

First of all, just by looking at the domains and codomains, we could note that the map ℓ' is not just the image of ℓ through the hom-tensor adjunction, the functor T is also involved. The same can be said about r' and r. Indeed, these maps can be seen as intertwining the hom-tensor adjunction with T. (Note how the domain and codomain of ℓ' and r' are similar to the ones of $\mathrm{ev} : [A, B] \otimes A \to B$.)

Here is how to obtain these maps. The map ℓ' is defined in terms of ℓ in the following way:

$$\ell' \quad := \quad [A, B] \otimes TA \xrightarrow{\ell_{[A,B],A}} T([A, B] \otimes A) \xrightarrow{T\mathrm{ev}_{A,B}} TB.$$

The map r' can be defined in terms of r in the following way:

$$r' \quad := \quad T[A, B] \otimes A \xrightarrow{r_{[A,B],A}} T([A, B] \otimes A) \xrightarrow{T\mathrm{ev}_{A,B}} TB.$$

Conversely, from these maps, one can reobtain ℓ and r, as follows:

$$\ell \quad = \quad A \otimes TB \xrightarrow{\mathrm{coev}_{A,B} \otimes TB} [B, A \otimes B] \otimes TB \xrightarrow{\ell'} T(A \otimes B)$$

$$r \quad = \quad TA \otimes B \xrightarrow{(T\mathrm{coev}_{A,B}) \otimes B} T[B, A \otimes B] \otimes B \xrightarrow{r'} T(A \otimes B).$$

Exercise 6.5.24. Show that this indeed recovers ℓ and r.

There is actually a bijection between strengths and maps in the form of ℓ' and r' that satisfy particular conditions, but that's beyond the scope of this book.[30]

Exercise 6.5.25. Prove that ℓ' and r', as above, are natural in B. If you can, show that they are also *extranatural* in A, in a similar way to ev and coev (Proposition 6.5.21).

Let's now look at what these maps do. The map $\ell' : [A, B] \otimes TA \to TB$ can be considered *an internal version of the action of* T *on morphisms*. In **Set**, suppose you have a function $A \to B$, a monad

[30]See again the series of papers [Koc70; Koc71; Koc72; Koc75] for more information (where functors are composed on the *right* and the internal hom is denoted by ⋔).

T, and an element of TA. Then, we can apply T to the function and get an element of TB. (Compare with how `fmap` is typed in Haskell, which can be considered as working in a cartesian-closed category similar to **Set**.) Let's see this through an example.

Example 6.5.26 (algebra, computer science). Consider the list monad L on $(\mathbf{Set}, \times, 1)$. Let's look a the map $\ell' : [A, B] \otimes TA \to TB$ in this case. (For consistency of notation with internal homs, we denote the set of functions $A \to B$ by $[A, B]$: It is not a list!) Given a function $f : A \to B$ and a list $[b_1, b_2, b_3]$ of elements of B, the map ℓ' works as follows (recall what the strength ℓ does for the monad L, see Example 6.4.12):

$$\begin{array}{ccccc} [A,B] \times LA & \xrightarrow{\ell_{[A,B],A}} & L([A,B] \times A) & \xrightarrow{L\mathrm{ev}_{A,B}} & LB \\ (f, [b_1, b_2, b_3]) & \longmapsto & [(f, b_1), (f, b_2), (f, b_3)] & \longmapsto & [f(b_1), f(b_2), f(b_3)]. \end{array}$$

That is, ℓ' applies a function to all items in a list of arguments, and gives a list with the results.

In other words, it is the action of the functor L on morphisms.

Exercise 6.5.27. Show this for other strong monads, such as the power set, the probability monad on **Set**, and the action monad of a monoid.

There is however more to this story, and it holds for general endofunctors, not just for monads. Let F be an endofunctor on a closed monoidal category **C**. Given objects A and B of **C**, the action of F on morphisms gives a function between the hom-sets

$$\begin{array}{ccc} \mathrm{Hom}_{\mathbf{C}}(A, B) & \longrightarrow & \mathrm{Hom}_{\mathbf{C}}(FA, FB) \\ f & \longmapsto & Ff. \end{array}$$

We know that $\mathrm{Hom}_{\mathbf{C}}(A, B)$ and $\mathrm{Hom}_{\mathbf{C}}(FA, FB)$ can be obtained by applying the forgetful functor $\mathrm{Hom}_{\mathbf{C}}(I, -)$ to the internal homs $[A, B]$ and $[FA, FB]$. One may then wonder if we can lift, or *strengthen*, the function $f \mapsto Ff$ to a morphism $[A, B] \to [FA, FB]$ of $\mathbf{C}$. Perhaps surprisingly, the map ℓ' does exactly this, and that's where the name "strength" comes from (it was first defined in the closed case).

Proposition 6.5.28. *Let (F, ℓ) be a strong endofunctor on a closed monoidal category $\mathbf{C}$. Construct the map ℓ' as above. Then, $(\ell')^\sharp : [A, B] \to [FA, FB]$ lifts the action of F on morphisms.*

This means that the following diagram commutes, where F_* denotes the action of F on morphisms:

$$\begin{array}{ccccc} \mathrm{Hom}_{\mathbf{C}}(I, [A,B]) & \xrightarrow{\cong} & \mathrm{Hom}_{\mathbf{C}}(I \otimes A, B) & \xrightarrow{\cong} & \mathrm{Hom}_{\mathbf{C}}(A, B) \\ {\scriptstyle (\ell')^\sharp \circ -}\downarrow & & & & \downarrow{\scriptstyle F_*} \\ \mathrm{Hom}_{\mathbf{C}}(I, [FA,FB]) & \xrightarrow{\cong} & \mathrm{Hom}_{\mathbf{C}}(I \otimes FA, FB) & \xrightarrow{\cong} & \mathrm{Hom}_{\mathbf{C}}(FA, FB). \end{array} \tag{6.5.6}$$

One can also express the proposition above in the language of *enriched category theory*, but this is beyond the scope of this book.[31]

Proof. First of all, we can use (6.5.4) to get the following explicit form of $(\ell')^\sharp = (F\mathrm{ev} \circ \ell)^\sharp$:

$$[A, B] \xrightarrow{\text{coev}} [FA, [A, B] \otimes FA] \xrightarrow{[FA,\ell]} [FA, F([A, B] \otimes A)] \xrightarrow{[FA,F\mathrm{ev}]} [FA, FB].$$

Let's now show that (6.5.6) commutes. We start with $f : I \to [A, B]$ at the top-left corner. Using (6.5.4) and the explicit form of $(\ell')^\sharp$, the top-right path in (6.5.6) gives the top-right path in the following diagram. Similarly, and using the triangle identities, the bottom-left path in (6.5.6) gives the bottom-left path in the following diagram:

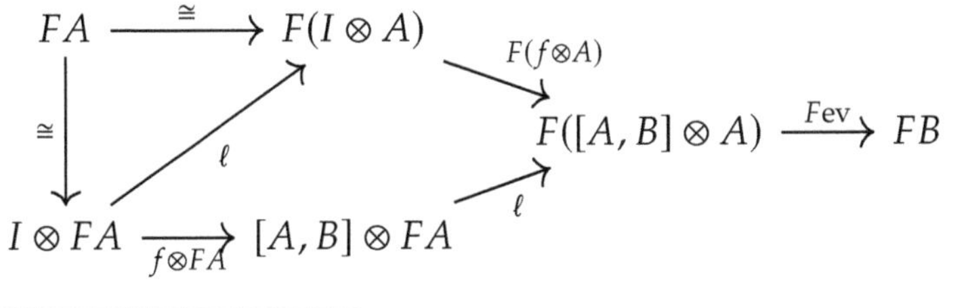

[31] See once again https://ncatlab.org/nlab/show/strong+monad.

which commutes since the strength ℓ is natural and compatible with unitors. □

Let's now turn to $r' : T[A,B] \otimes A \to TB$ and observe once again what it does for the list monad L.

Example 6.5.29 (algebra, computer science). Let $[f,g,h]$ be a length-three list of functions $A \to B$, i.e. an element of $L[A,B]$. Recalling what r does for the monad L, we have the following:

$$\begin{array}{ccccc} T[A,B]\times A & \xrightarrow{r_{[A,B],A}} & T([A,B]\times A) & \xrightarrow{T\mathrm{ev}_{A,B}} & TB \\ ([f,g,h],a) & \longmapsto & [(f,a),(g,a),(h,a)] & \longmapsto & [f(a),g(a),h(a)]. \end{array}$$

In other words, the map r' is *evaluating a list of functions all on the same argument and returning a list with the results.* (Note the duality with Example 6.5.26.)

Other monads behave in a similar way, as seen in the following example.

Exercise 6.5.30 (algebra). Define explicitly the map r' for the action monad induced by an internal monoid.

Exercise 6.5.31 (probability). Define explicitly the map r' for the probability monad on **Set**. (Warning: this does not extend to general measurable spaces since, as we remarked, the category of measurable spaces is not cartesian-closed.[32])

Once again, of course, the action on the internal hom $[A,B]$ has in general more information than what it does to the underlying hom-set (can you give an example?).

[32]While, as we have seen, there is a closed monoidal structure on Meas, the Giry monad is not strong for that structure, see [Sat18].

The map r' is particularly interesting for us if we look at it through the hom-tensor adjunction: It corresponds to a map $(r')^\sharp : T[A,B] \to [A,TB]$, which can be considered to be "like a strength for T in the variable B of $[A,B]$." Here is the precise statement.

Proposition 6.5.32. *The map $(r')^\sharp$ defined above makes the following diagrams commute:*

$$\begin{array}{ccc} & [A,B] & \\ {}^{\eta}\swarrow & & \searrow^{[A,\eta]} \\ T[A,B] & \xrightarrow{(r')^\sharp} & [A,TB] \end{array} \qquad \begin{array}{ccccc} TT[A,B] & \xrightarrow{T(r')^\sharp} & T[A,TB] & \xrightarrow{r} & [A,TTB] \\ \downarrow{\scriptstyle \mu} & & & & \downarrow{\scriptstyle [A,\mu]} \\ T[A,B] & & \xrightarrow{(r')^\sharp} & & [A,TB] \end{array}$$

Compare these diagrams with the ones in Definition 6.4.8.

Proof. We prove the case of the unit diagram; the multiplication one is proven similarly, see the following exercise. We can decompose the unit diagram as follows, using (6.5.4) to write $(r')^\sharp$ in terms of r', and then expressing it in terms of r:

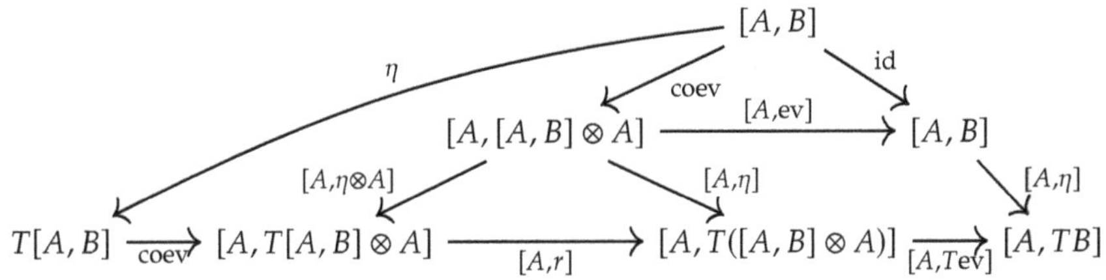

This diagram now commutes by the triangle identities, the unit condition for the strength r, and the naturality of η and coev. □

Exercise 6.5.33. Prove that the multiplication diagram commutes.

Here is an important consequence.

Proposition 6.5.34. *Let (T,μ,η) be a monad with right strength r on a closed monoidal category* **C**. *Let (A,e) be a T-algebra and X be any object. Then, the object $[X,A]$ has a canonical T-algebra structure given by the map*

$$T[X,A] \xrightarrow{(r')^\sharp} [X,TA] \xrightarrow{[X,a]} [X,A].$$

Proof. We can decompose the algebra diagrams as follows:

$$
\begin{array}{ccc}
[X,A] & \xrightarrow{\eta} & T[X,A] \\
 & \searrow^{[X,\eta]} & \downarrow{\scriptstyle (r')^\sharp} \\
 & {\scriptstyle \mathrm{id}} & [X,TA] \\
 & \searrow & \downarrow{\scriptstyle [X,a]} \\
 & & [X,A]
\end{array}
\qquad
\begin{array}{ccccc}
TT[X,A] & \xrightarrow{T(r')^\sharp} & T[X,TA] & \xrightarrow{T[X,a]} & T[X,A] \\
\Big\downarrow & & \downarrow{\scriptstyle (r')^\sharp} & & \downarrow{\scriptstyle (r')^\sharp} \\
{\scriptstyle \mu} & & [X,TTA] & \xrightarrow{[X,Ta]} & [X,TA] \\
\Big\downarrow & & \downarrow{\scriptstyle [X,\mu]} & & \downarrow{\scriptstyle [X,a]} \\
T[X,A] & \xrightarrow{(r')^\sharp} & [X,TA] & \xrightarrow{[X,a]} & [X,A]
\end{array}
$$

and so they commute by Proposition 6.5.32, the naturality of $(r')^\sharp$ (following from the naturality of r), and the algebra laws for (A, a'). □

Note that X in the proposition above did not have to be an algebra. The basic idea is that even if A is an algebra, then "things in A parametrized by an arbitrary X" are algebras too. Here is the prototypical example.

Example 6.5.35 (algebra, computer science). Let X be a set, and consider the monoid $(\mathbb{R}, +)$, which we view as an algebra of the list monad L. Then, the set $[X, \mathbb{R}]$ of functions $X \to \mathbb{R}$ is a monoid too due to Proposition 6.5.34. Let's first of all see what the map $(r')^\sharp$ does explicitly. Again, let $[f, g, h]$ be a length-three list of functions, this time, $X \to \mathbb{R}$. Using Examples (6.5.4) and 6.5.29, we get

$$
\begin{array}{ccccc}
L[X,\mathbb{R}] & \xrightarrow{\quad\mathrm{coev}\quad} & [X, L[X,\mathbb{R}] \times X] & \xrightarrow{\qquad [X,r'] \qquad} & [X, L\mathbb{R}] \\
[f,g,h] & \longmapsto & \big(x \mapsto ([f,g,h], x)\big) & \longmapsto & \big(x \mapsto [f(x), g(x), h(x)]\big).
\end{array}
$$

Therefore, the monoid structure on $[X, \mathbb{R}]$ is the following:

$$
\begin{array}{ccccc}
L[X,\mathbb{R}] & \xrightarrow{\qquad (r')^\sharp \qquad} & [X, L\mathbb{R}] & \xrightarrow{\qquad [X,a] \qquad} & [X,\mathbb{R}] \\
[f,g,h] & \longmapsto & \big(x \mapsto [f(x), g(x), h(x)]\big) & \longmapsto & f(x) + g(x) + h(x).
\end{array}
$$

In other words, it is the usual monoid (group) structure of real-valued functions, defined pointwise.

We can also say that $(r')^\sharp$ equips the monad L with a *pointwise structure*. For other monads, the situation is similar. (For example, what happens for internal modules?)

Exercise 6.5.36. Let T be a strong monad on a closed monoidal category $\mathbf{C}$:

- Given an object X of $\mathbf{C}$ and a morphism of T-algebras $f : (A, a) \to (B, b)$, show that the internal postcomposition
$$[X, A] \xrightarrow{[X,f]} [X, B]$$
is a morphism of T-algebras (for the pointwise algebra structures).
- Given a morphism $g : X \to Y$ of $\mathbf{C}$ and a T-algebra (A, a), show that the internal precomposition
$$[Y, A] \xrightarrow{[g,A]} [X, A]$$
is a morphism of T-algebras (for the pointwise algebra structures).

Instantiate these situations for the monoid of real-valued functions on a set.

To conclude this section, let's connect what we saw in Section 6.4 to this setting. Recall that we defined bimorphisms, or "T-morphisms in each variable separately," in terms of the strength (Definition 6.4.17). Here is an equivalent way of defining them, when the category is closed monoidal.

Exercise 6.5.37. Let T be a strong monad on a closed monoidal category. Let (A, a), (B, b), and (C, c) be T-algebras. Let $f : A \otimes B \to C$ be a morphism of $\mathbf{C}$, and denote by $f^\sharp : A \to [B, C]$ its curried version. Consider the following diagrams:

$$
\begin{array}{ccc}
TA \xrightarrow{T(f^\sharp)} T[B,C] \xrightarrow{(r')^\sharp} & [B,TC] \\
\big\downarrow{\scriptstyle a} & \big\downarrow{\scriptstyle [B,c]} \\
A \xrightarrow{\qquad\quad f^\sharp \qquad\quad} & [B,C]
\end{array}
\qquad
\begin{array}{ccc}
A \xrightarrow{f^\sharp} [B,C] \xrightarrow{(\ell')^\sharp} & [TB,TC] \\
\big\downarrow{\scriptstyle f^\sharp} & \big\downarrow{\scriptstyle [TB,c]} \\
[B,C] \xrightarrow{\qquad\quad [b,C] \qquad\quad} & [TB,C].
\end{array}
$$

Prove that f is a T-morphism in A if and only if the left diagram commutes and a T-morphism in B if and only if the right diagram commutes.

(*Hint*: Using Corollary 6.5.20 and (6.5.4), write ℓ' and ℓ as well as r' and r in terms of each other, and use the results of Section 6.5.3 and naturality when needed.)

A concrete interpretation for these diagrams will be given in the following section.

Finally, we can also equivalently state the commutativity condition for monads (Definition 6.4.25) in terms of internal homs, as follows.

Exercise 6.5.38. Show that a strong monad on a symmetric monoidal category is commutative if and only if the following diagram commutes for all objects A and B:

$$
\begin{array}{ccccc}
 & & T[A,B] & & \\
 & \swarrow{\scriptstyle T\left((\ell')^\sharp_{A,B}\right)} & & \searrow{\scriptstyle (r')^\sharp_{A,B}} & \\
T[TA,TB] & & & & [A,TB] \\
{\scriptstyle (r')^\sharp_{A,TB}}\big\downarrow & & & & \big\downarrow{\scriptstyle (\ell')^\sharp_{TA,TB}} \\
[TA,TTB] & & & & [TA,TTB] \\
 & \searrow{\scriptstyle [TA,\mu_B]} & & \swarrow{\scriptstyle [TA,\mu_B]} & \\
 & & [TA,TB] & &
\end{array}
\tag{6.5.7}
$$

6.5.5 The internal hom of algebras

In Section 6.5.1, we saw that vector spaces, modules, abelian groups, and other structures, which can be seen as algebras of commutative monads, all admit an internal hom. Just as the tensor products of all these structures can be seen as special cases of the same construction (Section 6.4.4), all these internal homs can be seen as instances of the same *internal hom of algebras of a commutative monad.*

Let's take as an example the case of vector spaces. Given vector spaces A and B, the internal hom $[A, B]$:

- as a set, is the set of all those maps $A \to B$ which are linear (not all functions);
- as a vector space, is equipped with the *pointwise* linear structure, induced by the one of B.

We can translate both points to the internal language of a closed monoidal category:

- The linearity condition means that we want our maps to be morphisms of algebras (in this case, of the vector space monad):

$$\begin{array}{ccc} TA & \xrightarrow{Tf} & TB \\ \downarrow{\scriptstyle a} & & \downarrow{\scriptstyle b} \\ A & \xrightarrow{f} & B. \end{array} \tag{6.5.8}$$

- The pointwise linear structure means that we are considering the internal hom $[A, B]$ equipped with the pointwise algebra structure given by r', as in the previous section.

We then have to form, using the language of $\mathbf{C}$, a *subobject*[33] of the internal hom $[A, B]$ which intuitively "does not contain *all* morphisms $A \to B$ of $\mathbf{C}$, but only the ones that make (6.5.8) commute." Moreover, we have to restrict the pointwise algebra structure to this subobject. This motivates the following definition.

Definition 6.5.39. Let (T, μ, η, ∇) be a commutative monad on a closed symmetric monoidal category with equalizers. Let (A, a) and

[33]By a *subobject* of X, here we mean an object S with a monic arrow $S \to X$ — think of subsets.

(B, b) be T-algebras. The **internal hom** between the algebras A and B, denoted by $[A, B]_T$, is the equalizer of the following diagram of **C**:

$$[A, B] \underset{[a,B]}{\overset{(\ell')^\sharp}{\rightrightarrows}} [TA, TB] \xrightarrow{[TA,b]} [TA, B]. \tag{6.5.9}$$

Denote also by $i : [A, B]_T \to [A, B]$ the equalizer map.

Let's interpret this definition by instantiating it for the case of sets, starting with $f : A \to B$. The upper arrow in (6.5.9) applies the functor T to f, as we saw in the previous section, and then postcomposes it with the algebra structure map b:

$$\begin{array}{ccccc} [A, B] & \xrightarrow{(\ell')^\sharp} & [TA, TB] & \xrightarrow{[TA,b]} & [TA, B] \\ f & \longmapsto & Tf & \longmapsto & b \circ Tf. \end{array}$$

The lower arrow in (6.5.9), instead, precomposes f with the algebra structure map a:

$$\begin{array}{ccc} [A, B] & \xrightarrow{[a,B]} & [TA, B] \\ f & \longmapsto & f \circ a. \end{array}$$

Therefore, f is in the equalizer if and only if $b \circ Tf = f \circ a$, i.e. if f makes the diagram (6.5.8) commute, which means exactly that f is a morphism of algebras.

Exercise 6.5.40. Show that the *points* of the objects $[A, B]_T$ are the T-morphisms $A \to B$.

Let's now equip the object $[A, B]_T$ with a T-algebra structure, in some sense inherited from the pointwise one of $[A, B]$. We use the following result, which can be seen as an instance of the more general fact that *monadic functors create limits* [Rie16, Theorem 5.6.5].

Lemma 6.5.41. *Let (A, a) and (B, b) be algebras of a monad T on a category* **C**, *and let $f, g : (A, a) \to (B, b)$ be T-morphisms. The equalizer*

E of f and g in $\mathbf{C}$, if it exists, has a unique T-algebra structure that makes the universal map $i : E \to A$ a T-morphism and hence an equalizer in $\mathbf{C}^T$ as well.

Proof. Consider the following diagram:

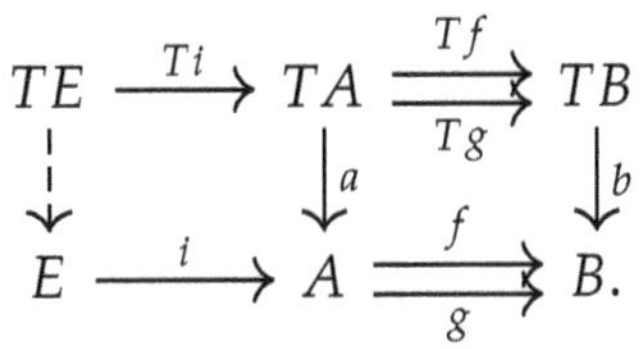

Since E is an equalizer, we have $f \circ i = g \circ i$, hence $Tf \circ Ti = Tg \circ Ti$. Moreover, since f and g are T-morphisms, $b \circ Tf = f \circ a$ and $b \circ Tg = g \circ a$. Therefore, the map $a \circ Ti : TE \to A$ satisfies

$$f \circ a \circ Ti = b \circ Tf \circ i = b \circ Tg \circ Ti = g \circ a \circ Ti.$$

Therefore, by the universal property of E, there exists a unique map $e : TE \to E$ such that $i \circ e = a \circ Ti$. (Why is e a T-algebra structure map?) □

Exercise 6.5.42 (linear algebra). Relate the lemma above, for the monad of vector spaces, to the fact that the solutions to a linear equation on a vector space form themselves a vector space. (Can you think of other examples with other algebraic structures?)

Therefore, to give a T-algebra structure on $[A, B]_T$, it suffices to show that the morphisms in (6.5.9) are T-morphisms. We already know that the lower morphism in (6.5.9), $[a, B]$, is a T-morphism by Exercise 6.5.36. For the upper one, see the following exercise.

Exercise 6.5.43 (important!). Show that the upper composite arrow in (6.5.9) is a T-morphism. (*Hint*: While $[TA, b]$ is a T-morphism by Exercise 6.5.36, $(\ell')^\sharp$ separately may not be on its own, so you have to prove something about the composite. Use the fact that T is commutative so that the diagram (6.5.7) commutes and the fact that B is a T-algebra.)

This completes the construction of the internal hom $[A, B]_T$.

One may ask, when the tensor product of algebras as in Section 6.4.4 exists, does this internal hom make the Eilenberg–Moore category closed monoidal?[34] In the case (say) of vector spaces, the answer is positive: Recall from Example 6.5.3 that for vector spaces, a linear map $f^\sharp : A \to [B, C]$ corresponds to a *bilinear* map $f : A \times B \to C$, where;

- f is linear in A since the map $f^\sharp : A \to [B, C]$ is linear (in A);
- f is also linear in B since for every $a \in A$, the map $f^\sharp_a \in [B, C]$ is linear as a map $B \to C$.

Moreover, as bilinear maps $f : A \times B \to C$ are in bijection with linear maps $A \otimes B \to C$, by composing these bijections, we have the usual hom-tensor adjunction of vector spaces:

$$\mathrm{Hom}_{\mathbf{Vect}}(A \otimes B, C) \xrightarrow{\cong} \mathrm{Bi}(A \otimes B; C) \xrightarrow{\cong} \mathrm{Hom}_{\mathbf{Vect}}(A, [B, C]).$$

Perhaps surprisingly, we can adapt the very same idea for generic algebras of a commutative monad. So, let T be a commutative monad on a closed monoidal category $\mathbf{C}$, let (A, a), (B, b), and (C, c) be T-algebras, and let $f : A \otimes B \to C$ be a (generic) morphism of $\mathbf{C}$. In Exercise 6.5.37, it was asked to show that f is a T-morphism in A if and only if the left diagram in the following commutes and a T-morphism in B if and only if the right diagram in the following commutes:

$$\begin{array}{ccccc} TA & \xrightarrow{T(f^\sharp)} & T[B,C] & \xrightarrow{(r')^\sharp} & [B,TC] \\ \downarrow{\scriptstyle a} & & & & \downarrow{\scriptstyle [B,c]} \\ A & & \xrightarrow{\quad f^\sharp \quad} & & [B,C] \end{array} \qquad \begin{array}{ccccc} A & \xrightarrow{f^\sharp} & [B,C] & \xrightarrow{(\ell')^\sharp} & [TB,TC] \\ \downarrow{\scriptstyle f^\sharp} & & & & \downarrow{\scriptstyle [TB,c]} \\ [B,C] & & \xrightarrow{\quad [b,C] \quad} & & [TB,C]. \end{array}$$

[34]Note that if $\otimes_T$ has a right adjoint, it preserves coequalizers and all colimits, and so $(\mathbf{C}^T, \otimes_T, TI)$ is a symmetric monoidal category, see Section 6.4.4.

Let's interpret this more concretely, keeping in mind what we have just said for vector spaces:

- The condition that the first diagram commutes states exactly that $f^\sharp : A \to [B, C]$ is a T-morphism for the pointwise algebra structure on $[B, C]$. This is the internal version of the fact that, say, for vector spaces, the assignment $a \mapsto f^\sharp_a$ is linear in a.
- The condition that the second diagram commutes states exactly that $f^\sharp$ makes the following compositions equal:

$$A \xrightarrow{f^\sharp} [B,C] \xrightarrow{(\ell')^\sharp} [TB,TC] \xrightarrow{[TB,c]} [TB,C], \qquad [B,C] \xrightarrow{[b,C]} [TB,C].$$

As we had remarked above, this can be interpreted as an internal version of "being a T-morphism"; in this case, it can be interpreted as the fact that "all the things parametrized by $f^\sharp : A \to [B,C]$ are T-morphisms $[B,C]$." Therefore, by the universal property of $[B,C]_T$ as an equalizer, $f^\sharp$ factors uniquely through $[B,C]_T$ via a T-morphism:

$$\begin{array}{ccccccc} A & & & & & & \\ {\scriptstyle \tilde f}\downarrow & \searrow^{f^\sharp} & & & & & \\ [B,C]_T & \xrightarrow{i} & [B,C] & \xrightarrow{(\ell')^\sharp} & [TB,TC] & \xrightarrow{[TB,c]} & [TB,C]. \end{array} \qquad [B,C] \xrightarrow{[b,C]} [TB,C]$$

In other words, we have the following.

Corollary 6.5.44. *There is a natural bijection between the T-bimorphisms $A \otimes B \to C$ and the T-morphisms $A \to [B,C]_T$:*

$$\begin{aligned} \mathrm{Bi}(A,B;C) &\xrightarrow{\cong} \mathrm{Hom}_{\mathbf{C}^T}(A,[B,C]_T) \\ f &\longmapsto \tilde f. \end{aligned} \tag{6.5.10}$$

Suppose now that the tensor product $A \otimes_T B$ exists, which is guaranteed for example if $\mathbf{C}$ has coequalizers (see Section 6.4.4).

Then, by definition, $A \otimes_T B$ represents the functor $\mathrm{Bi}(A, B; -)$, giving a natural bijection

$$\mathrm{Hom}_{\mathbf{C}^T}(A \otimes_T B, C) \cong \mathrm{Bi}(A, B; C).$$

Composing this bijection with the one in (6.5.10), we get, just as for vector spaces, the desired hom-tensor adjunction.

Let's state this in the form of a theorem — the last one of this book.

Theorem 6.5.45. *Let $(\mathbf{C}, \otimes, I)$ be a closed monoidal category with equalizers and coequalizers, and let T be a commutative monad on $\mathbf{C}$. Then, the tensor product of any two algebras $A \otimes_T B$ and their internal hom $[A, B]_T$ both exist, and we have a hom-tensor adjunction*

$$\mathrm{Hom}_{\mathbf{C}^T}(A \otimes_T B, C) \cong \mathrm{Hom}_{\mathbf{C}^T}(A, [B, C]_T)$$

that makes the category of T-algebras closed monoidal.

This can be seen as the abstract principle underlying the hom-tensor adjunction of

- vector spaces,
- modules over a commutative ring,
- abelian groups.

By means of this categorical construction, the same intuition can be used in other contexts as well, giving for example a hom-tensor adjunction for:

- commutative monoids,
- commutative rings,
- convex spaces,
- G-sets for an abelian group G,
- join-semilattices,

and many more.

Conclusion

This is the end! Hopefully, this book has sparked your interest in categories and their applications and has prepared you to get started on more advanced topics. If you want to deepen your understanding of category theory, my advice is to *reach out to other people*. Many members of the category theory community are happy to help.

Further reading: Where does it go from here?

- If you are interested in learning more pure category theory or in applying category theory to areas of pure math such as algebraic geometry or topology, the next step is a more advanced category theory textbook. You can use Emily Riehl's book [Rie16], as well as the classic texts in Refs. [Bor94] and [AHS90]. You should be ready to understand those books if you are motivated.
- If you are interested in algebraic topology or homotopy theory, you can read Emily Riehl's notes [Rie] and then her book [Rie14]. For algebraic topology from a categorical viewpoint, you can look at J. P. May's classic [May99]. I recommend, however, that before approaching those, you should learn a bit more category theory, for example from the references given in the point above. An excellent reference for classical algebraic topology [Hat02].
- If you are interested in learning category theory from a more application-oriented perspective, the next reading could be Brendan Fong and David Spivak's recent book [FS19]. It has an overview

of possible applications of category theory, from which you can choose where to focus in particular.

- If you want to learn more about monoidal categories and string diagrams, I recommend Noson Yanofsky's new introduction [Yanng], which should be in print by the time this book is out. Other introductory references are again [FS19] and Dan Marsden's book [Mar14].
- If you are interested in applications to quantum information theory, especially using string diagrams, check out Peter Selinger's work [Sel10]. I also recommend the book by Bob Coecke and Aleks Kissinger [CK17], which is (in their words) "not a book about category theory," but definitely of great interest for the categorically-minded.
- If you are interested in category theory in the context of computer science and functional programming, I recommend Bartosz Milewski's book [Mil19].

These are only a few references (as of 2023). The category theory literature is quite vast and still growing.

Online resources: There are a lot of category theory resources online as well. The category theory community is very open to sharing knowledge and very keen on large-scale community projects. Here are some of them (available at the time of writing in 2023):

- The *nLab* (http://ncatlab.org) is a wiki about category theory, higher category theory, and their applications. You can not only learn from it but also contribute yourself.
- The *nForum* (https://nforum.ncatlab.org/) is the forum of the nLab, where category-related questions and discussions are more than welcome.

- The *n-Category Café* (https://golem.ph.utexas.edu/category/) is a blog on mathematics with a focus on category theory. Many recent ideas in category theory and their applications were born in the comment threads of this blog.
- The *Applied Category Theory* page (https://www.appliedcategorytheory.org/) advertises events you can apply to, is about category theory and its applications, and displays a number of learning resources.
- The *Category Theory Zulip Server* (https://categorytheory.zulipchat.com/) is a forum for questions and general discussion about category theory and its applications. This forum is private — if you are interested, please contact either me (the author) or one of the server moderators for an invite link.
- The *Stacks Project* (https://stacks.math.columbia.edu/) is a wiki about algebraic geometry with in-depth material on categorical constructions.

There are also several YouTube channels with category theory content. Here are some of them: (At the time of writing, these YouTube contents are freely available.)

- The *Catsters* (http://www.simonwillerton.staff.shef.ac.uk/TheCatsters/) are several short videos by Eugenia Cheng and Simon Willerton that explain the basics of category theory.
- Bartosz Milewski's YouTube channel (https://www.youtube.com/playlist?list=PLbgaMIhjbmEnaH_LTkxLI7FMa2HsnawM_) has videos on category theory and programming.
- The Topos Institute's YouTube channel (https://www.youtube.com/playlist?list=PLhgq-BqyZ7i5lOqOqqRiS0U5SwTmPpHQ5) has videos of the lectures at MIT of the *Applied Category Theory* course.

- The *Applied Category Theory* YouTube channel (https://www.youtube.com/@appliedcategorytheory5517/playlists) has recordings of tutorials and conference talks on applied category theory.
- For Italian speakers, the *ItaCa project* (https://www.youtube.com/@ProgettoItaCaCT) has category theory video lectures in Italian as well as recordings of conference talks (in English).
- Finally, my YouTube channel (https://www.youtube.com/@PaoloPerroneMath) has some material on category theory, mostly in the context of probability and statistics, which is my main research topic at the time of writing.

Bibliography

[AHS90] Jiří Adámek, Horst Herrlich, and George Strecker. *Abstract and Concrete Categories: The Joy of Cats*. Wiley, 1990. Available at http://www.tac.mta.ca/tac/reprints/articles/17/tr17.pdf.

[AM10] M. Aguiar and S. Mahajan. *Monoidal Functors, Species and Hopf Algebras*. AMS, 2010.

[Aum61] Robert J. Aumann. "Borel structures for function spaces'p'. In: *Illinois Journal of Mathematics* 5 (1961), pp. 614–630.

[Bor94] Francis Borceux. *Handbook of Categorical Algebra*. Cambridge University Press, 1994.

[Bra14] Martin Brandenburg. *Tensor Categorical Foundations of Algebraic Geometry*. Available at http://arxiv.org/abs/1410.1716. PhD thesis. University of Münster, 2014.

[Che23] Eugenia Cheng. *The Joy of Abstraction. An Exploration of Math, Category Theory, and Life*. Cambridge University Press, 2023.

[CK17] Bob Coecke and Aleks Kissinger. *Picturing Quantum Processes. A First Course in Quantum Theory and Diagrammatic Reasoning*. Cambridge University Press, 2017.

[Dij72] Edsger W. Dijkstra. "The humble programmer." *Communications of the ACM* 15.10 (1972), p. 864.

[FS19] Brendan Fong and David I. Spivak. *An Invitation to Applied Category Theory: Seven Sketches in Compositionality*. Cambridge University Press, 2019. Available at https://arxiv.org/abs/1803.05316.

[GLN08] Jean Goubault-Larrecq, Sławomir Lasota, and David Nowak "Logical relations for monadic types." In: *Mathematical Structures in Computer Science* 18.6 (2006), pp. 1169–1217.

[Hat02] Allen Hatcher. *Algebraic Topology*. Cambridge University Press, 2002. Available at http://pi.math.cornell.edu/~hatcher/AT/ATpage.html.

[JY21] Niles Johnson and Donald Yau. *2-Dimensional Categories*. Oxford University Press, 2021.

[Kle14] Achim Klenke. *Probability Theory, A Comprehensive Course*. Second edition. Springer, 2014.

[Koc70] Anders Kock. "Monads on symmetric monoidal closed categories." In: *Archiv der Mathematik* 21 (1970), pp. 1–10.

[Koc71] Anders Kock. "Bilinearity and cartesian closed monads." In: *Mathematica Scandinavica* 29 (1971), pp. 161–174.

[Koc72] Anders Kock. "Strong functors and monoidal monads." In: *Archiv der Mathematik* 23 (1972), pp. 113–120.

[Koc75] Anders Kock. "Closed categories generated by commutative monads." In: *Journal of the Australian Mathematical Society* 12 (1975), pp. 405–424.

[Lac09] Stephen Lack. "A 2-category companion." In: *Towards Higher Categories*. Ed. by John C. Baez and J. Peter May. Springer, 2009.

[Lor21] Fosco Loregian. *(Co)end Calculus*. Cambridge University Press, 2021. Available at https://arxiv.org/abs/1501.02503.

[Mar14] Daniel Marsden. *Category Theory Using String Diagrams*. 2014. Available at https://arxiv.org/abs/1401.7220.

[May99] J. P. May. *A Concise Course in Algebraic Topology*. University of Chicago Press, 1999. Available at https://www.math.uchicago.edu/~may/CONCISE/ConciseRevised.pdf.

[Mil19] Bartosz Milewski. *Category Theory in Context*. Blurb, 2019.

[Per18] Paolo Perrone. "Categorical probability and stochastic dominance in metric spaces." PhD thesis. University of Leipzig, 2018. Available at http://www.paoloperrone.org/phdthesis.pdf.

[Rie] Emily Riehl. *A Leisurely Introduction to Simplicial Sets*. Available at http://www.math.jhu.edu/~eriehl/ssets.pdf.

[Rie14] Emily Riehl. *Categorical Homotopy Theory*. Cambridge University Press, 2014. Available at http://www.math.jhu.edu/~eriehl/cathtpy.pdf.

[Rie16] Emily Riehl. *Category Theory for Programmers*. Dover, 2016. Available at http://www.math.jhu.edu/~eriehl/context.pdf.

[Sat18] Tetsuya Sato. "The Giry monad is not strong for the canonical symmetric monoidal closed structure on Meas." In: *Journal of Pure and Applied Algebra* 222.10 (2018), pp. 2888–2896.

[Sea13] Gavin J. Seal. "Tensors, monads and actions." In: *Theory and Applications of Categories* 28.15 (2013). arxiv:1205.0101, pp. 403–434.

[Sel10] Peter Selinger. "A survey of graphical languages for monoidal categories." In: *New Structures for Physics*. Springer, 2010, pp. 289–355.

[Shu08] Michael A. Shulman. *Set Theory for Category Theory*. 2008. Available at https://arxiv.org/abs/0810.1279.

[Tru20] Luke Trujillo. *A Coherent Proof of Mac Lane's Coherence Theorem*. 2020. Available at https://scholarship.claremont.edu/hmc_theses/243/.

[Yanng] Noson S. Yanofsky. *Monoidal Category Theory. A Unifying Concept in Mathematics, Physics, and Computing*. MIT Press, Forthcoming.

Index

Mathematical Terms

B

C

I

J

K

L

N

O

T

U

Milton Keynes UK
Ingram Content Group UK Ltd.
UKHW021356120624
444122UK00006B/66

9 789811 286001